Harmal

The Genus *Peganum*

Traditional Herbal Medicines for Modern Times

Each volume in this series provides academia, health sciences, and the herbal medicines industry with in-depth coverage of the herbal remedies for infectious diseases, certain medical conditions, or the plant medicines of a particular country.

Series Editor: Dr. Roland Hardman

Volume 1
Shengmai San, edited by Kam-Ming Ko

Volume 2
Rasayana: Ayurvedic Herbs for Rejuvenation and Longevity, by H.S. Puri

Volume 3
Sho-Saiko-To: (Xiao-Chai-Hu-Tang) Scientific Evaluation and Clinical Applications, by Yukio Ogihara and Masaki Aburada

Volume 4
Traditional Medicinal Plants and Malaria, edited by Merlin Willcox, Gerard Bodeker, and Philippe Rasoanaivo

Volume 5
Juzen-taiho-to (Shi-Quan-Da-Bu-Tang): Scientific Evaluation and Clinical Applications, edited by Haruki Yamada and Ikuo Saiki

Volume 6
Traditional Medicines for Modern Times: Antidiabetic Plants, edited by Amala Soumyanath

Volume 7
Bupleurum *Species: Scientific Evaluation and Clinical Applications,* edited by Sheng-Li Pan

Volume 8
Herbal Principles in Cosmetics: Properties and Mechanisms of Action, by Bruno Burlando, Luisella Verotta, Laura Cornara, and Elisa Bottini-Massa

Volume 9
Figs: The Genus Ficus, by Ephraim Philip Lansky and Helena Maaria Paavilainen

Volume 10
Phyllanthus Species: Scientific Evaluation and Medicinal Applications, edited by Ramadasan Kuttan and K. B. Harikumar

Volume 11
Honey in Traditional and Modern Medicine, edited by Laïd Boukraâ

Volume 12
Caper: The Genus Capparis, by Ephraim Philip Lansky, Helena Maaria Paavilainen, and Shifra Lansky

Volume 13
Chamomile: Medicinal, Biochemical, and Agricultural Aspects, by Moumita Das

Harmal
The Genus *Peganum*

Ephraim Shmaya Lansky
Shifra Lansky
Helena Maaria Paavilainen

with a Preface by Andrew Weil

CRC Press
Taylor & Francis Group
Boca Raton London New York

CRC Press is an imprint of the
Taylor & Francis Group, an **informa** business

CRC Press
Taylor & Francis Group
6000 Broken Sound Parkway NW, Suite 300
Boca Raton, FL 33487-2742

First issued in paperback 2021

© 2017 by Taylor & Francis Group, LLC
CRC Press is an imprint of Taylor & Francis Group, an Informa business

No claim to original U.S. Government works

ISBN 13: 978-1-03-209623-0 (pbk)
ISBN 13: 978-1-4822-4956-9 (hbk)

Library of Congress Cataloging-in-Publication Data

Names: Lansky, Ephraim S., author. | Lansky, Shifra, 1991- author. | Paavilainen, Helena M., author.
Title: Harmal : the genus Peganum / author(s): Ephraim Shmaya Lansky, Shifra Lansky, and Helena Maaria Paavilainen.
Other titles: Traditional herbal medicines for modern times ; [v. 20]
Description: Boca Raton : Taylor & Francis, 2017. | Series: Traditional herbal medicines for modern times ; [v. 20] | Includes bibliographical references.
Identifiers: LCCN 2017015869 | ISBN 9781482249569 (hardback : alk. paper)
Subjects: LCSH: Peganum harmala.
Classification: LCC QK495.Z9 L36 2017 | DDC 583/.79--dc23
LC record available at https://lccn.loc.gov/2017015869

Visit the Taylor & Francis Web site at
http://www.taylorandfrancis.com

and the CRC Press Web site at
http://www.crcpress.com

For Peganum harmala and the science of healing.

E.S.L.

For my family.

S.L.

*In memory of my grandmother Eimi Lempinen,
always listening and always supporting.*

H.M.P.

Contents

List of Figures

List of Tables

Foreword

One might ask: from where came the idea to research *Peganum harmala*? The answer is that the idea found the researchers, as much as they found it. As children, one of them (ESL) treasured the little Golden Guides, concise, colorful packets of information and plenty of color with titles like *Mammals*, *Fishes*, *Birds*, *Reptiles and Amphibians* (the favorite), *Trees*, and *Rocks and Minerals*. It was nostalgic, then, to have also acquired at some point to the family home an additional Golden Guide, written by Professor Richard Schultes of Harvard, and titled simply *Hallucinogenic Plants*. *P. harmala* is depicted in that book with minimal text, apologizing that not much was known about this Old World hallucinogen, which seemed to no longer be in use. That guide joined the others in the series as valued possessions. ESL had earlier encountered harmala alkaloids in undergraduate Natural Products Chemistry at New College, Sarasota, Florida, with Prof. Rodger W. Griffin, who encouraged him to concentrate his interest in indolic chemistry for the sake of that seminar on the smaller indolics—lysergics would make the study unwieldy—it was to be restricted to simple substituted tryptamines and the harmala alkaloids, with harmala as the limit in chemical complexity. Only decades later while searching botanical candidates for putative complementary cancerology were harmala alkaloids encountered again, in what seemed a surprising context.

Some words of caution are in order. First, though *P. harmala* can be widely used safely in medicine, it *is* classified worldwide as a poisonous plant. Due to its monoamine oxidase inhibiting action, it will generally be contraindicated when SSRIs and other antidepressants or antipsychotic drugs are being used. It inhibits cytochrome P450 enzymes (Zhao et al. 2011), so when taken along with drugs dependent on P450 facilitated metabolism, e.g., statins, must be understood by the prescribing physician. One should be assiduously careful if taking other drugs at the same time, and always under strict and compassionate medical supervision.

However, the real and consistent danger underscoring all poisonings and fatalities from *P. harmala* described in the literature (Berdai et al. 2014; Mahmoudian et al. 2002; Moshiri et al. 2013) has been overdosing. A person either does not understand what harmal is (a poisonous plant with the potential to damage internal organs and kill), or otherwise, *does* know what harmal is and is overly enthusiastic to achieve some particular objective. There are fatalities reported every year from *P. harmala* poisoning worldwide. Please treat this plant with humility and respect. NOTHING CONTAINED WITHIN THESE PAGES REPRESENTS ANY TYPE OF PRESCRIPTION, ENDORSEMENT, OR RECOMMENDATION FOR TREATING ANY AILMENT. IMPROPER USE OF *P. HARMALA* IN ANY FORM FOR PERSONAL USE CAN RESULT IN MORBIDITY AND DEATH. DO NOT TRY THIS AT HOME!

In animals, harmal poisoning can be recognized by digestive and nervous syndromes—inebriation engenders a narcotic state interrupted by an occasional short period of excitement. Abortion is frequent in animals that digest this plant in a dry year. In humans, vomiting and diarrhea can occur together with tremors, ataxia, bradycardia, delirium, hallucinations, and functional paralysis.

Nevertheless, the worldwide popularity of *P. harmala* is great. In a survey in a small Pakistani community, the inhabitants were queried about their use of 82 different medicinal plants. The leading preference among the 82 was for *P. harmala*, the second for *Punica granatum* (pomegranate), according to a preference scale as follows: *P. harmala* (0.93), *Punica granatum* (0.91), *Thymus mongolicus* (0.90), *Chenopodium album* (0.89), *Coriandrum sativum* (0.87), *Mentha longifolia* (0.87), *Lactuca serriola* (0.87), and *Portulaca oleracea* (0.87) (Aziz et al. 2016).

One aspect of this study was its globality—most feasible in the wired world. The contribution of scientists from Iran is singular, especially since Persia has been the seat of harmal and of the devotional culture that likely developed around it for thousands of years, and modern Iranian scientists continually report creative new uses for *P. harmala* in medicine and public health.

In China, trials and quiet experiments continue to methodically advance the medical use of harmal at all levels: *in vitro*, *in vivo*, and clinically. Much has been discovered through Chinese efforts in recent years regarding *Peganum*'s anticancer properties. *Harmalogy* thrives in China.

Scientific communities in many other countries also contribute, notably in Algeria, Egypt, Jordan, Morocco, Tunisia, Pakistan, India, Saudi Arabia, Iraq, Spain, Italy, and Turkey, where ancient knowledge of indigenous plants as medicines continues in daily life. More work, especially for synthetic harmala analogues, is done in Russia, Germany, France, Belgium, and England. As seen within, it is an *extremely* important and highly *utilitarian* herb for public health concerns (see use against pests, bacteria, fungi, and parasites in Chapter 7).

Persons seeking *P. harmala* for "recreational" hedonistic "light shows" and similar pleasures (Burillo-Putze et al. 2013) should be forewarned: (1) the blur between recreation and healing allows for considerable overlap, but if it is truly for recreational use, it should actually recreate—in a safe, physician-supervised, setting; (2) if the individual is looking for psychedelic shows, disappointment may occur. The psychoactive effect of harmal could easily be more subtle than expected, and physical discomforts including vomiting, headache, and difficulty in walking may occur. Overdose may cause a life-threatening emergency. **The possibility for unfavorable interactions with prescription drugs is real, very serious, and must always be considered. The physician prescribing for other drugs or supplements needs to ensure appropriate supervision prior to the use of harmal.**

Like the authors' previous works by CRC, *Figs: The Genus* Ficus, and *Caper: The Genus* Capparis, the pages to follow aspire to integrate ancient and indigenous ethnopharmacology with modern preclinical and clinical research, and as well to present something of the basis for the research through an exposition of its chemistry. The present work includes examples of direct clinical experience, self-inebriations, and subjective elaboration of the psychoactive effects of harmal and its extracts. Chapter 9 offers a speculative hypothesis of why psychoactive phytochemicals ranging from harmala alkaloids to cannabinoids exert anticancer effects clinically, based on a putative *retuning* of serotonin receptors on both neurons and lymphocytes.

In the Conclusion, the goal is to integrate the divergent streams of investigations and to reach elaborate possibilities for future research and collaborations.

REFERENCES

Aziz, M.A., M. Adnan, A.H. Khan, A.U. Rehman, R. Jan, and J. Khan. 2016. Ethno-medicinal survey of important plants practiced by indigenous community at Ladha subdivision, South Waziristan agency, Pakistan. *J Ethnobiol Ethnomed* 12(1): 53.

Berdai, M.A., S. Labib, and M. Harandou. 2014. *Peganum harmala* L. intoxication in a pregnant woman. *Case Rep Emerg Med* 2014: 783236.

Burillo-Putze, G., E. López Briz, B. Climent Díaz, P. Munné Mas, S. Nogue Xarau, M.A. Pinillos, and R.S. Hoffman. 2013. Emergent drugs (III): hallucinogenic plants and mushrooms. *An Sist Sanit Navar* 36(3): 505–18.

Mahmoudian, M., H. Jalilpour, and P. Salehian. 2002. Toxicity of *Peganum harmala*: Review and a Case Report. *Iranian Journal of Pharmacology & Therapeutics* 1: 1–4. http://ijpt.iums.ac.ir.

Moshiri, M., L. Etemad, S. Javidi, and A. Alizadeh. 2013. *Peganum harmala* intoxication, a case report. *Avicenna J Phytomed* 3(3): 288–92.

Zhao, T., Y.Q. He, J. Wang, K.M. Ding, C.H. Wang, and Z.T. Wang. 2011. Inhibition of human cytochrome P450 enzymes 3A4 and 2D6 by β-carboline alkaloids, harmine derivatives. *Phytother Res* 25(11): 1671–7.

Series Preface

Global warming and global travel are contributing factors in the spread of infectious diseases such as malaria, tuberculosis, hepatitis B, and HIV. These are not well controlled by the present drug regimes. Antibiotics are also failing because of bacterial resistance. Formerly less well-known tropical diseases are reaching new shores.

A whole range of illnesses, such as cancer, occur worldwide. Advances in molecular biology, including methods of *in vitro* testing for a required medical activity, give new opportunities to draw judiciously on the use and research of traditional herbal remedies from around the world. The reexamining of the herbal medicines must be done in a multidisciplinary manner.

There have been 51 volumes published since the start of the book series Medicinal and Aromatic Plants—Industrial Profiles in 1997. The series continues.

The same series editor, Dr. Roland Hardman, is also covering a second series entitled Traditional Herbal Medicines for Modern Times. Each volume of this series reports on the latest developments and discusses key topics relevant to interdisciplinary health sciences, research by ethnobiologists, taxonomists, conservationists, agronomists, chemists, pharmacologists, clinicians, and toxicologists. The series is relevant to all these scientists and will enable them to guide business, government agencies, and commerce in the complexities of these matters. The background to the subject is outlined next.

Over many centuries, the safety and limitations of herbal medicines have been established by their empirical use by the "healers" who also took a holistic approach. The healers are aware of the infrequent adverse effects and often know how to correct contraindications when they occur. Consequently, and ideally, the preclinical and clinical studies of an herbal medicine need to be carried out with the full cooperation of the traditional healer. The plant composition of the medicine, the stage of the development of the plant material, when it is to be collected from the wild or from its cultivation, its postharvest treatment, the preparation of the medicine, the dosage and frequency, and much other essential information is required. A consideration of the intellectual property rights and appropriate models of benefit sharing may also be necessary.

Wherever the medicine is being prepared, the first requirement is a well-documented reference collection of dried plant material. Such collections are encouraged by organizations including the World Health Organization and the United Nations Industrial Development Organization. The Royal Botanic Gardens at Kew (United Kingdom) is now increasing its collection of traditional Chinese dried plant material relevant to its purchase and use by those who sell or prescribe traditional Chinese medicine in the United Kingdom.

In any country, the control of the quality of plant raw material, of its efficacy, and of its safety in use is essential. The work requires sophisticated laboratory equipment and highly trained personnel. This kind of "control" cannot be applied to the locally produced herbal medicines in the rural areas of many countries, on which millions of people depend. Local traditional knowledge of the healers has to suffice.

Conservation and protection of plant habitats are required, and breeding for biological diversity is important. Gene systems are being studied for medicinal exploitation. There can never be too many seed conservation "banks" to conserve genetic diversity. Unfortunately, such banks are usually dominated by agricultural and horticultural crops, with little space for medicinal plants. Developments, such as random amplified polymorphic DNA, enable the genetic variability of a species to be checked. This can be helpful in deciding whether specimens of close genetic similarity warrant storage.

From ancient times, a great deal of information concerning diagnosis and the use of traditional herbal medicines has been documented in the scripts of China, India, and elsewhere. Today, modern formulations of these medicines exist in the form of powders, granules, capsules, and tablets. They are prepared in various institutions, such as government hospitals in China and Korea and by companies such as the Tsumura Company of Japan, with good quality control. Similarly, products are produced by many other companies in India, the United States, and elsewhere with a varying degree of quality control. In the United States, the Dietary Supplement and Health Education Act of 1994 *recognized* the class of physiotherapeutic agents derived from medicinal and aromatic plants. Furthermore, under public pressure, the U.S. Congress set up an Office of Alternative Medicine. In 1994, this office assisted in the filing of several investigational new drug (IND) applications required for clinical trials of some Chinese herbal preparations. The significance of these applications was that each Chinese preparation involved several plants and yet was handled with a single IND. A demonstration of the contribution to efficacy, of *each* ingredient of *each* plant, was not required. This was a major step forward toward more sensible regulations with regard to phytomedicines.

The subject of Western herbal medicines is now being taught again to medical students in Germany and Canada. Throughout Europe, the United States, Australia, and other countries, pharmacy and health-related schools are increasingly offering training in phytotherapy. Traditional Chinese medicine clinics are now common outside of China. An Ayurvedic hospital now exists in London, with a BSc Honors degree course in Ayurvedic medicine being available: Professor Shrikala Warrier, Registrar/Dean, MAYUR, Ayurvedic University of Europe, 81 Wimpole Street, London, WIG 9RF, email sw@unifiedherbal.com. This is a joint venture with a university in Manipal, India.

The term *integrated medicine,* which selectively combines traditional herbal medicine with "modern medicine," is now being used. In Germany, there is now a hospital in which traditional Chinese medicine is integrated with Western medicine. Such co-medication has become common in China, Japan, India, and North America by those educated in both systems. Benefits claimed include improved efficacy, reduction in toxicity and the period of medication, as well as a reduction in the cost of the treatment. New terms, such as *adjunct therapy, supportive therapy,* and *supplementary medicine* now appear as a consequence of such comedication. Either medicine may be described as an adjunct to the other, depending on the communicator's view. Great caution is necessary when traditional herbal medicines are used by doctors not trained in their use and likewise when modern medicines are used by traditional herbal doctors. Possible dangers from drug interactions need to be stressed.

Roland Hardman, BPharm, BSc, PhD, FRPharmS
Head of Pharmacognosy (retired), School of Pharmacy and Pharmacology
University of Bath, United Kingdom

Preface

In this exhaustive, thoroughly researched, and beautifully illustrated book, Ephraim and Shifra Lansky and Helena Paavilainen describe all aspects of the genus *Peganum,* in particular the species *Peganum harmala*, a desert plant commonly known throughout the Middle East as harmal. Often regarded mainly as toxic, harmal is in fact a most interesting plant, with complex chemistry and pharmacological activity, a rich history in ethnomedicine in the Near and Far East, and many potential therapeutic applications in contemporary medicine. I have no doubt that *Harmal: The Genus* Peganum will be the definitive reference work on the subject.

I first learned about *Peganum* in 1960 as an undergraduate at Harvard University in a course titled "Plants and Human Affairs" taught by the late Richard Schultes, then director of the Harvard Botanical Museum, now known as the godfather of modern ethnobotany. He cited *Peganum* as a rare example of an Old World hallucinogen with some of the same alkaloids as *Banisteriopsis caapi,* the sacred vine of Amazonian Indians. Those natives learned to decoct the woody stems of the vine with leaves of plants containing dimethyltryptamine (DMT) to "make the visions brighter." The result is the powerful psychedelic or "entheogenic" brew, *ayahuasca,* little known outside the Amazon basin in 1960 that is now in widespread use throughout North and South America and elsewhere for religious, psychotherapeutic, and healing purposes.

Before ayahuasca became so available in the United States, people wanting to experience it came up with formulas for "ayahuasca analogs," mostly based on crushed seeds of *Peganum harmala* combined with DMT-containing leaves of species of *Psychotria* or *Acacia*. This made many more people aware of harmal, but in my experience, none of these experimenters knew it as anything other than a potential source of psychoactivity. As the authors of this book make clear, the pharmacological actions of *P. harmala* are much broader than that. The plant is, indeed, toxic, but it illustrates perfectly the truth that all drugs become toxic as dose is increased, and some poisons become useful therapeutic agents at appropriately low doses.

Harmal: The Genus Peganum documents the extensive history of use of these plants in traditional Arab and Iranian medicine, where their indications are numerous and diverse. Most exciting is the authors' review of ways the compounds in harmal might be used today to treat a range of serious diseases, from malaria to diabetes and, especially, cancer.

I congratulate the Lanskys and Helena Paavilainen on producing a monumental work on a most promising genus.

Andrew Weil, MD

Acknowledgments

E.S.L.: I wish to express my gratitude first to Dr. Eli Har Lev for wildcrafting and providing me with my first dried seed-bearing capsules of *Peganum harmala*, and for taking the initiative to produce our review on anticancer properties of desert plants, which included a discussion of *P. harmala*. Israel Prize winner Prof. Emeritus Eviatar Nevo helped provide a biological context for this work by frequently supporting and endorsing my explorations in field pharmacognosy through the Institute of Evolution at Haifa University, which he founded in the 1960s, and Prof. Solomon Wasser, also of that institution, provided me with biomass from medicinal mushrooms to evaluate as possible components and as correctives for practical treatments. Dr. Tomas Pavlicek supported me on many occasions in obtaining supplies of *P. harmala* seeds for extractions. Prof. Bashar Saad and Dr. Omar Said of The Galilee Society R&D Center (affiliated with Haifa University) shared generously with me their time, knowledge, and experience regarding the use of harmal in traditional Arab medicine. Prof. Adam Halberstadt of the University of California, San Diego, was instrumental in clarifying aspects of the psychopharmacology of harmala alkaloids. Distinguished Prof. Mark Geyer, also of UCSD, and Dr. Stephen Fulder guided me to help find the right guides. Dr. S.W. Kaplan and Dr. Ofer Finkelstein were always there to discuss the manuscript and future directions of the work. Shelley Chen-Aridor of the Haifa University Library was, as ever, my enthusiastic, industrious, and indefatigable research assistant. My family, Dr. Shen Yu and PhD students, visual artists, and musicians Zipora and Shifra Lansky, provided incalculable assistance for getting it all done. Tikun Olam Ltd. of Tel Aviv kindly supplied the medical *Cannabis* used as a corrective in the cases described in Chapter 8. Rabbi Dr. Joel Covitz of Brookline, Massachusetts, enabled me to place the healing work in its proper archetypal, personal, and historical contexts and Rav Avidan Chazani and the faithful of Haifa's Sephardis Synagogue of Ahuza, provided the context and venue for patent intercessory support, the secret "corrective" to the treatments described. Neurologist, Dr. Jonathan Grunfeld provided kindness too numerous to count, and Prof. J. Moshe Gomori and Dr. David Michaeli helped in clarifying the possible effect of harmal on brain tumors. Prof. Pinchas Mandel, Dr. Martin Goldman, and Dr. Daniel Rubin offered their friendship and support. Without the ballast and practicality of my coauthor, Dr. Helena Maaria Paavilainen, it is unlikely that this book would ever have achieved fruition. Of course, the greatest debt is to my patients with cancer seeking complementary and alternative herbal treatments for their diseases, without whom the impetus for this work would not have been, and to the memory of my parents, Atty. Sidney and Edith Lansky, who laid the earliest groundwork for the start that would come. Roland Hardman was the best and most helpful editor that could ever be imagined. Dr. Samuel Epstein was my incredibly generous and intuitive mentor in medicine and biology from my earliest maturity.

H.M.P.: My greatest thanks go to my coauthor, Dr. Ephraim Lansky, who introduced me to the hero of this book, *Peganum harmala*, and helped me by his enthusiasm to go on. Without him this book would never have been started—and definitely not brought to completion. I am indebted also to my other coauthor, Shifra Lansky, for her painstaking and detail-oriented work on the chemical aspects of harmal and especially for drawing the chemical formulas. Thank you for making this book possible.

Special thanks are due to my friends Sasha Sedan and Brian Bush who both took me in their cars to the desert to search for the plant that I had just heard "grew in the parking lot of the Inn of the Good Samaritan" (thank you also, kind gardener in the Botanical Garden of the Hebrew University in Mount Scopus, for guiding me there and for the wonderful, detailed advice that actually helped me to find it!). Without you two, a great part of the photographs would not be in the book.

I am also grateful to John Sulzycki for accepting the manuscript for publication and for his and his staff's great patience with delays, and for Dr. Roland Hardman for his support and unselfish generosity in supplying us with information.

There are four persons who have a special part in my life among ancient herbs. Dr. Paget Stanfield was instrumental in creating my first professional contact with Israeli medicinal plant researchers. Thank you for your encouragement!

Prof. Shmuel Kottek has supported and guided me from the beginning of my studies in history of medicine, encouraging me in things I thought impossible. Thank you for believing in me and helping me achieve.

A special thanks to my sister, Kaarina Paavilainen, for being a safety net for me—the person to whom to turn to find security in the complications of life (such as writing a book).

And last but not least, I want to thank my mother, Maija Paavilainen, for her support for and interest in this project—to the extent of growing *Peganum harmala* from seed in her garden in cold Finland, and with remarkable success! There are not many who would have listened to me with such patience.

Authors

Ephraim Shmaya Lansky is a University of Pennsylvania trained integrative medical doctor concentrating in complementary cancer care. He is a licensed practitioner of medical hypnosis in Israel. He teaches Kokikai Aikido once a week, in which he holds a second-degree black belt. His work with pomegranate fruits and their value-added healthcare products yielded several patents, a PhD in pharmacognosy from Leiden University in the Netherlands, and an MBA from Bradford University in England. He is a coauthor with Helena Paavilainen and Shifra Lansky of *Caper: The Genus* Capparis, and with Helena Paavilainen of *Figs: The Genus* Ficus, both books published by CRC Press. He is also a coauthor with Robert A. Newman, formerly of M.D. Anderson Cancer Center in Houston, Texas, of *Pomegranate: The Most Medicinal Fruit* (Basic Books).

Shifra Lansky earned a BSc degree in chemistry from Hebrew University in Jerusalem, where she is presently pursuing her graduate studies. Her focus is on the biochemical and structural characterization of bacterial proteins that may be used for the production of biofuel. Shifra also enjoys playing the violin, reading, and skiing in her spare time.

Helena Maaria Paavilainen is a researcher at the Hadassah Medical School, Hebrew University of Jerusalem, Israel. Her main research interests are ethnomedicine, historical ethnopharmacology, and the history of pharmacology, especially the Hebrew, Arabic, and Latin traditions. She wrote her PhD thesis (published as *Medieval Pharmacotherapy: Continuity and Change; Case Studies from Ibn Sina and Some of His Late Medieval Commentators*, Leiden: Brill, 2009) on the development of medical drug therapy in medieval times and on the potential validity of medieval herbal treatments. She also coauthored with Dr. Lansky the monograph *Figs: The Genus* Ficus (Boca Raton, FL: CRC Press, 2010), and with Dr. Lansky and Shifra Lansky the sequel *Caper: The Genus* Capparis in 2014. She currently works as a freelance consultant bioprospecting ancient and medieval herbal texts for practical applications in medicine, functional nutrition, and agriculture.

Zipora Lansky, illustrator, holds a BSc in Biology from the Technion-Israel Institute of Technology, and is currently pursuing her PhD in the Department of Chemistry under Prof. Deborah Fass at the Weizmann Institute of Rehovot, Israel, where she is concentrating on the extracellular matrix. She provided the paintings for the covers of *Figs: The Genus* Ficus and *Caper: The Genus* Capparis. More of her work can be viewed at www.ziporalanskyart.com.

1 Introduction

PEGANUM HARMALA: FROM TOXIC TO HEALING

We begin here with *Peganum harmala*, the "flagship species" of the genus *Peganum*. *P. harmala,* known as harmal throughout the Middle East, where it may also be labeled, especially in small herb shops, *esfand*, is officially known in English as "Syrian rue," or "wild rue" (Figure 1.1). It must be noted at the outset that the seeds of this tough little desert shrub are most notably toxic. Although usually the final outcomes of *P. harmala* poisoning are benign, fatalities occur (Achour et al. 2012a,b; Berdai et al. 2014; Moshiri et al. 2013) and call to mind the power of, and respect due, *Peganum*. There is apparently no antidote to the poisoning, with supportive treatment only being indicated (Sadr Mohammadi et al. 2016) (Figure 1.2).

Wild rue, almost always the seeds, is not only consumed by chewing or by imbibing an aqueous decoction. Seeds may also be smoked, or scattered on red-hot charcoal as incense, and may be used in this or a similar manner as a fumigant (Koyuncu et al. 2008). And the first medicinal uses to come to attention are for external salves (Figure 1.3).

The toxicity of harmal owes to alkaloids, specifically β-carboline, indole alkaloids, and most specifically the harmala alkaloids, named for the plant, *P. harmala*. These compounds, including **harmine** (Patel et al. 2012), **harmalol**, and **harmaline** (Khan et al. 2013; Zhao et al. 2012), bind strongly to nucleic acids (Nafisi et al. 2010a,b), owing to their long bulky side chains, and thus can affect cells at their "cores." Practically, this toxicity is utilizable for its cytotoxic (anticancer) (Chabir et al. 2014; Dastagir and Hussain 2014; El Gendy and El-Kadi 2013; Hamsa and Kuttan 2011; Lamchouri et al. 2013), antiprotozoal (Khoshzaban et al. 2014; Rahimi-Moghaddam et al. 2011; Tanweer et al. 2014), antibacterial (Ali et al. 2011; Irshaid et al. 2014), antiviral (Ma et al. 2013), antifungal (Hashem 2011; Nenaah 2010), and insecticidal effects (Rehman et al. 2009; Zeng et al. 2010), and even for its toxic effects on fellow plant species (Deng et al. 2014; Shao et al. 2013). Humans need to be careful with it, and usually they have been, using it particularly as part of sacramental rites. Indeed, harmal is the most revered plant in the Koran, and credibly the "burning bush" of the Torah (Shanon 2008). This is the toxicity of inebriation, or frank entheogenesis, where use of harmal extracts as bright red dyes for Oriental carpets led to careless licking of many a finger, sometimes with resultant perceived "magic carpet rides" (Figures 1.4 through 1.7 and cover painting).

The human mind has intuited and proven that in the midst of this toxic milieu also lie *entourage* (Ben-Shabat et al. 1998; Russo 2011) proteins (Shi et al. 2013; Soliman et al. 2013), polysaccharides (Golovchenko et al. 2012), and complex alkaloid superstructures (Wang et al. 2014) with immunostimulatory, antitoxic (Soliman and Fahmy 2011), cytoprotective (Singh et al. 2013), antiaging (Mohammadirad et al. 2013), sedating (Moloudizargari et al. 2013), and diabetes-benefiting (Naresh et al.

FIGURE 1.1 *Peganum harmala* blossoms in the Judaean Desert in May. The flowers develop slowly and last long. (5/12/2015, Judaean Desert, by the side of the Inn of the Good Samaritan, Israel: by Helena Paavilainen.)

FIGURE 1.2 Bundle of dried ripe harmal, seeds still inside the capsules. (3/11/2015, Jerusalem, Israel: by Helena Paavilainen.)

FIGURE 1.3 *P. harmala* L. Ripening capsules containing the medicinally important seeds are slowly starting to open. An oasis in the middle of a desert is an ideal place for these plants that can stand the drought but thrive with adequate moisture. (7/29/2008, Tadmur, Homs, Syria: by Rafael Medina via www.flickr.com.)

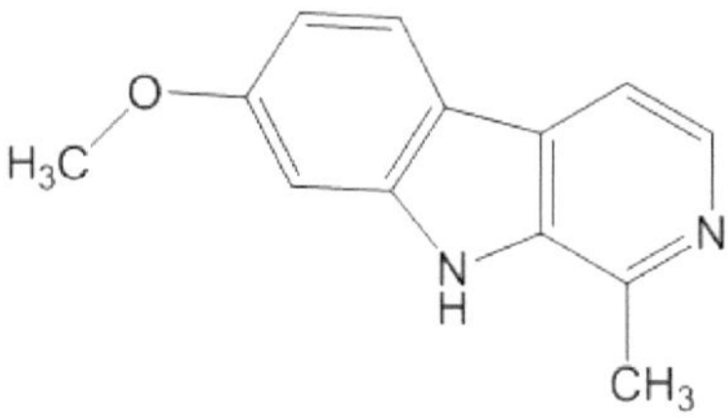

FIGURE 1.4 Harmine.

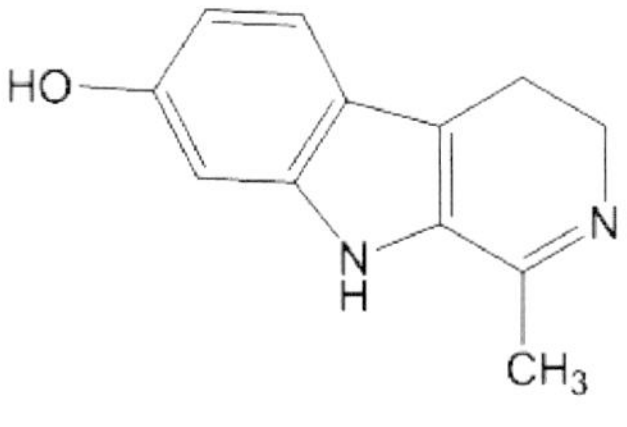

FIGURE 1.5 Harmalol.

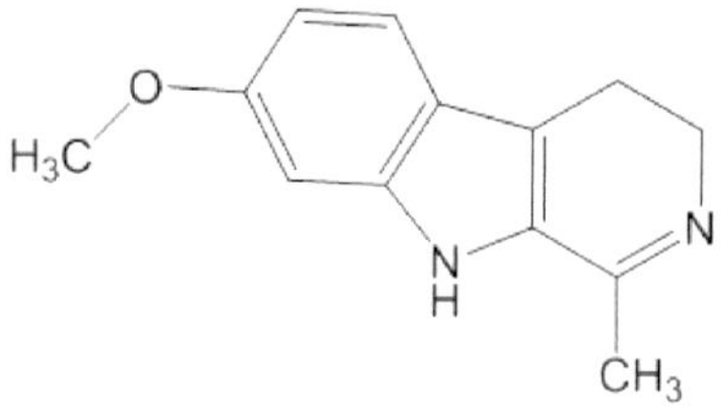

FIGURE 1.6   Harmaline.

FIGURE 1.7   *P. harmala* L. The ripe seeds are still attached to the dry capsule. (July 2011, Flix, Ribera d'Ebre, Catalonia: by Ferran J. Lloret, in F.J. Lloret. 2007. Galanthus: Web de botanica, http://www.galanthus.cat, retrieved 3/16/2015.)

2012; Rahimifard et al. 2014; Singh et al. 2012) properties. Combined with another herb indigenous to Iran, *Dracocephalum kotschyi* (Talari et al. 2014), *P. harmala* is the most important traditional Persian herb for treating cancer (Sobhani et al. 2002). Its particular benefit as a putative neuro-oncologic is increasingly recognized (Liu et al. 2013), and its anti-inflammatory (Bensalem et al. 2014; Khlifi et al. 2013) value against brain deterioration and senility (Ali et al. 2013; Biradar et al. 2013; Grabher et al. 2012) is appreciated.

OVERVIEW OF THE BOOK

This volume follows our earlier works published by CRC Press, namely *Figs: The Genus* Ficus and *Caper: The Genus* Capparis, in connecting traditional uses of *P. harmala* in different times and cultures with modern research, especially in the realms of cancer, infectious disease, diabetology, and brain science. Also, as then,

FIGURE 1.8 *P. harmala* L. growing in Ladakh's high altitude desert area. Genus *Peganum* has adapted well to extreme conditions. Its strong smell and taste protect it from grazing animals even in areas where green plants are scarce. (Ladakh, Jammu-Kashmir, India: © Andrés V. Pérez Latorre 2012, in Valle de Nubra. [Blog.] http://valledenubra.blogspot.co .il/p/floravegetacion.html.)

we link the pharmacological power of the genus to its evolution in the context of species survival and adaptation to environmental stress and niche (Figure 1.8).

The next chapter expounds briefly on the botanical placement of *P. harmala* within the *Peganum* genus as a whole, including other species of note. The chapters to follow will traverse in more depth the thrust of the supporting ethnography and scientific work that has provided a practical scaffold for ongoing clinical and veterinary applications (Bahmani and Eftekhari 2012).

REFERENCES

Achour, S., N. Rhalem, A. Khattabi, H. Lofti, A. Mokhtari, A. Soulaymani, A. Turcant et al. 2012a. *Peganum harmala* L. poisoning in Morocco: About 200 cases. *Therapie* 67(1): 53–8.

Achour, S., H. Saadi, A. Turcant, A. Banani, A. Mokhtari, A. Soulaymani, and R. Soulaymani Bencheikh. 2012b. *Peganum harmala* L. poisoning and pregnancy: Two cases in Morocco. *Med Sante Trop* 22(1): 84–6.

Ali, N.H., S. Faizi, and S.U. Kazmi. 2011. Antibacterial activity in spices and local medicinal plants against clinical isolates of Karachi, Pakistan. *Pharm Biol* 49(8): 833–9.

Ali, S.K., A.R. Hamed, M.M. Soltan, U.M. Hegazy, E.E. Elgorashi, I.A. El-Garf, and A.A. Hussein. 2013. *In vitro* evaluation of selected Egyptian traditional herbal medicines for treatment of Alzheimer disease. *BMC Complement Altern Med* 13: 121.

Bahmani, M., and Z. Eftekhari. 2012. An ethnoveterinary study of medicinal plants in treatment of diseases and syndromes of herd dog in southern regions of Ilam province, Iran. *Comp Clin Path* 22(3): 403–7.

Ben-Shabat, S., E. Fride, T. Sheskin, T. Tamiri, M.H. Rhee, Z. Vogel, T. Bisogno et al. 1998. An entourage effect: Inactive endogenous fatty acid glycerol esters enhance 2-arachidonoyl-glycerol cannabinoid activity. *Eur J Pharmacol* 353(1): 23–31.

Bensalem, S., J. Soubhye, I. Aldib, L. Bournine, A.T. Nguyen, M. Vanhaeverbeek, A. Rousseau et al. 2014. Inhibition of myeloperoxidase activity by the alkaloids of *Peganum harmala* L. (*Zygophyllaceae*). *J Ethnopharmacol* 154(2): 361–9.

Berdai, M.A., S. Labib, and M. Harandou. 2014. *Peganum harmala* L. intoxication in a pregnant woman. *Case Rep Emerg Med* 2014: 783236.

Biradar, S.M., H. Joshi, and K.C. Tarak. 2013. Cerebroprotective effect of isolated harmine alkaloids extracts of seeds of *Peganum harmala* L. on sodium nitrite-induced hypoxia and ethanol-induced neurodegeneration in young mice. *Pak J Biol Sci* 16(23): 1687–97.

Chabir, N., H. Ibrahim, M. Romdhane, A. Valentin, B. Moukarzel, M. Mars, and J. Bouajila. 2014. Seeds of *Peganum harmala* L.: Chemical analysis, antimalarial and antioxidant activities, and cytotoxicity against human breast cancer cells. *Med Chem* 11(1): 94–101.

Dastagir, G., and F. Hussain. 2014. Cytotoxic activity of plants of family *Zygophyllaceae* and *Euphorbiaceae*. *Pak J Pharm Sci* 27(4): 801–5.

Deng, C., H. Shao, X. Pan, S. Wang, and D. Zhang. 2014. Herbicidal effects of harmaline from *Peganum harmala* on photosynthesis of *Chlorella pyrenoidosa*: Probed by chlorophyll fluorescence and thermoluminescence. *Pestic Biochem Physiol* 115: 23–31.

El Gendy, M.A., and A.O. El-Kadi. 2013. Harmine and harmaline downregulate TCDD-induced Cyp1a1 in the livers and lungs of C57BL/6 mice. *Biomed Res Int* 258095.

Golovchenko, V.V., D.S. Khramova, A.S. Shashkov, D. Otgonbayar, A. Chimidsogzol, and Y.S. Ovodov. 2012. Structural characterisation of the polysaccharides from endemic Mongolian desert plants and their effect on the intestinal absorption of ovalbumin. *Carbohydr Res* 356: 265–72.

Grabher, P., E. Durieu, E. Kouloura, M. Halabalaki, L.A. Skaltsounis, L. Meijer, M. Hamburger et al. 2012. Library-based discovery of DYRK1A/CLK1 inhibitors from natural product extracts. *Planta Med* 78(10): 951–6.

Hamsa, T.P., and G. Kuttan. 2011. Harmine activates intrinsic and extrinsic pathways of apoptosis in B16F-10 melanoma. *Chin Med* 6: 11.

Hashem, M. 2011. Antifungal properties of crude extracts of five Egyptian medicinal plants against dermatophytes and emerging fungi. *Mycopathologia* 172(1): 37–46.

Irshaid, F.I., K.A. Tarawneh, J.H. Jacob, and A.M. Alshdefat. 2014. Phenol content, antioxidant capacity and antibacterial activity of methanolic extracts derived from four Jordanian medicinal plants. *Pak J Biol Sci* 17(3): 372–9.

Khan, F.A., A. Maalik, Z. Iqbal, and I. Malik. 2013. Recent pharmacological developments in β-carboline alkaloid "harmaline." *Eur J Pharmacol* 721(1–3): 391–4.

Khlifi, D., R.M. Sghaier, S. Amouri, D. Laouini, M. Hamdi, and J. Bouajila. 2013. Composition and anti-oxidant, anti-cancer and anti-inflammatory activities of *Artemisia herba-alba*, *Ruta chalpensis* L. and *Peganum harmala* L. *Food Chem Toxicol* 55: 202–8.

Khoshzaban, F., F. Ghaffarifar, and H.R. Jamshidi Koohsari. 2014. *Peganum harmala* aqueous and ethanol extracts effects on lesions caused by *Leishmania major* (MRHO/IR/75/ER) in BALB/c mice. *Jundishapur J Microbiol* 7(7): e10992.

Koyuncu, O., D. Ozturk, I. Potoglu Erkera, and A. Kaplan. 2008. Anatomical and palynological studies on economically important *Peganum harmala* L. (*Zygophyllaceae*). *Biol Diver Conserv* 20: 108–15, www.biodicon.com.

Lamchouri, F., M. Zemzami, A. Jossang, A. Abdellatif, Z.H. Israili, and B. Lyoussi. 2013. Cytotoxicity of alkaloids isolated from *Peganum harmala* seeds. *Pak J Pharm Sci* 26(4): 699–706.

Liu, H., D. Han, Y. Liu, X. Hou, J. Wu, H. Li, J. Yang et al. 2013. Harmine hydrochloride inhibits Akt phosphorylation and depletes the pool of cancer stem-like cells of glioblastoma. *J Neurooncol* 112(1): 39–48.

Ma, X., D. Liu, H. Tang, Y. Wang, T. Wu, Y. Li, J. Yang et al. 2013. Purification and characterization of a novel antifungal protein with antiproliferation and anti-HIV-1 reverse transcriptase activities from *Peganum harmala* seeds. *Acta Biochim Biophys Sin (Shanghai)* 45(2): 87–94.

Mohammadirad, A., F. Aghamohammadali-Sarraf, S. Badiei, Z. Faraji, R. Hajiaghaee, M. Baeeri, M. Gholami et al. 2013. Anti-aging effects of some selected Iranian folk medicinal herbs-biochemical evidences. *Iran J Basic Med Sci* 16(11): 1170–80.

Moloudizargari, M., P. Mikaili, S. Aghajanshakeri, M.H. Asghari, and J. Shayegh. 2013. Pharmacological and therapeutic effects of *Peganum harmala* and its main alkaloids. *Pharmacogn Rev* 7(14): 199–212.

Moshiri, M., L. Etemad, S. Javidi, and A. Alizadeh. 2013. *Peganum harmala* intoxication, a case report. *Avicenna J Phytomed* 3(3): 288–92.

Nafisi, S., Z.M. Malekabady, and M.A. Khalilzadeh. 2010a. Interaction of β-carboline alkaloids with RNA. *DNA Cell Biol* 29(12): 753–61.

Nafisi, S., M. Bonsaii, P. Maali, M.A. Khalilzadeh, and F. Manouchehri. 2010b. Beta-carboline alkaloids bind DNA. *J Photochem Photobiol B* 100: 84–91.

Naresh, G., N. Jaiswal, P. Sukanya, A.K. Srivastava, A.K. Tamrakar, and T. Narender. 2012. Glucose uptake stimulatory effect of 4-hydroxypipecolic acid by increased GLUT 4 translocation in skeletal muscle cells. *Bioorg Med Chem Lett* 22(17): 5648–51.

Nenaah, G. 2010. Antibacterial and antifungal activities of (beta)-carboline alkaloids of *Peganum harmala* (L) seeds and their combination effects. *Fitoterapia* 81(7): 779–82.

Patel, K., M. Gadewar, R. Tripathi, S.K. Prasad, and D.K. Patel. 2012. A review on medicinal importance, pharmacological activity and bioanalytical aspects of beta-carboline alkaloid "Harmine." *Asian Pac J Trop Biomed* 2(8): 660–4.

Rahimifard, M., M. Navaei-Nigjeh, N. Mahroui, S. Mirzaei, Z. Siahpoosh, A. Nili-Ahmadabadi, A. Mohammadirad et al. 2014. Improvement in the function of isolated rat pancreatic islets through reduction of oxidative stress using traditional Iranian medicine. *Cell J* 16(2): 147–163.

Rahimi-Moghaddam, P., S.A. Ebrahimi, H. Ourmazdi, M. Selseleh, M. Karjalian, G. Haj-Hassani, M.H. Alimohammadian et al. 2011. *In vitro* and in vivo activities of *Peganum harmala* extract against *Leishmania major*. *J Res Med Sci* 16(8): 1032–9.

Rehman, J.U., X.G. Wang, M.W. Johnson, K.M. Daane, G. Jilani, M.A. Khan, and F.G. Zalom. 2009. Effects of *Peganum harmala* (*Zygophyllaceae*) seed extract on the olive fruit fly (*Diptera*: *Tephritidae*) and its larval parasitoid *Psyttalia concolor* (*Hymenoptera*: *Braconidae*). *J Econ Entomol* 102(6): 2233–40.

Russo, E.B. 2011. Taming THC: Potential cannabis synergy and phytocannabinoid–terpenoid entourage effects. *Br J Pharmacol* 163(7): 1344–64.

Sadr Mohammadi, R., R. Bidaki, F. Mirdrikvand, S.N.M. Yazdi, and P.Y. Anari. 2016. *Peganum harmala* (*Aspand*) intoxication; A case report. *Emerg (Tehran)* 4(2): 106–7.

Shanon, B. 2008. Biblical entheogens: A speculative hypothesis. *Time and Mind: The Journal of Archaeology, Consciousness and Culture* 1(1): 51–74.

Shao, H., X. Huang, Y. Zhang, and C. Zhang. 2013. Main alkaloids of *Peganum harmala* L. and their different effects on dicot and monocot crops. *Molecules* 18(3): 2623–34.

Shi, Z., Z.J. Wang, H.L. Xu, Y. Tian, X. Li, J.K. Bao, S.R. Sun et al. 2013. Modeling, docking and dynamics simulations of a non-specific lipid transfer protein from *Peganum harmala* L. *Comput Biol Chem* 47: 56–65.

Singh, A.B., T. Khaliq, J.P. Chaturvedi, T. Narender, and A.K. Srivastava. 2012. Anti-diabetic and anti-oxidative effects of 4-hydroxypipecolic acid in C57BL/KsJ-db/db mice. *Hum Exp Toxicol* 31(1): 57–65.

Singh, V.K., V. Mishra, S. Tiwari, T. Khaliq, M.K. Barthwal, H.P. Pandey, G. Palit et al. 2013. Anti-secretory and cyto-protective effects of peganine hydrochloride isolated from the seeds of *Peganum harmala* on gastric ulcers. *Phytomedicine* 20(13): 1180–5.

Sobhani, A.M., S.A. Ebrahimi, and M. Mahmoudian. 2002. An *in vitro* evaluation of human DNA topoisomerase I inhibition by *Peganum harmala* L. seeds extract and its beta-carboline alkaloids. *J Pharm Pharm Sci* 5(1): 19–23.

Soliman, A.M., H.S Abu-El-Zahab, and G.A. Alswiai. 2013. Efficacy evaluation of the protein isolated from *Peganum harmala* seeds as an antioxidant in liver of rats. *Asian Pac J Trop Med* 6(4): 285–95.

Soliman, A.M., and S.R. Fahmy. 2011. Protective and curative effects of the 15 KD isolated protein from the *Peganum harmala* L. seeds against carbon tetrachloride induced oxidative stress in brain, tests and erythrocytes of rats. *Eur Rev Med Pharmacol Sci* 15(8): 888–99.

Talari, M., E. Seydi, A. Salimi, Z. Mohsenifar, M. Kamalinejad, and J. Pourahmad. 2014. *Dracocephalum*: Novel anticancer plant acting on liver cancer cell mitochondria. *Biomed Res Int* 2014: 892170.

Tanweer, A.J., N. Chand, U. Sanddigue, C.A. Bailey, and R.U. Khan. 2014. Antiparasitic effect of wild rue (*Peganum harmala* L.) against experimentally induced coccidiosis in broiler chicks. *Parasitol Res* 113(8): 2951–60.

Wang, K.B., Y.T. Di, Y. Bao, C.M. Yuan, G. Chen, D.H. Li, J. Bai et al. 2014. Peganumine A, a β-carboline dimer with a new octacyclic scaffold from *Peganum harmala*. *Org Lett* 16(15): 4028–31.

Zeng, Y., Y. Zhang, Q. Weng, M. Hu, and G. Zhong. 2010. Cytotoxic and insecticidal activities of derivatives of harmine, a natural insecticidal component isolated from *Peganum harmala*. *Molecules* 15(11): 7775–91.

Zhao, T., S.S. Zheng, B.F. Zhang, Y.Y. Li, S.W. Bligh, C.H. Wang, and Z.T. Wang. 2012. Metabolic pathways of the psychotropic-carboline alkaloids, harmaline and harmine, by liquid chromatography/mass spectrometry and NMR spectroscopy. *Food Chem* 134(2): 1096–105.

2 The Genus *Peganum*

OVERVIEW OF GENUS *PEGANUM*

Peganum (L.) is a smallish genus consisting of four to seven species, mostly in the arid and semiarid regions of North Africa and Eurasia, from the Mediterranean to Mongolia, and in North America, from Texas to Mexico (Figure 2.1) (Ronse De Craene et al. 1996). As an introduced species, it also can be found in Australia and South Africa. It belongs to the family of Nitrariaceae (Lindl.), a family of three to four genera and some 16 species in the arid regions mostly from North Africa to East Asia, but also in North America and Australia (Encyclopedia of Life n.d.; Liu and Zhou 2008; Porter 2013; Sheahan and Chase 1996; Stevens 2001–; The Plant List 2013; Zohary and Feinbrun-Dothan 1966–2004). Formerly, the genus was included in the family Zygophyllaceae (Cronquist 1981; Zohary 1989). Molecular evidence and its morphological differences from the Zygophyllaceae, however, support its recognition as a separate family (Table 2.1) (Kubitzki 2011; Liu and Zhou 2008; Muellner et al. 2007; Savolainen et al. 2000; Sheahan and Chase 1996).

Concisely, the genus *Peganum* could be described as follows (Danin 2003–; Ghafoor 1985; Komarov et al. 1934–1964; Liu and Zhou 2008; Porter 2013; Stevens 2001–; Täckholm 1974; USDA, NRCS 2015; Vail and Rydberg 1907; Zohary and Feinbrun-Dothan 1966–2004):

- Perennial herbs or shrubs, under 1 m tall, prostrate, spreading, or erect, with erect, corymbosely branched, glabrous or puberulous stems rising from a woody rhizome (Figure 2.2).
- Dense, glabrous or hispid leaves and branches (Figure 2.3).
- Leaves simple, alternate, sessile, nonsheathing, entire or divided into irregular linear lobes, which are acute to abruptly soft pointed. Stipules, when present, are intrapetiolar, minute, distinct, setaceous, and caducous (Figure 2.4).
- Peduncles 1-flowered, opposite the leaves.
- Flowers hermaphrodite, solitary, axillary or terminal, or forming terminal raceme-like dichasia.
- Sepals 3 to 5, foliaceous, linear or lobes linear, acute or obtuse, entire to irregularly, deeply pinnately lobed, persistent.
- Petals 4 or 5, white to yellow, oblong or oblanceolate, spreading, subequal, entire, imbricate.
- Stamens 10–15, distinct, three-whorled, in groups of three opposite the sepals or paired opposite the petals; filaments linear, dilated at base, inserted in two rows at the base of the annular or cup-shaped disc; anthers linear.

FIGURE 2.1 *Peganum harmala* L. bush blooming in April in typical arid surroundings. (4/21/2006, Range Rd 7, WSMR: by Patrick Alexander, in SEINet. Southwest Environmental Information Network, SEINet—Arizona Chapter. 2015. http://swbiodiversity.org/seinet /index.php, retrieved 3/16/2015.)

TABLE 2.1
Classification of Genus *Peganum* L.

- Plantae
  - Viridiplantae
    - Streptophyta
      - Embryophyta
        - Tracheophyta
          - Spermatophytina
            - Magnoliopsida
              - Rosanae
                - Sapindales
                  - Nitrariaceae
                    - Peganum L.

Source: Retrieved from the Integrated Taxonomic Information System (ITIS) (http:// www.itis.gov).

- Ovary two- to three-lobed, spheric, two- to four-celled, many-ovuled, subsessile; one style rising from the center of the lobes, triquetrous, somewhat twisted, two- to three-carinate above the middle. Keels of style stigmatose. Styles apically three-angulate (Figure 2.5).
- Fruit is a capsule or berry, globose, leathery, fleshy or not fleshy, usually three-locular, loculidal dehiscent, or indehiscent and berry-like. The

FIGURE 2.2 Beginning of the growing season when new stems grow from the woody rhizome. (3/11/2015, Mount Scopus Botanical Garden, Jerusalem, Israel: by Helena Paavilainen.)

capsule is slightly compressed at both ends, opening by three valves at the apex to release numerous dark triangular seeds.

- The seeds are 10–100 per fruit, angular; testa pitted; endosperm fleshy; embryo curved (Figure 2.6).

The differences between the species of the genus *Peganum* are insignificant, except for size, habit, and fruit morphology (Gray and Wright 1852–1853; Ronse De Craene et al. 1996). Their chemical profiles seem to correlate with each other, though only *P. harmala* has been adequately studied (Cheng et al. 2010; Hegnauer 1973; Umadevi et al. 1990). In the following, one can note some of the similarities and differences

FIGURE 2.3 Morphological drawing of *Peganum harmala* L. (From Engler, A. and Prantl, K., 3, Fig. 58, 1897. [© 2006 by L.H. Bailey Hortorium, ref. DOL1951] via PlantSystematics .org http://www.plantsystematics.org.)

in the external appearance of the species (Figures 2.7a–d through 2.9) and some examples of their typical habitats (Figure 2.10a–h).

As seen, the genus *Peganum* has adapted itself efficiently to the difficult natural surroundings in which it has found itself. At the same time, different cultures have found a nearly endless variety of uses for this intriguing and beautiful plant. In the continuation of this chapter, we will see some of them through a short overview of the ethnopharmacological uses of *Peganum harmala*, the genus that has received the most attention from researchers. We will also show, in a tabular form, the other spe- cies of *Peganum* species highlighted during the writing of this book. This species- by-species description will encompass the sadly limited data on the ethnography, distribution, and local uses, both common and medical, of each of them (Tables 2.2 through 2.5 and Figures 2.11 through 2.15). As a neutral and random means of arranging the species, alphabetical order is used. Description of the geographical

FIGURE 2.4   *Peganum harmala* L., leaf structure. (By Thiaudière, J.-C., in Peltier, J.P., *Plant Biodiversity of South-Western Morocco*, 2006. http://www.teline.fr, retrieved 3/16/2015.)

distribution of the species is based on *World Geographical Scheme for Recording Plant Distributions*, 2nd ed. (Brummitt et al. 2001).

TECHNICAL NOTES ON *PEGANUM* SPECIES OTHER THAN *P. HARMALA*

PEGANUM MULTISECTUM

Liu (2011) isolated six alkaloids from *P. multisectum* seeds: deoxyvasicinone, l-vasicinone, harmine, vasicine, evodiamine, and fagomine. It was the first time the latter two compounds were identified in *P. multisectum* and possibly ever in any *Peganum* species (Figures 2.13 through 2.16).

Evodiamine is known from its occurrence in the evodia fruit, *Tetradium ruticarpum* (= *Evodia rutaecarpa*), a warming bitter used in traditional Chinese medicine (TCM), (*wuzhuyu*). Evodiamine is used as an anti-obesity drug with anti-allergenic, analgesic, antitumor, anti-ulcerogenic, and neuroprotective properties (Tan and Zhang 2016). Evodiamine has been shown to be a potent promoter of apoptosis in human cancer cells of the kidney (Wu et al. 2016), lung (Lin et al. 2016), and ovary (Wei et al. 2016).

(a)

(b)

FIGURE 2.5 *Peganum harmala* L. has white or yellowish white flowers, 2–2.5 cm in diameter. (a) South-Western Morocco. (By Fouad Msanda, in Peltier, J.P., *Plant Biodiversity of South-Western Morocco*, 2006. http://www.teline.fr, retrieved 16 March 2015.) (b) 4/27/2014, Parque Natural Cabo de Gata-Níjar, Almeria, Spain. (From Colomer, E.M., *Flores silvestres del Mediterraneo*, 2012. http://floressilvestresdelmediterraneo.blogspot.co.il, retrieved 3/16/ 2015.)

(a)

(b)

FIGURE 2.6 From a ripening capsule to seed. (a) *Peganum harmala* L. capsules ripening to orange-brown color. (South-Western Morocco. By Fouad Msanda, in Peltier, J.P., *Plant Biodiversity of South-Western Morocco*, 2006. http://www.teline.fr, retrieved 3/16/2015.) (b) *P. harmala* L., dehiscing capsule. (7/12/2009, Hwy 180 at Ridge Road [Grant County Airport Road], New Mexico, USA. From Kleinman, R., n.d. http://wnmu.edu/academic /nspages/gilaflora/, retrieved 3/16/2015.)   *(Continued)*

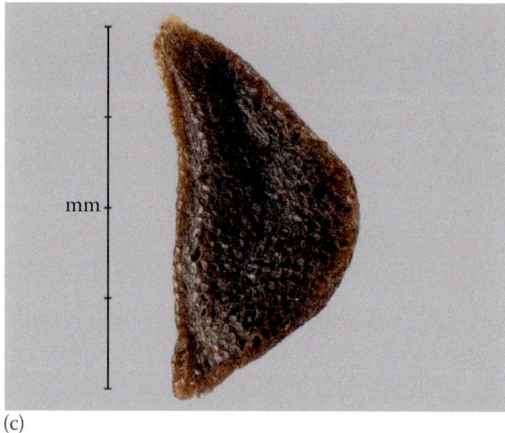

(c)

(d)

FIGURE 2.6 (CONTINUED) From a ripening capsule to seed. (c) *P. harmala* L., dark brown triangular seed with spongy coat. (From CFIA, n.d. http://www.inspection.gc.ca, retrieved 3/16/2015.) (d) *P. nigellastrum* Bunge, seeds still attached to capsule. (8/25/2007, Khovd, Mongolia. By Schnittler, M., in FloraGREIF, 2010–. http://greif.uni-greifswald.de/floragreif/, 2010–. Retrieved 3/16/2015.)

Fagomine is named for the genus from which it was first identified and isolated, namely *Fagopyrum esculentum*, i.e., buckwheat. Fagomine is an iminosugar—a sugar in which the endocyclic oxygen is replaced by a nitrogen. When given in Barcelona with either sucrose or starch to rats, "D-fagomine lowered blood glucose in a dose-dependent manner without stimulating insulin secretion" and modulated bacterial adhesion in gut

FIGURE 2.7 Examples of the variability in the appearance of the leaves and stems within genus *Peganum*. (a) *Peganum mexicanum*. Herb from a central root with prostrate branches. (October 2007, Hudspeth County, Texas, USA. From Spjut, R.W., *The World Botanical Associates*, 2004. http://www.worldbotanical.com, retrieved 3/16/2015.) (b) *Peganum harmala*, young leaves. (3/11/2015, Mount Scopus Botanical Garden, Jerusalem, Israel: by Helena Paavilainen.) (*Continued*)

(c)

(d)

FIGURE 2.7 (CONTINUED)   Examples of the variability in the appearance of the leaves and stems within genus *Peganum*. (c) *Peganum harmala* L. (October 2007, Hudspeth County, Texas, USA. From Spjut, R.W., *The World Botanical Associates*, 2004. http://www.world botanical.com, retrieved 7/18/2017.) (d) *P. nigellastrum* Bunge growing in the Gobi Desert. (6/10/2014, Bayanzag, Bulgan, Mongolia: by Timmermann U., in FloraGREIF 2010–. http://greif.unigreifswald.de/floragreif/, retrieved 7/18/2017.)

(a)

(b)

FIGURE 2.8 Extreme case of influence of surroundings on *Peganum harmala*: a *Peganum* plant growing outside in the cool, rainy climate of Southern Finland. The plant was grown from seed. Notice the abundance of the leaves. At least partly due to the relative shortness of Finnish summer, the plant did not blossom. (8/26/2016, Loimaa, Finland: by Helena Paavilainen.)

FIGURE 2.9   Flowering *Peganum harmala* L. in the desert. (South-Western Morocco. By Thiaudière, J.-C., in Peltier, J.P., *Plant Biodiversity of South-Western Morocco*, 2006. http://www.teline.fr, retrieved 3/16/2015.)

(Gómez et al. 2012). The beneficial effects were replicated in rats fed with high fat *and* high sucrose diets (Molinar-Toribio et al. 2015). Furthermore, the amount of fagomine in a human diet containing buckwheat products is sufficient to account for many of buckwheat's health benefits such as helping to prevent or attenuate metabolic syndrome (Amézqueta et al. 2013). Fagomine inhibits glycogen phosphorylase, the enzyme catalyzing breakdown of liver glycogen to glucose units (Jakobsen et al. 2001), and potentiates glucose-induced insulin secretion (Taniguchi et al. 1998). Fagomine is antihyperglycemic in streptozotocin induced diabetic mice (Nojima et al. 1998), and both fagomine and its derivatives are cytotoxic to cancer cell lines, especially when the length of their added aliphatic side chains is increased (Padró et al. 2010). Other important natural sources of fagomine are mulberry leaves (Asano et al. 1994), the droppings of the silkworm (Zhou et al. 2007) that consumes mulberry leaves, and the silkworms themselves (Nakagawa et al. 2010). The traditional therapeutic benefit of mulberry leaves, Chinese *sang ye*, for diabetes is related to the presence of fagomine (Figure 2.17) (Hunyadi et al. 2013).

Cheng et al. (2010) used chromatographic methods for creating HPLC fingerprints to distinguish seeds from three *Peganum* species, *P. harmala* Linn., *P. multisectum* (Maxim) Bobr., and *P. nigellastrum* Bunge, though the seeds from the three species are not distinguishable by gross physical assessment. *P. multisectum* is a host to the parasitic medicinal plant *Cynomorium songaricum* (Wang et al. 2011), and on its leaves, to a recently discovered new powdery species of fungus, *Erysiphe alashanensis* (Figure 2.18) (Liu and Shang 2008).

(a)

(b)

FIGURE 2.10 Examples of the habitats to which species of the genus *Peganum* have adapted. (a) *P. harmala* L. on Zagreb Mountains, Central Western Iran. (April–May 2008: by Marijn van den Brink. photos.v-d-brink.eu/keyword/zygophyllaceae/i-PZ882Nh.) (b) *P. harmala* L. on the Skardu-Askole area of Baltistan. (Summer 2008, Karakoram, Pakistan: by Marijn van den Brink photos.v-d-brink.eu/keyword/zygophyllaceae/i-sG8gDkr.) (*Continued*)

(c)

(d)

FIGURE 2.10 (CONTINUED)   Examples of the habitats to which species of the genus *Peganum* have adapted. (c) *P. harmala* L. on Elburz Mountains. (May 2010, Sendan Mountains west of Sefid Rud Reservoir, Qazvin Province, Iran: by Marijn van den Brink photos.v-d-brink .eu/keyword/zygophyllaceae/i-jdPqtFc.) (d) *P. nigellastrum* Bunge on the Flaming Cliffs of the Gobi Desert. (6/10/2014, Bayanzag, Bulgan, Mongolia. By Timmermann, U., in *FloraGREIF*, 2010–. http://greif.uni-greifswald.de/floragreif/, retrieved 3/16/2015.)   (*Continued*)

(e)

(f)

FIGURE 2.10 (CONTINUED)   Examples of the habitats to which species of the genus *Peganum* have adapted. (e) *P. harmala* L. on semiarid areas of Central and South Alicante. (From Torregrosa, S.G., *Apatita: Flora de Alicante*, n.d. http://www.apatita.com/herbario /floradealicante.html, retrieved 3/16/2015.) (f) *P. harmala* L. in the vicinity of River Ebro. (July 2011, Flix, Ribera d'Ebre, Catalonia. From Lloret, F.J., *Galanthus: Web de botanica*, 2007. http://www.galanthus.cat, retrieved 3/16/2015.)   (*Continued*)

(g)

(h)

FIGURE 2.10 (CONTINUED)   Examples of the habitats to which species of the genus *Peganum* have adapted. (g) *P. harmala* L. in the plains in Oregon. (9/10/2008, Harney County, Oregon, USA. From Oregon Department of Agriculture, http://www.streamwebs.org /invasive-species-list/african-rue-peganum-harmala retrieved 3/16/2015.) (h) *P. harmala* L. in the arid semidesert of Morocco. (South-Western Morocco. By Philippe Geniez, in Peltier, J.P., *Plant Biodiversity of South-Western Morocco*, 2006. http://www.teline.fr, retrieved 3/16/2015.)

TABLE 2.2
Ethnography, Distribution, and Local Uses of *Peganum harmala*

Peganum harmala		References
Synonyms	*Peganum harmala* L.:	African Plant Database n.d.; Ghafoor
	Harmala multifida All.	1985; Hassler 2014; Tela Botanica
	H. peganum Crantz	1999; The Plant List 2013; Tropicos
	H. syriaca Bubani	2013
	Peganon harmalum (L.) St.-Lag.	
	Peganum crithmifolium Retz.	
	P. dauricum L.	
	P. dauricum Pall.	
	P. harmala var. *garamantum* Maire	
	P. harmala var. *rotschildianum*	
	(Buxb.) Maire	
	P. rothschildianum F.Buxbaum	
	P. rotschildianum Buxb. (1927)	
	Maire	
	Infraspecific taxons:	
	Peganum harmala var. *grandiflorum*	
	Hadidi	
	Peganum harmala var. *stenophyllum*	
	Boiss.	
Common names	Arabic: 'alqat al-dib, ḥarjal, ḥarmal, ḥre-milan, ḥuraymilān, legherma, mogannanna, saḍab-šami, saḍab-bari	Aldén et al. 2009; Bailey and Danin 1981; Bedevian 1936; Burkill 1985; Chopra et al. 1965; Clauson 1972;
	Armenian: khowniv, sbant, vahri-pegenay	Danin 2003–; Ducros 1930; efloraofindia 2007; Flattery
	Azerbaijani: üzərlik	and Schwartz 1989; Foucauld 1951;
	Bengali: ispand	Granot 1994; GRIN 1994; Hanelt et al.
	Berber: bender-tifré	2001; Hooper and Field 1937; ITIS
	Catalan: harmala, harmalà, ruda borda	n.d.; Kaganoff 1977; Liu and Zhou
	Chinese: luo tuo peng	2008; Lokotsch 1927; Makowiecki
	Dutch: harmala	1936; Marzell 1943–1979; Masclans
	English: African rue, foreign henna, harmal, harmal peganum, harmal shrub, harmel, harmel peganum, isband, ozallaik, peganum, rue, ruin weed, Syrian rue, wild rue	and Girvès 1954; Mayer 1909; MEDUSA n.d.; Palevitch and Yaniv 1991; Parsa 1960; Pedrotti and Bertoldi 1930; Rehm 1994; Safina et al. 1970; Sorūšīān 1956; SRTD, AMS n.d.;
	Finnish: pilviharmikki, syyrianängelmä	Steingass 1892; Takhtadzhian
	French: harmal, rue sauvage	1954–1966; Tela Botanica 1999;
	German: Harmalkraute, Harmal-Raute, Harmel, Harmelkraute, Harmelraute, Harmelstaude, Steppenraute, wilde Raute	Townsend and Guest 1966–1980; USDA Forest Service 2014; USDA, NRCS 2015; Watson 1868; Watt et al. 1889–1896; Wild Flowers
	Gujarati: harmal, hurmaro, ispun	of Israel. n.d.; ZipcodeZoo 2004

(Continued)

TABLE 2.2 (CONTINUED)
Ethnography, Distribution, and Local Uses of *Peganum harmala*

Peganum harmala		References
	Hebrew: shabar lavan	
	Hindi: harmal, harmul, isband lahouri, isbund-lahouri, kaladana, lahouri harmal, lahouri-hurmul	
	Hungarian: szíriai rutafü	
	Italian: armora, pégano, ruta selvatica	
	Kazakh: adiraspan, adraspan	
	Kermani: espend, sepen, svon	
	Kurdish: aspand	
	Morocco: madjouna	
	Pashto: spandah, spilanai	
	Persian: asbatān, fāšarsīn, isfanj, ispand, sifand, sipand	
	Polish: poganek rutowaty	
	Portuguese: harmale	
	Punjabi: hurmul	
	Russian: garmala obiknovennaya, mogil'nik	
	Sindi: hurmur, spelani	
	Spanish: alfarma, alharma, amargaza, armalá, gamarza, harma, hármaga	
	Svenska: harmelbuske	
	Tamil: simaiyaravandi, simayaiyalavinai	
	Tamachek: harmel	
	Telugu: simagoranta	
	Turkish: ilezik, uzarih, üzerlik	
	Ukrainan: rebryk, ruta-dyka, smerdioxa, sobakarne	
	Uzbek: isirik	
	Yiddish: bibek	
Habitat and distribution	Dry places, such as roadsides and abandoned fields, slightly saline sands near oases, dry grasslands and steppes in desert and semidesert regions. Grows best on sites that receive some runoff water. Altitude 0–3600 m.	Bézanger-Beauquesne et al. 1980; Burkill 1985; Chopra et al. 1960; Danin 2003–; Ghafoor 1985; Granot 1994; GRIN 1994; Hammiche and Merad 1997; Hanelt et al. 2001; Hassler 2014; IPNI 2012; ITIS n.d.; Liu and Zhou 2008; MEDUSA n.d.; Ozenda 1977; Paris and Dilleman 1960; Porter 2013; Quezel and Santa 1963; USDA Forest Service 2014; USDA, NRCS 2015; Wild Flowers of Israel n.d.; ZipcodeZoo 2004; Zohary and Feinbrun-Dothan 1966–2004
	Wild distribution: Southeastern Europe to Mediterranean region, N Africa till Mali, Southwest, Middle, and Central Asia till Siberia, India. Introduced into North America, Australia, and South Africa.	

(Continued)

TABLE 2.2 (CONTINUED)
Ethnography, Distribution, and Local Uses of *Peganum harmala*

Peganum harmala	References
Northern Africa: Algeria; Egypt; Libya; Morocco; Tunisia	
West Tropical Africa: Mali, Mauritania	
Southern Africa: South Africa (introduced)	
Arabian Peninsula: Kuwait; Saudi Arabia	
Western Asia: Afghanistan; Cyprus; Egypt; Iran; Iraq; Israel; Jordan; Lebanon; Syria; Turkey	
Caucasus: Armenia; Azerbaijan; Georgia; Russian Federation (Ciscaucasia, Dagestan)	
Middle Asia: Kazakhstan; Kyrgyzstan; Tajikistan; Turkmenistan; Uzbekistan	
Mongolia: Mongolia	
China: Gansu, W Hebei, W Nei Mongol, Ningxia, Qinghai, N Shanxi, Xinjiang, Xizang, Zizhiqu, Hui, Uygur; Tibet	
Indian Subcontinent: India, Pakistan	
East Europe: Moldova; Russian Federation, European part; Ukraine	
Southeastern Europe: Bulgaria; Greece; Italy; Romania; Serbia	
Southwestern Europe: Spain; Andorra; France (introduced)	
Middle Europe: Germany (introduced); Hungary (introduced)	
Canada: Canada (introduced)	
Southwestern USA: Arizona (introduced), California (introduced), Nevada (introduced)	
Northwestern USA: Colorado (introduced), Montana (introduced), Oregon (introduced), Washington (introduced)	
South-Central USA: New Mexico (introduced), Texas (introduced)	
Mexico: Mexico (introduced)	
Australia: New South Wales (introduced)	

(Continued)

TABLE 2.2 (CONTINUED)
Ethnography, Distribution, and Local Uses of *Peganum harmala*

Peganum harmala		References
Economic importance	Leaf: medicinal; social: religious uses, smoking materials, incense, narcotics; insectifuge	Abdel Aziz et al. 2010; Abrol and Chopra 1962; Abu-Irmaileh and Afifi 2003; Ahmad 1957; Akhtar et al. 2000; Alami et al. 1976; Al-Awdat and Laham 1994; Al-Khalil 1995; Alkofahi et al. 1996; Al-Qura'n 2008; Al-Rawi and Chakravarty 1964; Al-Yahya 1986; Ayensu 1978; Bailey and Danin 1981; Baillon 1871–1888; Bellakhdar 1997; Bellakhdar et al. 1991; Bhattacharyya 1991; Blatter et al. 1919; Boukef et al. 1982; Boulos 1983; Bown 1995; Bremner et al. 2009; Brobst et al. 2009; Burkill 1909, 1985; Chopra 1933; Chopra et al. 1949; Çitoğlu 1997; Dragendorff 1898; Duke 1992–2016, 2002; Emboden 1979; Ertug 2000; Facciola 1990; Farouk et al. 2008; Flattery and Schwartz 1989; Font Quer 1979; Friedman et al. 1986; Frison et al. 2008; Ghafoor 1985; Goodman and Ghafoor 1992; Gorsi and Miraj 2002; Granot 1994; Hammiche and Merad 1997; Hamsa and Kuttan 2011; Hanelt et al. 2001; Hartwell 1967–1971; Hassan 1967; Herraiz et al. 2010; Honda et al. 1996; Hooper and Field 1937; Hussain and Tobji 1997; Ivanovska et al. 1997; Kakrani and Saluja 1993; Kamboj 1988; Karaki et al. 1986; Kunkel 1984; Lamchouri et al. 1999; Leporatti and Ghedira 2009; Leporatti and Lattanzi 1994; Lewis and Elvin-Lewis 1977; Liu and Zhou 2008; Malhi and Trivedi 1972; McLean and Ivimey-Cook 1956; MEDUSA n.d.; Merzouki et al. 2000; Mirzaei et al. 2007; Mohagheghzadeh et al. 2006; Nafisi et al. 2010b; Navchoo and Buth 1990; Nayar 1954, 1955; Pelt 1971; Pennacchio et al. 2010; PFAF 1996; Phillips and Rix 1991; Pieroni et al 2005; Prashanth and John 1999; Rashan et al. 1989; Razzack 1980
	Branches: medicinal	
	Stem: medicinal; products: dyes, stains, inks, tattoos, and mordants	
	Root: medicinal; insecticide; phytochemistry: alkaloids; products: dyes, stains, inks, tattoos, mordants	
	Flower: medicinal	
	Fruit: medicinal; air purifier; for making oil; magic	
	Seed: trade item; food: edible oil, condiment; medicinal; veterinary; phytochemistry: alkaloids; insecticide; disinfectant; cosmetics; social: religious uses, smoking materials, incense, narcotics, hallucinogen, intoxicant, magic; possible candidate for the Zoroastrian drug "haoma"; products: stains, inks, tattoos, mordants, dyes: red color for dyeing tarboosh hats and carpets	
	Whole plant: medicinal; mosquito repellant; social: religious uses, smoking materials, narcotics, magic; agricultural: fodder for donkeys	
	Part not mentioned: medicinal; disinfectant; insecticide, insect repellant; drink: alcoholic, stimulant; dye; tannin/red dyestuff; fodder; weed (poisonous or repellent)	

(Continued)

TABLE 2.2 (CONTINUED)
Ethnography, Distribution, and Local Uses of *Peganum harmala*

Peganum harmala		References
		Ross et al. 1980; Saadabi 2006; Saha et al. 1961; Said 1984; Schipper and Volk 1960; Sezik et al. 1997, 2004; Shah 1982; Shapira et al. 1989; Singh 1995; Singh and Kachroo 1976; Singh and Pandey 1998; Singh et al. 1996a,b; Sokmen et al. 1999; Steinmetz 1957; Tabata et al. 1994; Tahraoui et al. 2007; Uphof 1959, 1968; USDA Forest Service 2014; USDA, NRCS 2015; Usher 1974; van Wyk and Wink 2004; Wild Flowers of Israel n.d.; Winkler 1912
Local medicinal uses	Leaf: narcotics; antiseptic; wounds; abscesses; throat inflammation; rheumatism; facilitating birth Branches: counterirritant Stem: facilitating birth Root: alkaloids; pediculoside; gingivitis; toothache; hemorrhoids Flower: abscesses; throat inflammation Fruit: common cold; epilepsy; mental and nervous illnesses; digestive; stomach complaints; pregnancy/birth Seed: alkaloids; panacea; alterative; disinfectant; antiseptic; bactericide; antimicrobial; anti-inflammatory; analgesic; antispasmodic; antihypertensive, hypotonic; stimulant; psychoactive; narcotic; hallucinogen; CNS stimulant; hypnotic; sedative; soporific; sudorific; relaxant; purifying; blood purifier; cicatrizing; wounds; neoplasms; subcutaneous tumors; ulcers; rheumatism; sciatica; back pain; muscle pain; joint pain; arthritis; for baldness; strengthens hair; skin diseases; common cold; rhinitis; laryngitis; bronchitis; asthma; bronchodilator, expectorant; toothache; suppresses saliva during sleep; eye diseases; otitis; epilepsy; hysteria; neuralgia; parkinsonism;	Abdel Aziz et al. 2010; Abrol and Chopra 1962; Abu-Irmaileh and Afifi 2003; Ahmad 1957; Akhtar et al. 2000; Alami et al. 1976; Al-Awdat and Laham 1994; Al-Khalil 1995; Alkofahi et al. 1996; Al-Qura'n 2008; Al-Rawi and Chakravarty 1964; Al-Yahya 1986; Ayensu 1978; Bailey and Danin 1981; Baillon 1871–1888; Bellakhdar 1997; Bellakhdar et al. 1991; Blatter et al. 1919; Boukef et al. 1982; Boulos 1983; Bown 1995; Bremner et al. 2009; Brobst et al. 2009; Burkill 1909, 1985; Chopra 1933; Chopra et al. 1949; Çitoğlu 1997; Dragendorff 1898; Duke 1992–2016, 2002; Emboden 1979; Ertug 2000; Facciola 1990; Farouk et al. 2008; Flattery and Schwartz 1989; Font Quer 1979; Friedman et al. 1986; Frison et al. 2008; Ghafoor 1985; Goodman and Ghafoor 1992; Granot 1994; Hammiche and Merad 1997; Hamsa and Kuttan 2011; Hanelt et al. 2001; Hartwell 1967–1971; Hassan 1967; Herraiz et al. 2010; Honda et al. 1996; Hooper and Field 1937; Hussain and Tobji 1997; Ivanovska et al. 1997; Kakrani and Saluja 1993; Kamboj 1988; Karaki et al. 1986; MEDUSA n.d.; Nayar 1955; Pelt 1971; Pennacchio et al. 2010; PFAF 1996; Phillips and Rix 1991;

(Continued)

TABLE 2.2 (CONTINUED)
Ethnography, Distribution, and Local Uses of *Peganum harmala*

Peganum harmala	References
paralysis; asthenia; nervous illness; mental illness; cardiac disease; lactagogue; diabetes, hypoglycemic; emetic; digestive; indigestion: stomachache, intestinal pain, intestinal complaints, colic; gas; diarrhea; jaundice; diuretic; dysuria; involuntary urination in prostatitis; urinary disorders; anthelmintic, against tapeworms, vermifuge; leishmaniasis; hemorrhoids; aphrodisiac; menstrual problems; emmenagogue; sexual disorders; impotence; sexual stimulant; against infertility; abortifacient; antifertility agent; pregnancy/birth; uterine pain during pregnancy; prolapse of uterus; syphilis; fever; malaria; as frictions and applications; veterinary: diuretic for camels	Pieroni et al. 2005; Prashanth and John 1999; Rashan et al. 1989; Razzack 1980; Ross et al. 1980; Saadabi 2006; Saha et al. 1961; Said 1984; Schipper and Volk 1960; Sezik et al. 1997, 2004; Shah 1982; Shapira et al. 1989; Singh 1995; Singh and Pandey 1998; Singh et al. 1996b; Sokmen et al. 1999; Steinmetz 1957; Tabata et al. 1994; Tahraoui et al. 2007; Uphof 1959, 1968; Usher 1974; van Wyk and Wink 2004; Wild Flowers of Israel n.d.
Whole plant: depurative; anticonvulsant; sedative; narcotic; analgetic; anti-inflammatory; vulnerary; skin disease; purulent conjunctivitis; toothache; headache; arthritis, rheumatism; neurotic pains; arterial hypertension; common cold; antitussive; emetic; diuretic; hemorrhoids; abortifacient; emmenagogue; leishmaniasis;	
Part not mentioned: panacea; alterative; for general weakness; sudorific; depurative; astringent; hypotensive; stimulant; CNS stimulant; CNS depressant; narcotic; intoxicant; soporific; antispasmodic; analgesic; antiseptic; antibacterial; antifungal; antiparasitic; pediculicide; wounds; burns; skin diseases; dermatitis; folliculitis; strengthens hair; joint pains; rheumatism; neonatal tetanus; alopecia; improves eyesight; eye diseases; eye inflammation; suppurative conjunctivitis; blepharitis;	

(Continued)

TABLE 2.2 (CONTINUED)
Ethnography, Distribution, and Local Uses of *Peganum harmala*

Peganum harmala		References
	toothache; for nerves; neuralgia; Parkinsonism; nervous breakdown; hysteria; psychiatric conditions; depression; lactagogue; antitussive, asthma, laryngitis; hypertension; dropsy in the feet; diabetes; jaundice; spleen disease; hiccup; emetic; digestive disorders; colic; anthelmintic; diuretic; calculus; hemorrhoids; aphrodisiac; against female infertility; emmenagogue; dysmenorrhea; abortifacient; uterine cancer; fever; malaria; mumps; poisonous; against poisons and snake venoms	
Toxicity	Leaf: poisonous, but the percentage of alkaloids in the leaf (0.52%) is much lower than in the seeds (3–4%)	Abrol and Chopra 1962; Al-Rawi and Chakravarty 1964; Bellakhdar 1997; Bellakhdar et al. 1991; Ben Salah et al. 1986; Duke 1992–2016; Granot 1994; Hammiche and Merad 1997; Nayar 1955; Kamel et al. 1970; MEDUSA n.d.; Merzouki et al. 2000; Nayar 1955; Schipper and Volk 1960; USDA Forest Service 2014; Wild Flowers of Israel n.d.
	Stem: poisonous, but the percentage of alkaloids in the stem (0.36%) is much lower than in the seeds (3–4%)	
	Root: pediculoside; root is poisonous but the percentage of alkaloids in the root is much lower than in the seeds	
	Flower: claimed to be toxic	
	Fruit: poisonous, causes dysfunction of digestive system; causes convulsions and death especially in camels	
	Seed: toxic; insecticide; causes digestive disorders: nausea, vomiting; cardiac disorders: bradycardia, circulatory disorders, hypotension; dyspnoea; neurological disorders: intoxication, vertigo, stimulation, euphoria, tremors, convulsions, intense sensation of heat, hypothermia; paresthesia, paralysis, visual and auditive hallucinations, furiosity, deep sleep, CNS depression; nephrological disorders: anuria, uremia; paralysis of the respiratory system and death. Main toxins are alkaloids whose chemical structure combines an indole ring and a	

(Continued)

TABLE 2.2 (CONTINUED)
Ethnography, Distribution, and Local Uses of *Peganum harmala*

Peganum harmala	References
pyridine ring: harman, harmine, harmaline, harmalol (= harmol). The percentage of alkaloids in the seeds (3–4%) is much higher than in the rest of the plant. Alkaloid content rises sharply in the summer during the ripening phase of the fruit. Poisonings are mainly due to overdose or confusion with the seeds of *Conium maculatum* L.	
Whole plant: all plant parts are toxic; however, livestock seldom consumes them because of their bitter taste. Digestive disorders: nausea, vomiting; cardiac disorders: bradycardia, hypotension; neurological disorders: euphoria, visual hallucinations, tremors, convulsions, paralysis, CNS depression; dyspnoea; hypothermia; death. Main toxins are alkaloids whose chemical structure combines an indole ring and a pyridine ring: harman, harmine, harmaline, harmalol (= harmol). Alkaloid content rises sharply in the summer during the ripening phase of the fruit.	
Part not mentioned: poisonous to mammals; insecticide, pediculicide; causes weakness and paralysis of the hindquarters in guinea pigs. Alkaloids: low doses have a calming effect on the nervous system, high doses depress the CNS. Tannins: poisonous, causing hallucinations, convulsions and paralysis.	

Javzan et al. (2015) examined aerial parts of *P. multisectum* (Maxim) Bobrov growing in Mongolia using capillary gas chromatography–mass spectrometry (GC-MS) and column chromatography (CC), and noted **2-methylquinoline, 9-amino-2,3,5,6,7,8-hexahydro-1H-cyclopenta [b] quinoline, vasicinone** and harmine by GC-MS, and **peganine, deoxypeganine, deoxyvasicinone,** and **harmane** elucidated by MS and 1H and 13C NMR (Figures 2.19 through 2.24).

TABLE 2.3
Ethnography, Distribution, and Local Uses of *Peganum mexicanum*

Peganum mexicanum		References
Synonyms	*Peganum mexicanum* A. Gray: *P. texanum* M. E. Jones	GBIF 2001; The Plant List 2013; USDA, NRCS 2015; Vail and Rydberg 1907; ZipcodeZoo 2004
Common names	Spanish (Mexico): garbancillo, garbanzilla, limoncillo, romero del campo	Gray and Wright 1852–1853; ITIS n.d.; USDA, NRCS 2015
Habitat and distribution	Southwestern Texas to northeastern Mexico *South-Central USA*: Southwestern Texas *Mexico*: Coahuila, Chihuahua, San Luis Potosi	IPNI 2012; ITIS n.d.; Kearney and Peebles 1960; USDA, NRCS 2015; Vail and Rydberg 1907
Economic importance	Part not specified: medicinal	Baillon 1871–1888; Gray and Wright 1852–1853
Local medicinal uses	Part not specified: depurative; gonorrhea	Baillon 1871–1888, Gray and Wright 1852–1853
Toxicity	Part not specified: said to be poisonous to cattle	Gray and Wright 1852–1853

TABLE 2.4
Ethnography, Distribution, and Local Uses of *Peganum multisectum*

Peganum multisectum		References
Synonyms	*Peganum multisectum* (Maxim.) Bobrov: *P. harmala* var. *multisectum* Maxim.	Liu 2011; Liu and Zhou 2008; The Plant List 2013
Common names	Chinese: duo lie luo tuo peng	Liu and Zhou 2008
Habitat and distribution	Sandy areas, wastelands in semidesert areas; 1700–3900 m. *China*: Gansu, Nei Mongol, Ningxia, Qinghai, N Shaanxi, Xinjiang, Xizang.	Liu and Zhou 2008
Economic importance	–	
Local medicinal uses	–	
Toxicity	–	

TABLE 2.5
Ethnography, Distribution, and Local Uses of *Peganum nigellastrum*

Peganum nigellastrum		References
Synonyms	*Peganum nigellastrum* Bunge	GBIF 2001; IPNI 2012; Liu and Zhou 2008; The Plant List 2013
Common names	Chinese: luo tuo hao	Liu and Zhou 2008; Zheng et al. 2009
Habitat and distribution	Dry grasslands, hilly slopes, sandy and gravelly areas, semidesert and steppe areas. From eastern Siberia to China. *Siberia*: Eastern Siberia *Mongolia*: Mongolia *China*: Gansu, Hebei, Henan, Nei Mongol Zizhiqu, Ningxia Hui, Shaanxi, Shanxi, Xinjiang Uygur	Encyclopedia of Life n.d.; GBIF 2001; GRIN 1994; IPNI 2012; Liu and Zhou 2008
Economic importance	Part not specified: medicinal	Ma et al. 1997, 2000
Local medicinal uses	Part not specified: abscess; rheumatism; inflammatory diseases	Ma et al. 1997, 2000
Toxicity	Part not specified: effect on dopamine processes	Purevdorjiin and Gotovyn 1994

FIGURE 2.11 Ripening fruit of *P. harmala* L. (July 2011, Flix, Ribera d'Ebre, Catalonia. From Lloret, F.J., *Galanthus*: *Web de botanica*, 2007. http://www.galanthus.cat, retrieved 3/16/2015.)

FIGURE 2.12 *Peganum mexicanum* A. Gray. (October 2007, Hudspeth County, Texas, USA. From Spjut, R.W., *The World Botanical Associates*, 2004. http://www.worldbotanical.com, retrieved 3/16/2015.)

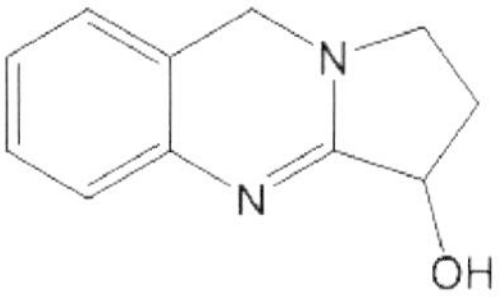

FIGURE 2.13 Vasicine.

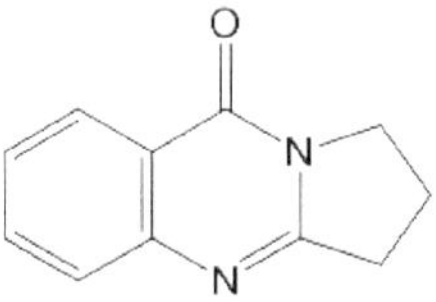

FIGURE 2.14 Deoxyvasicinone.

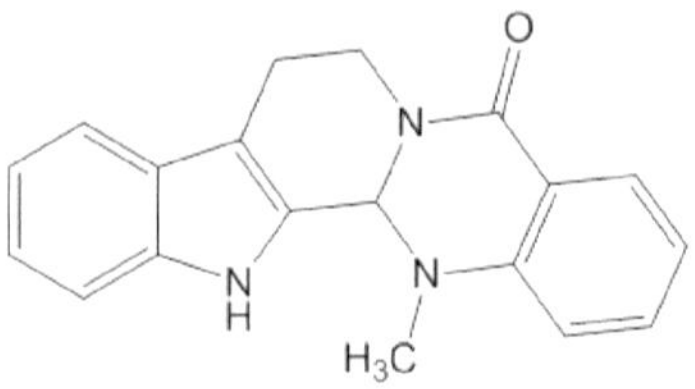

FIGURE 2.15   Evodiamine.

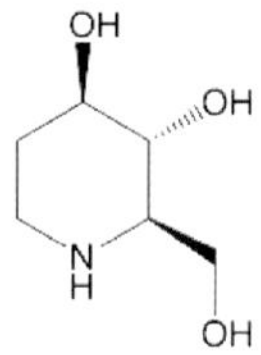

FIGURE 2.16   Fagomine.

PEGANUM NIGELLASTRUM

Peganum nigellastrum, in Chinese *luo tuo hao* or "little Peganum," is known in TCM to dispel damp and resolve toxin, quicken blood, relieve pain, and suppress cough. Indications for TCM use of *P. nigellastrum* include arthritis, menstrual disorder, bronchitis, and headache (Figure 2.25) (Zhou et al. 2011).

Jing Lu Liang and coworkers at Yeungnam University, Korea, found aerial parts of *P. nigellastrum* to contain a possible species-unique set of anticancer phytochemicals that they refer to as **luotonins A, B, C, D,** and **E (pyrroloquinazolino-quinoline alkaloids, canthin-6-one alkaloids,** and a **4(3H)-quinazolinone alkaloid)** that are all cytotoxic to P-388 human leukemia cells, with luotonin A the most deadly owing to its modulation of topoisomerase I-dependent DNA-cleavage in the leukemic cells (Liang et al. 2011). The mechanism for this effect is related to the fact that luotonin A "stabilizes the human DNA topoisomerase I-DNA covalent binary complex, affording the same pattern of cleavage as the structurally related topoisomerase I inhibitor camptothecin" (Cagir et al. 2003). Luotonin A also mediated topoisomerase I-dependent cytotoxicity toward *Saccharomyces cerevisiae* lacking yeast topoisomerase I, but possessing a plasmid with the human topoisomerase I gene under the aegis of a galactose promoter (Figures 2.26 through 2.30).

Zheng et al. (2009), screening for acetylcholinesterase inhibitors (AChEIs) as putative therapy for Alzheimer's disease, found two new alkaloids in *P. nigellastrum*

(a)

(b)

FIGURE 2.17 Mulberry leaves are used as a source of nutrition to silkworms. Both the leaves and the worms contain antidiabetic fagomine (Asano et al. 1994; Nakagawa et al. 2010; Hunyadi et al. 2013). (a) *Morus* sp. leaves and fruit. (Photo by Alexxx Maler (Alexxx1979) via Wikimedia Commons, https://commons.wikimedia.org/wiki/File:Morus_IMG_8131_1725.jpg.) (b) *Bombyx mori*, silkworms, on mulberry leaves. (6/26/2016 silk factory in Beijing: by Stephen Birch (sgbirch) via http://www.Flickr.com.)

FIGURE 2.18 *Cynomorium songaricum*, parasite of *Peganum multisectum*, is also a medicinal plant. (6/22/2015: by Paul Laney via http://www.flickr.com.)

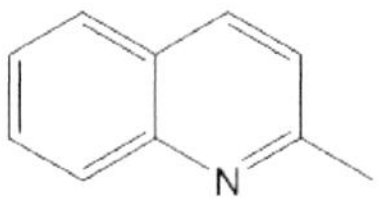

FIGURE 2.19   2-Methylquinoline.

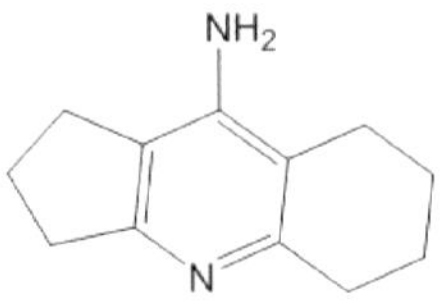

FIGURE 2.20   9-Amino-2, 3,5,6,7,8-hexahydro-1H-cyclopenta (b) quinoline.

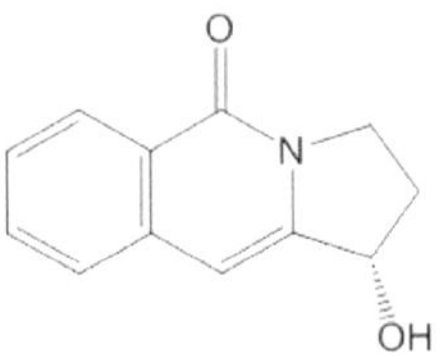

FIGURE 2.21   Vasicinone.

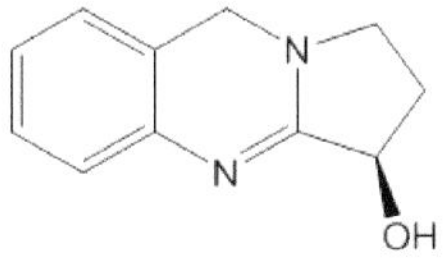

FIGURE 2.22   Peganine.

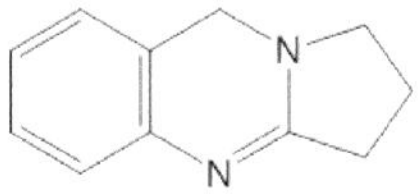

FIGURE 2.23   Deoxypeganine.

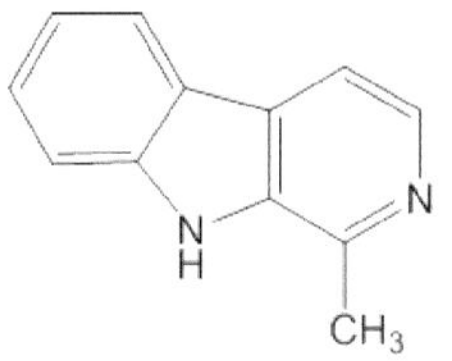

FIGURE 2.24   Harmane.

seeds, namely **nigellastrine I** and **nigellastrine II**, in addition to known alkaloids (e.g., known from *Peganum* sp.), vasicine, vasicinone, harmine, deoxyvasicinone, deoxyvasicine, harmaline, **harmol**, and harman. Potency of these *Peganum* alkaloids against acetylcholinesterase and as putative Alzheimer's symptoms therapy was comparable to that of galanthine, a natural product obtainable from plants such as *Narcissus pseudonarcissus* (Fraser et al. 2016), and currently in use both as a natural product and as a synthetic product for clinical treatments (Figures 2.31 through 2.33) (Galimberti and Scarpini 2016).

Novel triterpenoids from the roots of *P. nigellastrum* have been identified as **3alpha-acetoxy-27-trans-caffeoyloxyolean-12-en-28-oic acid methyl ester** and **3-oxotirucalla-7, 24-dien-21-oic acid** (Ma et al. 2007). Earlier, the same workers also isolated from the aerial parts *P. nigellastrum* two phenylpropanoids, **dihydrosinapyl ferulate** and **dihydroconiferyl ferulate**, and alkaloids, harmine, **3-phenylquinoline**, **3-(4-hydroxyphenyl)quinoline**, and **3-(1H-indol-3-yl)quinoline** (Ma et al. 2000) (Figures 2.34 through 2.40).

FIGURE 2.25 Aerial parts of *Peganum nigellastrum* contain anticancer phytochemicals (Liang et al. 2011). (7/7/2012, Gusinoye Ozero, Republic of Buryatia, Russia: by Daba Chimitov, http://www.plantarium.ru/page/image/id/145898.html.)

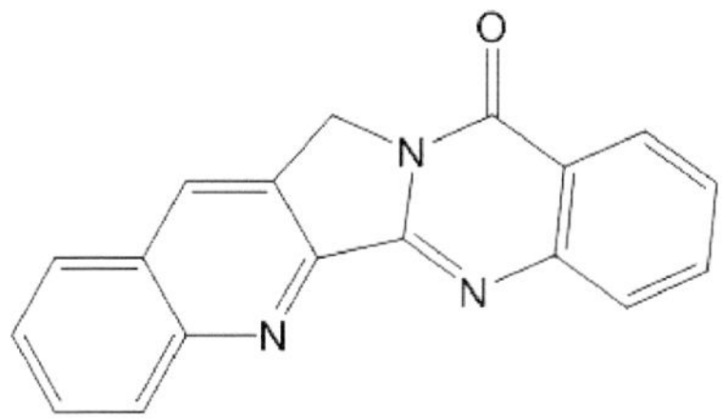

FIGURE 2.26 Luotonin A. (From Liang, J.L. et al., *Molecules*, 16(6): 4861–83, 2011.)

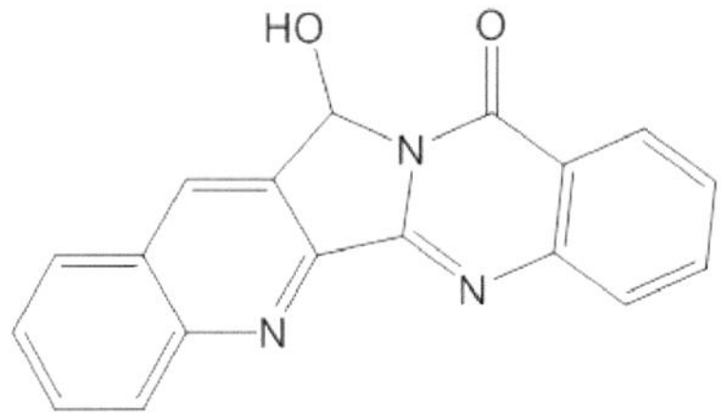

FIGURE 2.27 Luotonin B. (From Liang, J.L. et al., *Molecules*, 16(6): 4861–83, 2011.)

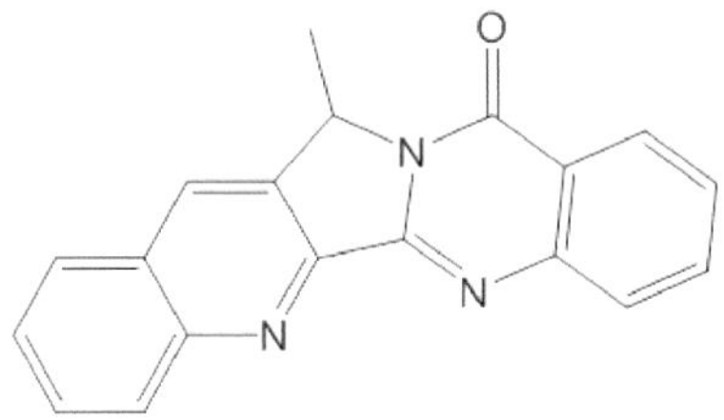

FIGURE 2.28 Luotonin C. (From Liang, J.L. et al., *Molecules*, 16(6): 4861–83, 2011.)

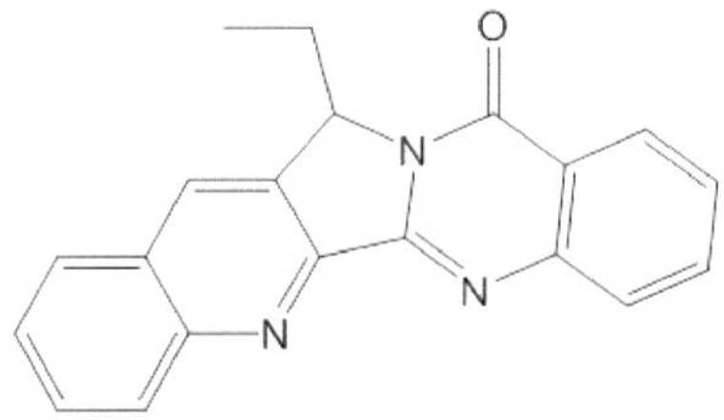

FIGURE 2.29 Luotonin D. (From Liang, J.L. et al., *Molecules*, 16(6): 4861–83, 2011.)

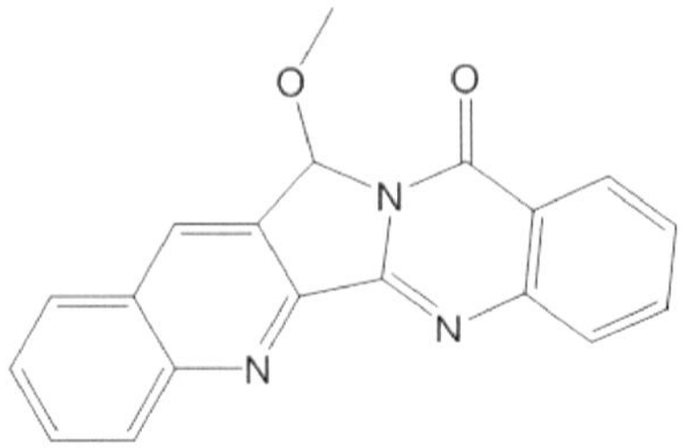

FIGURE 2.30   Luotonin E. (From Liang, J.L. et al., *Molecules*, 16(6): 4861–83, 2011.)

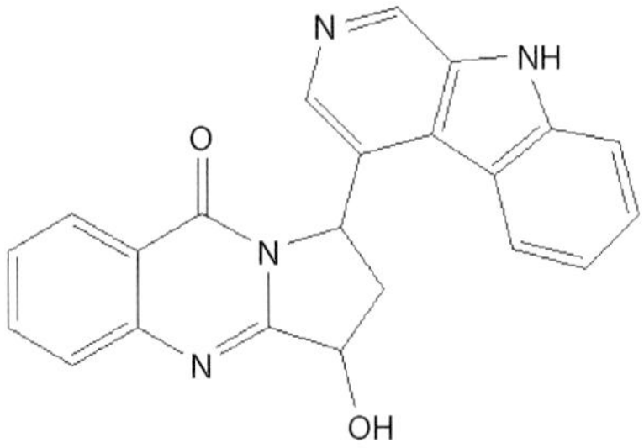

FIGURE 2.31   Nigellastrine I. (From Zheng, X.Y. et al., *Arch Pharm Res*, 32(9): 1245–51, 2009.)

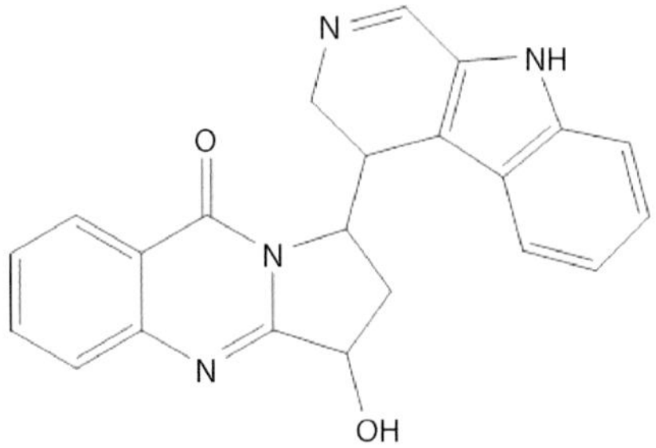

FIGURE 2.32   Nigellastrine II. (From Zheng, X.Y. et al., *Arch Pharm Res*, 32(9): 1245–51, 2009.)

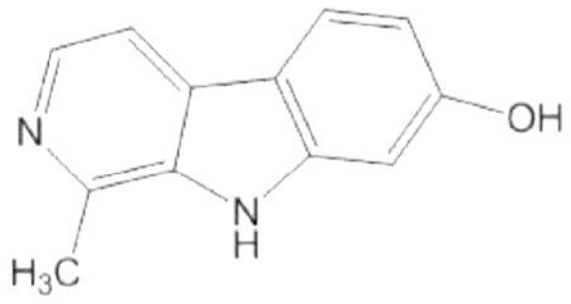

FIGURE 2.33   Harmol.

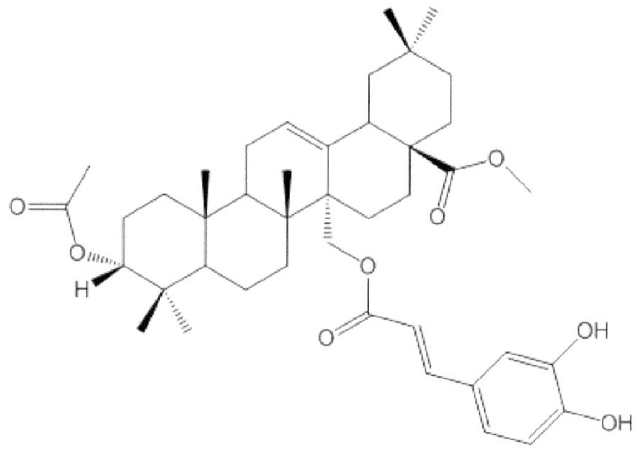

FIGURE 2.34 3alpha-acetoxy-27-trans-caffeoyloxyolean-12-en-28-oic acid methyl ester. (From Ma, Z.Z. et al., *J Asian Nat Prod Res*, 9(6–8): 575–8, 2007.)

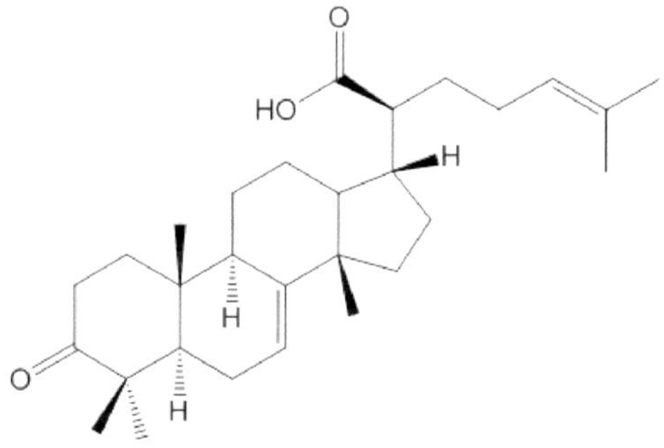

FIGURE 2.35 3-Oxotirucalla-7, 24-dien-21-oic acid. (From Ma, Z.Z. et al., *J Asian Nat Prod Res*, 9(6–8): 575–8, 2007.)

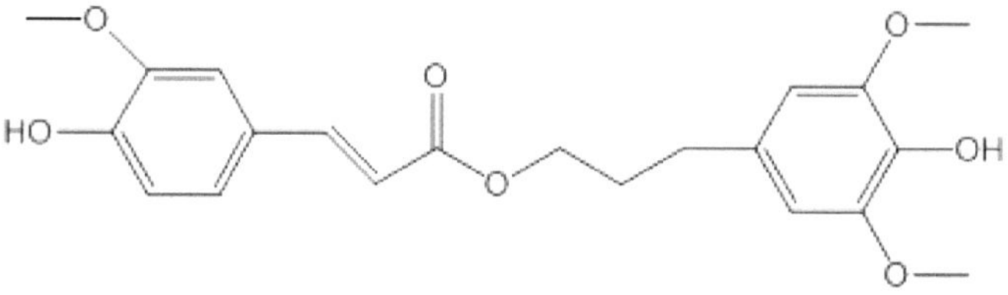

FIGURE 2.36 Dihydrosinapyl ferulate. (From Ma, Z.Z. et al., *Phytochemistry*, 53: 1075–8, 2000.)

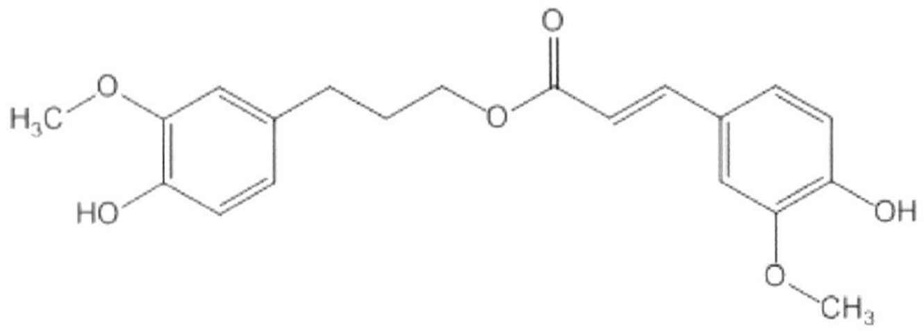

FIGURE 2.37   Dihydroconiferyl ferulate.

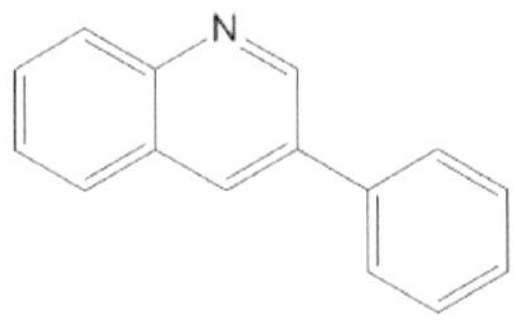

FIGURE 2.38   3-Phenylquinoline.

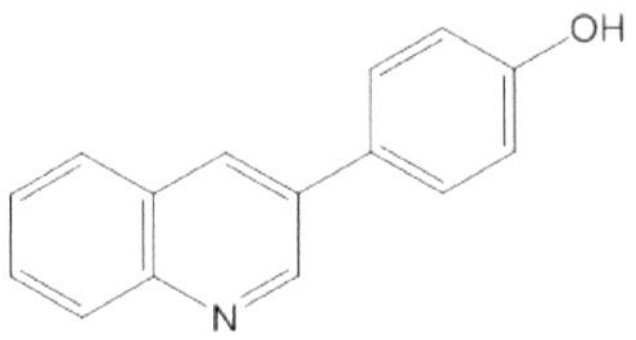

FIGURE 2.39   3-(4-Hydroxyphenyl)quinoline.

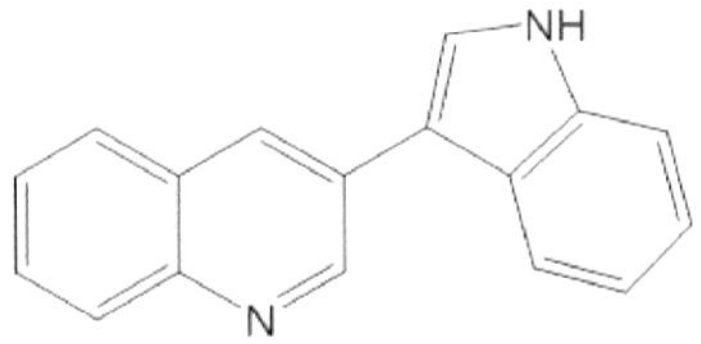

FIGURE 2.40   3-(1H-Indol-3-yl)quinoline.

REFERENCES

Abdel Aziz, N.G., S.M. Abdel Kader, M.M. El-Sayed, E.A. El-Malt, and E.S. Shaker (eds.). 2010. Novel carboline alkaloid from *Peganum harmala* as antibacterial agent. *Proceedings of the Tenth Radiation Physics and Protection Conference.* 27–30 November, Cairo, Egypt. pp. 359–68.

Abrol, B.K., and I.C. Chopra. 1962. Some vegetable drug resources of Ladakh (Little Tibet). Part I. *Curr Sci* 31: 324–6.

Abu-Irmaileh, B.E., and F.U. Afifi. 2003. Herbal medicine in Jordan with special emphasis on commonly used herbs. *J Ethnopharmacol* 89: 193–7.

African Plants Database (version 3.4.0). Conservatoire et Jardin botaniques de la Ville de Genève and South African National Biodiversity Institute, Pretoria. [n.d.] http://www .ville-ge.ch/musinfo/bd/cjb/africa/ (accessed 15 January 2015).

Ahmad, Y.S. 1957. *A Note on the Plants of Medicinal Value Found in Pakistan.* Karachi: Government of Pakistan Press.

Akhtar, M.S., Z. Iqbal, M.N. Khan, and M. Lateef. 2000. Anthelmintic activity of medicinal plants with particular reference to their use in animals in the Indo-Pakistan subcontinent. *Small Rumin Res* 38: 99–107.

Alami, R., A. Macksad, and A.R. El-Gindy. 1976. *Medicinal Plants in Kuwait.* Kuwait: Al-Assiriya Printing Press.

Al-Awdat, M., and G. Laham. 1994. *Al-Nabatat al-tibbiyah wa-sti'malatuha. Medicinal Plants and Their Uses.* Damascus: Al-Ahally Publ.

Aldén, B., S. Ryman, and M. Hjertson. 2009. *Våra kulturväxters namn: Ursprung och användning.* Stockholm: Formas. USDA, ARS, National Genetic Resources Program. Germplasm Resources Information Network (GRIN) [Online database.] National Germplasm Resources Laboratory, Beltsville, MD. http://www.arsgrin.gov/cgi-bin /npgs/html/stdlit.pl?Vara%20kulturvaxt%20namn (accessed 1 July 2014).

Al-Khalil, S. 1995. A survey of plants used in Jordanian traditional medicine. *Int J Pharmacog* 33(4): 317–23.

Alkofahi, A., H. Masaadeh, and S. Al-Khalil. 1996. Antimicrobial evaluation of some plant extracts of traditional medicine of Jordan. *Alex J Pharm Sci* 10(2): 123–6.

Al-Qura'n, S. 2008. Taxonomical and pharmacological survey of therapeutic plants in Jordan. *J Nat Prod* 1: 10–26.

Al-Rawi, A., and H.L. Chakravarty. 1964. *Medicinal Plants of Iraq.* Min Agr Techn Bull no 15. Baghdad: Govt Press.

Al-Yahya, M.A. 1986. Phytochemical studies of the plants used in traditional medicine of Saudi Arabia. *Fitoterapia* 57(3): 179–82.

Amézqueta, S., E. Galán, I. Vila-Fernández, S. Pumarola, M. Carrascal, J. Abian, L. Ribas-Barba, L. Serra-Majem, and J.L. Torres. 2013. The presence of D-fagomine in the human diet from buckwheat-based foodstuffs. *Food Chem* 136(3–4): 1316–21.

Asano, N., K. Oseki, E. Tomioka, H. Kizu, and K. Matsui. 1994. N-containing sugars from *Morus alba* and their glycosidase inhibitory activities. *Carbohydr Res* 259(2): 243–55.

Ayensu, E.S. 1978. *Medicinal Plants of West Africa.* Algonac: Reference Publications.

Bailey, C., and A. Danin. 1981. Bedouin plant utilization in Sinai and the Negev. *Econ Bot* 35(2): 145–62.

Baillon, H. 1871–1888. *The Natural History of Plants.* London: L. Reeve & Co.

Bedevian, A.K. 1936. *Illustrated Polyglottic Dictionary of Plant Names in Latin, Arabic, Armenian, English, French, German, Italian, and Turkish Languages including Economic, Medicinal, Poisonous and Ornamental Plants, and Common Weeds.* Cairo: Argus & Papazian Presses.

Bellakhdar, J. 1997. *La pharmacopée traditionnelle marocaine: Médecine arabe ancienne et savoirs populaires.* Paris: Ibis Press.

Bellakhdar, J., R. Claisse, J. Fleurentin, and C. Younos. 1991. Repertory of standard herbal drugs in the Moroccan pharmacopoeia. *J Ethnopharmacol* 35(2): 123–43.

Ben Salah, N., M. Amamou, Z. Jerbi, F. Ben Salah, and M. Yacoub. 1986. Aspects cliniques, pharmacologiques et toxicologiques du surdosage par une plante médicinale: le Harmel. *Essaydali scientifique* 21: 13–8.

Bézanger-Beauquesne, L., M. Pinkas, M. Torck, and F. Trotin. 1980. *Plantes médicinales des régions tempérées*. Paris: Ed. Maloine.

Bhattacharyya, A. 1991. Ethnobotanical observations in the Ladakh region of Northern Jammu and Kashmir State, India. *Econ Bot* 45(3): 305–8.

Blatter, E., P.F. Hallberg, and C. McCann. 1919. Contributions towards a flora of Baluchistan. *J Indian Bot* 1(2): 54–178.

Boukef, K., H.R. Souissi, and G. Balansard. 1982. Contribution to the study on plants used in traditional medicine in Tunisia. *Plant Med Phytother* 16(4): 260–79.

Boulos, L. 1983. *Medicinal Plants of North Africa*. Algonac: Reference Publications.

Bown, D. 1995. *Encyclopaedia of Herbs and their Uses*. London: Dorling Kindersley.

Bremner, P., D. Rivera, M.A. Calzado, C. Obón, C. Inocencio, C. Beckwith, B.L. Fiebich, E. Muñoz, and M. Heinrich. 2009. Assessing medicinal plants from South-Eastern Spain for potential anti-inflammatory effects targeting nuclear factor-Kappa B and other pro-inflammatory mediators. *J Ethnopharmacol* 124: 295–305.

Brobst, A., J. Lewis, B. Klett, C. Haustein, and J. Shriver. 2009. The free base extraction of harmaline from *Peganum harmala*. *Am J Undergrad Res* 8: 2–3.

Brummitt, R.K., F. Pando, S. Hollis, N.A. Brummitt et al. 2001. *World Geographical Scheme for Recording Plant Distributions*. 2nd ed. Pittsburgh, PA: Published for the International Working Group on Taxonomic Databases for Plant Sciences (TDWG) by the Hunt Institute for Botanical Documentation Carnegie Mellon University.

Burkill, H.M. 1985. *The Useful Plants of West Tropical Africa*. Vol. 5. Kew, UK: Royal Botanic Gardens.

Burkill, I.H. 1909. *A Working List of the Flowering Plants of Baluchistan*. Calcutta: Superintendent Government Printing.

Cagir, A., S.H. Jones, R. Gao, B.M. Eisenhauer, and S.M. Hecht. 2003. Luotonin A: A naturally occurring human DNA topoisomerase I poison. *J Am Chem Soc* 125(45): 13628–9.

CFIA. Canadian Food Inspection Agency. Government of Canada. [n.d.] http://www.inspection .gc.ca (accessed March 16 2015).

Cheng, X.-M., T. Zhao, T. Yang, C.-H. Wang, S.W.A. Bligh, and Z.-T. Wang. 2010. HPLC fingerprints combined with principal component analysis, hierarchical cluster analysis and linear discriminant analysis for the classification and differentiation of *Peganum* sp. indigenous to China. *Phytochem Anal* 21(3): 279–89.

Chopra, I.C., B.K. Abral, and K.L. Handa. 1960. *Les plantes médicinales des régions arides considérées surtout du point de vue botanique*. Ed. UNESCO.

Chopra, R.N. 1933. *Indigenous Drugs of India: Their Medical and Economic Aspects*. Calcutta, India: The Art Press.

Chopra, R.N., R.L. Badhwar, and S. Ghosh. 1949. *Poisonous Plants of India*. Calcutta: Manager of Publications, Government of India Press. Volume 1.

Chopra, R.N., R.L. Badhwar, and S. Ghosh. 1965. *Poisonous Plants of India*. 2nd rev. and enl. ed. New Delhi: Indian Council of Agricultural Research.

Çitoğlu, G. 1997. The public use of *Peganum harmala* in Turkey. *Hamdard Medicus* 40(3): 60–1.

Clauson, G. 1972. *An Etymological Dictionary of Pre-Thirteenth-Century Turkish*. Oxford: Clarendon Press.

Colomer, E.M. 2012. Flores silvestres del Mediterraneo. [Blog.] http://floressilvestresdel mediterraneo.blogspot.co.il (accessed March 16 2015).

Cronquist, A. 1981. *An Integrated System of Classification of Flowering Plants*. New York: Columbia University Press.

Danin, A. (ed.). 2003 (continuously updated). Flora of Israel online. Hebrew University of Jerusalem, Jerusalem, Israel. http://flora.huji.ac.il/browse.asp (accessed June 20 2014).

Dragendorff, G. 1898. *Die Heilpflanzen der verschiedenen Volker und Zeiten*. Stuttgart: F. Enke.

Ducros, A.H. 1930. *Essai sur le droguier populaire arabe de l'Inspectorat des pharmacies du Caire*. Le Caire: Impr. de l'Institut français d'archéologie orientale.

Duke, J.A. 1992–2016. Dr. Duke's Phytochemical and Ethnobotanical Databases. [Online database.] U.S. Department of Agriculture, Agricultural Research Service. Home Page, http://phytochem.nal.usda.gov/, http://dx.doi.org/10.15482/USDA.ADC/1239279 (accessed March 10 2017).

Duke, J.A. 2002. *Handbook of Phytochemical Constituents of GRAS Herbs and Other Economic Plants*. Boca Raton: CRC Press.

efloraofindia. 2007. [Online discussion group and database.] https://sites.google.com/site /efloraofindia/ (accessed January 3 2013).

Emboden, W. 1979. *Narcotic Plants*. New York: Macmillan.

Encyclopedia of Life. [n.d.] [Online database.] http://www.eol.org (accessed January 15 2014).

Engler, A., and K. Prantl. 1897. *Die natürlichen Pflanzenfamilien*. Vol. 3. In PlantSystematics. org, http://www.plantsystematics.org (accessed March 16 2015).

Ertug, F. 2000. An ethnobotanical study in Central Anatolia (Turkey). *Econ Bot* 54(2): 155–82.

Facciola, S. 1990. *Cornucopia—A Source Book of Edible Plants*. Vista: Kampong Publication.

Farouk, L., A. Laroubi, R. Aboufatima, A. Benharref, and A. Chait. 2008. Evaluation of the analgesic effect of alkaloid extract of *Peganum harmala* L.: Possible mechanisms involved. *J Ethnopharmacol* 115(3): 449–54.

Flattery, D.S., and M. Schwartz. 1989. *Haoma and Harmaline: The Botanical Identity of the Indo-Iranian Sacred Hallucinogen "Soma" and Its Legacy in Religion, Language, and Middle-Eastern Folklore*. Berkeley: University of California Press.

FloraGREIF. FloraGREIF—Virtual Flora of Mongolia. 2010– (continuously updated). University of Greifswald, Institute of Botany and Landscape Ecology, Institute of Geography and Geology, Computer Centre. http://greif.uni-greifswald.de/floragreif/ (accessed March 16 2015).

Font Quer, P. 1979. *Plantas medicinales: el Dioscórides renovado*. Barcelona: Editorial Labor.

Foucauld, C.E. 1951. *Dictionnaire touareg-français, dialecte de l'Ahaggar*. Paris.

Fraser, M.D., J.R. Davies, and X. Chang. 2016. New gold in them thar hills: Testing a novel sup-ply route for plant-derived galanthamine. *J Alzheimers Dis* Nov 9 [Epub ahead of print.]

Friedman, J., Z. Yaniv, A. Dafni, and D. Palevitch. 1986. A preliminary classification of the healing potential of medicinal plants, based on a rational analysis of an ethnopharma-cological field survey among Bedouins in the Negev Desert, Israel. *J Ethnopharmacol* 16(2–3): 275–87.

Frison, G., D. Favretto, F. Zancanaro, G. Fazzin, and S.D. Ferrara. 2008. A case of beta-carboline alkaloid intoxication following ingestion of *Peganum harmala* seed extract. *Forensic Sci Int* 179: e37–43.

Galimberti, D., and E. Scarpini. 2016. Old and new acetylcholinesterase inhibitors for Alzheimer's disease. *Expert Opin Investig Drugs* 25(10): 1181–7.

GBIF. Global Biodiversity Information Facility. 2001. Home page. http://www.gbif.org (accessed December 6 2014).

Ghafoor, A. 1985. Zygophyllaceae. In *Flora of Pakistan, eFloras*. St. Louis: Missouri Botanical Garden; Cambridge, MA: Harvard University Herbaria. http://www.efloras .org (accessed July 24 2014).

Gómez, L., E. Molinar-Toribio, M.Á. Calvo-Torras, C. Adelantado, M.E. Juan, J.M. Planas, X. Cañas, C. Lozano, S. Pumarola, P. Clapés, and J.L. Torres. 2012. D-Fagomine lowers postprandial blood glucose and modulates bacterial adhesion. *Br J Nutr* 107(12): 1739–46.

Goodman, S.M., and A. Ghafoor. 1992. *The Ethnobotany of Southern Balochistan, Pakistan: With Particular Reference to Medicinal Plants.* Chicago: Field Museum of Natural History.

Gorsi, M.S., and Sh. Miraj. 2002. Ethnomedicinal survey of plants of Khanabad village and its allied areas, District Gilgit. *Asian J Plant Sci* 1: 604–15.

Granot, Y. 1994. *Medical Plants of the Negev.* Midreshet Sde Boker. Interdisciplinary Education Center. http://www.boker.org.il/learning/?http://www.boker.org.il/meida/negev/meida.htm.

Gray, A., and C. Wright. 1852–1853. *Plantae Wrightianae Texano-neo-mexicanae: An Account of a Collection of Plants Made by Charles Wright in an Expedition from Texas to New Mexico, in the Summer and the Autumn of 1849.* Washington, DC: Smithsonian Institution.

GRIN. Germplasm Resources Information Network. 1994. [Online database.] USDA, ARS, National Genetic Resources Program. National Germplasm Resources Laboratory, Beltsville, MD. http://www.ars-grin. gov/npgs/ (accessed July 22 2014).

Hammiche, V., and R. Merad. 1997. *Peganum harmala.* International Programme on Chemical Safety Poisons Information Monograph 402. http://www.inchem.org/documents/pims/plant/pim402fr.htm [In French.]

Hamsa, T.P., and G. Kuttan. 2011. Harmine activates intrinsic and extrinsic pathways of apoptosis in B16F-10 melanoma. *Chin Med* 6: 11.

Hanelt, P., R. Büttner, and R. Mansfeld, eds. 2001. *Mansfeld's Encyclopedia of Agricultural and Horticultural Crops (except Ornamentals).* Berlin: Springer.

Hartwell, J.L. 1967–1971. Plants used against cancer: A survey. *Lloydia* 32(1): 78–107; 32(2): 153–205; 32(3): 247–96; 33(1): 97–194; 33(3): 288–392; 34(1): 103–60; 34(2): 204–55; 34(3): 310–61; 34(4): 386–425.

Hassan, I. 1967. Some folk uses of *Peganum harmala* in India and Pakistan. *Econ Bot* 21: 284.

Hassler, M. 2014. World Plants: Synonymic Checklists of the Vascular Plants of the World (version Mar 2014). In: *Species 2000 & ITIS Catalogue of Life, 2014 Annual Checklist.* Roskov Y., T. Kunze, T. Orrell, L. Abucay, L. Paglinawan, A. Culham, N. Bailly, P. Kirk, T. Bourgoin, G. Baillargeon, W. Decock, A. De Wever, and V. Didžiulis, (eds). Digital resource at http://www.catalogueoflife.org/annual-checklist/2014. Species 2000: Naturalis, Leiden, the Netherlands. (Accessed January 17 2015.)

Hegnauer, R. 1973. *Chemotaxonomie der Pflanzen.* Vol. 6. *Dicotyledoneae: Rafflesiaceae–Zygophyllaceae.* Basel: Birkhäuser.

Herraiz, T., D. González, C. Ancín-Azpilicueta, V.J. Arán, and H. Guillén. 2010. beta-Carboline alkaloids in *Peganum harmala* and inhibition of human monoamine oxidase (MAO). *Food Chem Toxicol* 48: 839–45.

Honda, G., E. Yesilada, M. Tabata, E. Sezik, T. Fujita, Y. Takeda, Y. Takaishi, and T. Tanaka. 1996. Traditional medicine in Turkey. VI. Folk medicine in West Anatolia: Afyon, Kutahya, Denizli, Mugla, Aydin Provinces. *J Ethnopharmacol* 53: 75–87.

Hooper, D., and H. Field. 1937. *Useful Plants and Drugs of Iran and Iraq.* Chicago: Field Museum of Natural History.

Hunyadi, A., E. Liktor-Busa, A. Márki, A. Martins, N. Jedlinszki, T.J. Hsieh, M. Báthori, J. Hohmann, and I. Zupkó. 2013. Metabolic effects of mulberry leaves: exploring potential benefits in type 2 diabetes and hyperuricemia. *Evid Based Complement Alternat Med* 948627.

Hussain, H., and R.S. Tobji. 1997. Antibacterial screening of some Libyan medicinal plants. *Fitoterapia* 68(5): 467–70.

IPNI. The International Plant Names Index. 2012. Home page. http://www.ipni.org (accessed September 5 2014).

ITIS. Integrated Taxonomic Information System. [n.d.] [Online database.] http://www.itis.gov/ (accessed May 12 2014).

Ivanovska, N., S. Philipov, and R. Istatkova. 1997. Evaluation of anti-inflammatory activity of plants used in Bulgarian folk medicines. *Fitoterapia* 68(5): 417–22.

Jakobsen, P., J.M. Lundbeck, M. Kristiansen, J. Breinholt, H. Demuth, J. Pawlas, M.P. Candela, B. Andersen, N. Westergaard, K. Lundgren, and N. Asano. 2001. Iminosugars: Potential inhibitors of liver glycogen phosphorylase. *Bioorg Med Chem* 9(3): 733–44.

Javzan, S., D. Selenge, N. Amartuvshin, D. Nedelcheva, V. Christov, and S. Philipov. 2015. Alkaloids from Mongolian species of *Peganum multisectum* (Maxim) Bobrov. *Mongolian J Chem* 16(42): 48–53.

Kaganoff, B.C. 1977. *A Dictionary of Jewish Names and Their History*. New York: Schocken Books.

Kakrani, H.K., and A.K. Saluja. 1993. Traditional treatment through herbal drugs in Kutch District, Gujarat State, India. Part I. Uterine disorders. *Fitoterapia* 65(5): 463–5.

Kamboj, V.P. 1988. A review of Indian medicinal plants with interceptive activity. *Indian J Med Res* 1988(4): 336–55.

Kamel, S.H., T.M. Ibrahim, A.A. Afifi, and S.M. Hamza. 1970. Chemical studies on the Egyptian plant *Peganum harmala*. *U A R J Vet Sci* 7(1): 61; *Biological Abstracts* 54 9840.

Karaki, H., T. Kishimoto, H. Ozaki, K. Sakata, H. Umeno, and N. Urakawa. 1986. Inhibition of calcium channels by harmaline and other harmala alkaloids in vascular and intestinal smooth muscles. *Br J Pharmacol* 89: 367–75.

Kearney, T.H., and R.H. Peebles. 1960. *Arizona Flora: With Supplement by John Thomas Howell, Elizabeth McLintock and Collaborators*. Berkeley: Univ. of California Press.

Kleinman, R. [n.d.] Vascular Plants of the Gila Wilderness. Dale A. Zimmerman Herbarium. Western New Mexico University Department of Natural Sciences. http://wnmu.edu/academic/nspages/gilaflora/ (accessed March 16 2015).

Komarov, V.L., B.K Schischkin, and E. Bobrow. 1934–1964. *Flora SSSR*. Moskow: Izdatel'stvo Akademii Nauk SSSR. USDA, ARS, National Genetic Resources Program. Germplasm Resources Information Network (GRIN) [Online database.] National Germplasm Resources Laboratory, Beltsville, MD. http://www.ars-grin.gov/cgi-bin/npgs/html/stdlit.pl?F%20USSR (accessed March 4 2014).

Kubitzki, K. 2011. *Flowering Plants, Eudicots: Sapindales, Cucurbitales, Myrtaceae*. Heidelberg; New York: Springer.

Kunkel, G. 1984. *Plants for Human Consumption: An annotated checklist of the edible phanerogams and ferns*. Koenigstein: Koeltz Scientific Books.

Lamchouri, F., A. Settaf, Y. Cherrah, M. Zemzami, B. Lyoussi, A. Zaid, N. Atif, and M. Hassar. 1999. Antitumour principles from *Peganum harmala* seeds. *Therapie* 54: 753–8.

Leporatti, M.L., and K. Ghedira. 2009. Comparative analysis of medicinal plants used in traditional medicine in Italy and Tunisia. *J Ethnobiol Ethnomed* 5: 31.

Leporatti, M.L., and E. Lattanzi. 1994. Traditional phytotherapy on coastal areas of Makran (Southern Pakistan). *Fitoterapia* 65(2): 158–61.

Lewis, W.H., and M.P.F. Elvin-Lewis. 1977. *Medical Botany*. New York: Wiley-Interscience.

Liang, J.L., H.C. Cha, and Y. Jahng. 2011. Recent advances in the studies on luotonins. *Molecules* 16(6): 4861–83.

Lin, L., L. Ren, L. Wen, Y. Wang, and J. Qi. 2016. Effect of evodiamine on the proliferation and apoptosis of A549 human lung cancer cells. *Mol Med Rep* 14(3): 2832–8.

Liu, B. 2011. Study on chemical constituents of *Peganum multisectum*. *Zhong Yao Cai* 34(11): 1719–21.

Liu, T.Z., and Y.Z. Shang. 2008. *Erysiphe alashanensis* sp. nov. from Inner Mongolia, China. *Nova Hedwigia (Stuttgart)* 86(1–2): 255–8.

Liu, Y., and L. Zhou. 2008. Nitrariaceae, in Z.Y. Wu, P.H. Raven, and D.Y. Hong, eds., *Flora of China. Vol. 11, Oxalidaceae through Aceraceae.* Beijing: Science Press, and St. Louis, MO: Missouri Botanical Garden Press. Via eFloras [Online Database.] http://www.efloras.org (accessed September 5 2014).

Lloret, F.J. 2007. Galanthus: Web de botanica, http://www.galanthus.cat (accessed March 16 2015).

Lokotsch, K. 1927. *Etymologisches Wörterbuch der europäischen (germanischen, romanischen und slavischen) Wörter orientalischen Ursprungs.* Heidelberg: C. Winter.

Ma, Z.Z., Y. Hano, T. Nomura, and Y.J. Chen. 1997. Two new pyrroloquinazolinoquinoline alkaloids from *Peganum nigellastrum*. *Heterocycles* 46(1): 541–6.

Ma, Z.Z., Y. Hano, T. Nomura, Y.J. Chen. 2000. Alkaloids and phenylpropanoids from *Peganum nigellastrum*. *Phytochemistry* 53: 1075–8.

Ma, Z.Z., Y. Hano, F. Qiu, G. Shao, Y.J. Chen, and T. Nomura. 2007. Triterpenoids from *Peganum nigellastrum*. *J Asian Nat Prod Res* 9(6–8): 575–8.

Makowiecki, S. 1936. *Słownik botaniczny łacińsko-małoruski, zebrał i ułożył w latach 1877–1932.* Kraków: Polska Akademia Umiejętności.

Malhi, B.S., and V.P. Trivedi. 1972. Vegetable antifertility drugs of India. *Q J Crude Drug Res* 12: 1922.

Marzell, H., W. Pfeifer, H. Paul, and W. Wissmann. 1943–1979. *Wörterbuch der deutschen Pflanzennamen.* Leipzig: S. Hirzel.

Masclans i Girvès, F. 1954. *Els noms vulgars de les plantes a les terres catalanes.* Barcelona: Institut d'Estudis Catalans.

Mayer, T.J.L. 1909. *English Biluchi Dictionary.* Lahore: Punjab Government Press.

McLean, R.C., and W.R. Ivimey-Cook. 1956. *Textbook of Theoretical Botany.* London: Longmans, Green and Co.

MEDUSA. The Medusa Database (http://medusa.maich.gr) and references contained therein. [n.d.] Home page (accessed November 12 2012).

Merzouki, A., F. Ed Derfoufi, and J.M. Mesa. 2000. Hemp (*Cannabis sativa* L.) and abortion. *J Ethnopharmacol* 73: 501–3.

Mirzaei, M., S.J. Nosratabadi, A. Derakhshanfar, and I. Sharifi. 2007. Antileishmanial activity of *Peganum harmala* extract on the *in vitro* growth of *Leishmania major* promastigotes in comparison to a trivalent antimony drug. *Veterinarski Arhiv* 77: 365–75.

Mohagheghzadeh, A., P. Faridi, M. Shams-Ardakani, and Y. Ghasemi. 2006. Medicinal smokes. *J Ethnopharmacol* 108: 161–84.

Molinar-Toribio, E., J. Pérez-Jiménez, S. Ramos-Romero, L. Gómez, N. Taltavull, M.R. Nogués, A. Adeva, O. Jáuregui, J. Joglar, P. Clapés, and J.L. Torres. 2015. D-Fagomine attenuates metabolic alterations induced by a high-energy-dense diet in rats. *Food Funct* 6(8): 2614–9.

Muellner, A.N., D.D. Vassiliades, and S.S. Renner. 2007. Placing *Biebersteiniaceae*, a herbaceous clade of Sapindales, in a temporal and geographic context. *Plant Syst Evol* 266: 233–52.

Nafisi, S., M. Bonsaii, P. Maali, M.A. Khalilzadeh, and F. Manouchehri. 2010b. Beta-carboline alkaloids bind DNA. *J Photochem Photobiol B* 100: 84–91.

Nakagawa, K., K. Ogawa, O. Higuchi, T. Kimura, T. Miyazawa, and M. Hori. 2010. Determination of iminosugars in mulberry leaves and silkworms using hydrophilic interaction chromatography-tandem mass spectrometry. *Anal Biochem* 404(2): 217–22.

Navchoo, I.A., and G.M. Buth. 1990. Ethnobotany of Ladakh, India: Beverages, narcotics, foods. *Econ Bot* 44(3): 318–21.

Nayar, S.L. 1954. Poisonous seeds of India. Part II. *J Bombay Nat Hist Soc* 52(2–3): 1–18.

Nayar, S.L. 1955. Vegetable insecticides. *Bull Natl Inst Sci India* 1955(4): 137–45.

Nojima, H., I. Kimura, F.J. Chen, Y. Sugihara, M. Haruno, A. Kato, and N. Asano. 1998. Antihyperglycemic effects of N-containing sugars from *Xanthocercis zambesiaca*, *Morus bombycis*, *Aglaonema treubii*, and *Castanospermum australe* in streptozotocin-diabetic mice. *J Nat Prod* 61(3): 397–400.

Ozenda, P. 1977. *Flore du Sahara*, Ed. du CNRS.

Padró, M., J.A. Castillo, L. Gómez, J. Joglar, P. Clapés, and C. de Bolós. 2010. Cytotoxicity and enzymatic activity inhibition in cell lines treated with novel iminosugar derivatives. *Glycoconj J* 27(2): 277–85.

Palevitch, D., and Z. Yaniv. 1991. Tzimḥey ha-marpe shel Eretz Yisra'el. *Medicinal Plants of the Holy Land*. Tel Aviv: Tamuz-Modan.

Paris, R., and G. Dillemann. 1960. *Les plantes médicinales des régions arides considérées surtout du point de vue pharmacologique*. Ed. UNESCO.

Parsa, A. 1960. Medicinal plants and drugs of plant origin in Iran. IV. *Plant Foods Human Nutr* 7(1): 65–136.

Pedrotti, G., and V. Bertoldi. 1930. *Nomi dialettali delle plante indigene del Trentino e della Ladinia Dolomitica*. Trento: G.B. Monauni.

Pelt, J.M. 1971. *Drogues et plantes magiques*. Paris: Horizons de France.

Peltier, J.P. 2006. Plant Biodiversity of South-Western Morocco. http://www.teline.fr (accessed March 16 2015).

Pennacchio, M., L.V. Jefferson, and K. Havens. 2010. *Uses and Abuses of Plant-Derived Smoke: Its Ethnobotany as Hallucinogen, Perfume, Incense, and Medicine*. Oxford: Oxford University Press.

PFAF. Plants for a Future Database. 1996. [Online database.] Plants for a Future, Lostwithiel, Cornwall, England. http://www.pfaf.org/ (accessed July 12 2014).

Phillips, R., and M. Rix. 1991. *Perennials*. Vols. 1–2. London: Pan Books.

Pieroni, A., H. Muenz, M. Akbulut, K.H. Başer, and C. Durmuşkahya. 2005. Traditional phytotherapy and trans-cultural pharmacy among Turkish migrants living in Cologne, Germany. *J Ethnopharmacol* 102: 69–88.

Porter, D.M. 2013. Peganum. Jepson Flora Project (eds.). *Jepson eFlora*. http://ucjeps.berkeley .edu/cgi-bin/get_IJM.pl?tid=36658 (accessed January 5 2015).

Prashanth, D., and S. John. 1999. Antibacterial activity of *Peganum harmala*. *Fitoterapia* 70: 438–9.

Purevdorjiin, T., and C. Gotovyn. 1994. Evaluation of *Peganum nigellastrum* effect on dopamine processes. *Vet Human Toxicol* 36(4): 351.

Quezel, P., and S. Santa. 1963. *Nouvelle flore de l'Algérie et des régions désertiques méridionales*. Vol. 2. Ed. du CNRS.

Rashan, L.J., M.H. Adaay, and A.L.T. Al-Khazraji. 1989. *In vitro* antiviral activity of the aqueous extract from the seeds of *Peganum harmala*. *Fitoterapia* 60(4): 365–7.

Razzack, H.M.A. 1980. The concept of birth control in Unani medical literature. Unpublished manuscript of the author.

Rehm, S. 1994. *Multilingual Dictionary of Agronomic Plants*. Dordrecht, Netherlands: Kluwer Academic. USDA, ARS, National Genetic Resources Program. Germplasm Resources Information Network (GRIN) [Online database.] National Germplasm Resources Laboratory, Beltsville, MD. http://www.ars-grin.gov/cgi-bin/npgs/html /stdlit.pl?Dict%20Rehm (accessed July 8 2014).

Ronse De Craene, L.P., J. De Laet, and E. Smets. 1996. Morphological studies in Zygophyllaceae. II. The floral development and vascular anatomy of *Peganum harmala*. *Am J Botany* 83: 201–15.

Ross, S.A., S.E. Megalla, D.W. Bishay, and A.H. Awad. 1980. Studies for determining anti-biotic substances in some Egyptian plants, Part II. Antimicrobial alkaloids from the seeds of *Peganum harmala* L. *Fitoterapia* 51(6): 309–12.

Saadabi, A.M. 2006. Antifungal activity of some Saudi plants used in traditional medicine. *Asian J Plant Sci* 5: 907–9.

Safina, L.K., R.G. Medvedeva, and L.D. Bryzgalova. 1970. Dinamika alkaloidov v gar-male obyknovennoj—*Peganum harmala* L. *Trudy Instituta Botaniki Akademij Nauk Kazakskoj SSR* 28: 226–35.

Saha, J.C., E.C. Savini, and S. Kasinathan. 1961. Ecbolic properties of Indian medicinal plants. Part 1. *Indian J Med Res* 49: 130–51.

Said, M. 1984. Potential of herbal medicines in modern medical therapy. *Ancient Sci Life* 4(1): 36–47.

Savolainen, V., M.F. Fay, D.C. Albach, A. Backlund, M. van der Bank, K.M. Cameron, S.A. Johnson, M.D. Lledó, J.-C. Pintaud, M. Powell, M.C. Sheahan, D.E. Soltis, P.S. Soltis, P. Weston, W.M. Whitten, K.J. Wurdack, and M.W. Chase. 2000. Phylogeny of the eudicots: A nearly complete familial analysis based on *rbc*L gene sequences. *Kew Bull* 55: 257–309.

Schipper, A., and O.H. Volk. 1960. The alkaloids of *Peganum harmala*. *Dtsch Apoth Ztg* 100: 255.

SEINet. Southwest Environmental Information Network, SEINet—Arizona Chapter. 2015. http://swbiodiversity.org/seinet/index.php (accessed March 16 2015).

Sezik, E., E. Yeşilada, H. Shadidoyatov, Z. Kulivey, A.M. Nigmatullaev, H.N. Aripov, Y. Takaishi, Y. Takeda, and G. Honda. 2004. Folk medicine in Uzbekistan. I. Toshkent, Djizzax, and Samarqand provinces. *J Ethnopharmacol* 92(2–3): 197–207.

Sezik, E., E. Yeşilada, M. Tabata, G. Honda, Y. Takaishi, T. Fujita, T. Tanaka, and Y. Takeda. 1997. Traditional medicine in Turkey. VIII. Folk medicine in East Anatolia: Erzurum, Erzincan, Agri, Kars, Igdir Provinces. *Econ Bot* 51(3): 195–211.

Shah, N.C. 1982. Herbal folk medicines in Northern India. *J Ethnopharmacol* 6(3): 293–301.

Shapira, Z., J. Terkel, Y. Egozi, A. Nyska, and J. Friedman. 1989. Abortifacient potential for the epigeal parts of *Peganum harmala*. *J Ethnopharmacol* 27(3): 319–25.

Sheahan, M.C., and M.W. Chase. 1996. A phylogenetic analysis of Zygophyllaceae R.Br. based on morphological, anatomical and *rbc*L DNA sequence data. *Bot J Linnean Soc* 122: 279–300.

Singh, B., O.P. Chaurasia, and K.L. Jadhav. 1996a. An ethnobotanical study of Indus Valley (Ladakh). *J Econ Taxon Bot Add Series* 12: 92–101.

Singh, G., and P. Kachroo. 1976. *Forest Flora of Srinagar*. Dehra Dun: Bishen Singh Mahendra Pal Singh.

Singh, V. 1995. Traditional remedies to treat asthma in North West and Trans-Himalayan Region in J. & K. State. *Fitoterapia* 65(6): 507–9.

Singh, V., and R.P. Pandey. 1998. *Ethnobotany of Rajasthan, India*. Jodhpur, India: Scientific Publishers.

Singh, V., B.K. Kapahi, and T.N. Srivastava. 1996b. Medicinal herbs of Ladakh especially used in home remedies. *Fitoterapia* 67(1): 38–48.

Sokmen, A., B.M. Jones, and M. Erturk. 1999. The *in vitro* antibacterial activity of Turkish medicinal plants. *J Ethnopharmacol* 67: 79–86.

Sorūšīān, J.S. 1956. *Farhang-e behdīnān*. Tehran.

Spjut, R.W. 2004. The World Botanical Associates. http://www.worldbotanical.com (accessed March 16 2015).

SRTD, AMS. Seed Regulatory and Testing Division, Agricultural Marketing Service, U.S.D.A. [n.d.] *State noxious-weed seed requirements recognized in the administration of the Federal Seed Act (updated annually)*. USDA, ARS, National Genetic Resources

Program. Germplasm Resources Information Network—(GRIN) [Online Database.] National Germplasm Resources Laboratory, Beltsville, Maryland. URL: http://www.ars-grin.gov.4/cgi-bin/npgs/html/stdlit.pl?State%20Noxweed%20Seed (accessed July 8 2014).

Steingass, F.J. 1892. *Persian-English dictionary*. London: Routledge and Kegan Paul Ltd.

Steinmetz, E.F. 1957. *Codex vegetabilis*. Amsterdam: [s.n.].

Stevens, P.F. 2001–. Angiosperm Phylogeny Website. Version July 12 2012 [and more or less continuously updated since.] http://www.mobot.org/MOBOT/research/APweb/ (accessed September 5 2014).

Tabata, M., E. Sezik, G. Honda, E. Yesilada, H. Fukui, K. Goto, and Y. Ikeshiro. 1994. Traditional medicine in Turkey III. Folk medicine in East Anatolia, Van and Bitlis Provinces. *Int J Pharmacog* 32(1): 3–12.

Täckholm, V. 1974. *Students' Flora of Egypt*. 2nd ed. Cairo: Cairo University.

Tahraoui, A., J. El-Hilaly, Z.H. Israili, and B. Lyoussi. 2007. Ethnopharmacological survey of plants used in the traditional treatment of hypertension and diabetes in south-eastern Morocco (Errachidia province). *J Ethnopharmacol* 110: 105–17.

Takhtadzhian, A.L. 1954–1966. *Flora Armenii*. Erevan: Izd-vo Akademii nauk Armianskoi SSR.

Tan, Q., and J. Zhang. 2016. Evodiamine and its role in chronic diseases. *Adv Exp Med Biol* 929: 315–28.

Taniguchi, S., N. Asano, F. Tomino, and I. Miwa. 1998. Potentiation of glucose-induced insulin secretion by fagomine, a pseudo-sugar isolated from mulberry leaves. *Horm Metab Res* 30(11): 679–83.

Tela Botanica: Le réseau de la botanique francophone. 1999. Association Tela Botanica. http://www.tela-botanica.org (accessed August 14 2014).

The Plant List. 2013. Home page. Version 1.1. http://www.theplantlist.org/ (accessed November 12 2014).

Torregrosa, S.G. [n.d.] Apatita: Flora de Alicante http://www.apatita.com/herbario/flora dealicante.html (accessed March 16 2015).

Townsend, C.C. and E. Guest, (eds). 1966–1980. *Flora of Iraq*. Baghdad, Iraq: Ministry of Agriculture and Agrarian Reform. USDA, ARS, National Genetic Resources Program. Germplasm Resources Information Network—(GRIN) [Online Database.] National Germplasm Resources Laboratory, Beltsville, Maryland. URL: http://www.ars-grin.gov.4/cgi-bin/npgs/html/stdlit.pl?F%20Iraq (accessed July 1 2014).

Tropicos®. Tropicos, botanical information system at the Missouri Botanical Garden. 2013. [Online database.] http://www.tropicos.org (accessed January 3 2014).

Umadevi, I., M. Daniel, and S.D. Sabnis. 1990. Chemotaxonomy of some Rutaceae. *Indian J Bot* 13: 23–8.

Uphof, J.C.Th. 1959. *Dictionary of Economic Plants*. Weinheim: Kramer.

Uphof, J.C.Th. 1968. *Dictionary of Economic Plants*. 2nd ed. Lehre: Kramer.

USDA Forest Service. 2014. *Field Guide for Managing African Rue in the Southwest*. United States Department of Agriculture. http://www.fs.usda.gov/detailfull/r3/forest-grassland health/invasivespecies/?cid=stelprdb5410084&width=full (accessed July 15 2015).

USDA, NRCS. 2015. The PLANTS Database. National Plant Data Team, Greensboro, NC. http://plants.usda.gov (accessed February 2 2015).

Usher, G. 1974. *A Dictionary of Plants Used by Man*. London: Constable.

Vail, A.M., and P.A. Rydberg. 1907. Zygophyllaceae. *N Am Flora* 25: 103–16.

van Wyk, B.-E., and M. Wink. 2004. *Medicinal Plants of the World*. Portland: Timber Press.

Wang, J., G. Luo, Y. Chen, Y. Zheng, and T. Zu. 2011. New record for China of host plant of *Cynomorium songaricum*: *Peganum multisectum*. *Zhongguo Zhong Yao Za Zhi* 36(23): 3244–6.

Watson, J.F. 1868. *Index to the Native and Scientific Names of Indian and other Eastern Economic Plants and Products.* London: Trubner & Co. USDA, ARS, National Genetic Resources Program. Germplasm Resources Information Network—(GRIN) [Online Database.] National Germplasm Resources Laboratory, Beltsville, Maryland. URL: http://www.ars-grin.gov.4/cgi-bin/npgs/html/stdlit.pl?Names%20Watson (accessed July 1 2014).

Watt, G., E. Thurston, and T.N. Mukharji. 1889–1896. *A Dictionary of the Economic Products of India.* Calcutta: Supt. of Govt. Print.

Wei, L., X. Jin, Z. Cao, and W. Li. 2016. Evodiamine induces extrinsic and intrinsic apoptosis of ovarian cancer cells via the mitogen-activated protein kinase/phosphatidylinositol -3-kinase/protein kinase B signaling pathways. *J Tradit Chin Med* 36(3): 353–9.

Wild Flowers of Israel. [n.d.] Home page. http://www.wildflowers.co.il/hebrew/plant.asp?ID=387 (accessed January 3 2015).

Winkler, H. 1912. *Botanisches Hilfsbuch für Pflanzer, Kolonialbeamte, Tropenkaufleute und Forschungsreisende.* Wismar: Hinstorffsche Verlagsbuchhandlung.

Wu, W.S., C.C. Chien, Y.C. Chen, and W.T. Chiu. 2016. Protein kinase RNA-like endoplasmic reticulum kinase-mediated Bcl-2 protein phosphorylation contributes to evodiamine-induced apoptosis of human renal cell carcinoma cells. *PLoS One* 11(8): e0160484.

Zheng, X.Y., Z.J. Zhang, G.X. Chou, T. Wu, X.M. Cheng, C.H. Wang, and Z.T. Wang. 2009. Acetylcholinesterase inhibitive activity-guided isolation of two new alkaloids from seeds of *Peganum nigellastrum* Bunge by an *in vitro* TLC-bioautographic assay. *Arch Pharm Res* 32(9): 1245–51.

Zhou, G.X., J.W. Ruan, M.Y. Huang, W.C. Ye, and Y.W. He. 2007. Alkaloid constituents from silkworm dropping of *Bombyx mori*. *Zhong Yao Cai* 30(11): 1384–5.

Zhou, J., G. Xie, and X. Yan. 2011. *Encyclopedia of Traditional Chinese Medicines: Molecular Structures, Pharmacological Activities, Natural Sources and Applications. Vol. 2: Isolated Compounds D–G.* London, UK: Springer Science and Business Media.

ZipcodeZoo. ZipcodeZoo.com. 2004. [Online Database.] BayScience Foundation. http://zipcodezoo.com http://zipcodezoo.com/Plants/P/Peganum_harmala/ (accessed July 22 2014).

Zohary, M. 1989. *Magdir ḥadaš le-tzimḥey Yisra'el. A New Analytical Flora of Israel.* Tel Aviv: Am Oved.

Zohary, M., and N. Feinbrun-Dothan. 1966–2004. *Flora Palaestina.* Jerusalem: Israel Academy of Sciences and Humanities.

3 Entheogenesis and Entheogenic Employment of Harmal

Entheogenesis may be both literally and liberally defined as "the creation, or invocation, of the deity, or Deity, within." It is a very ancient quest, and in one way or another, the basis of shamanism, animism, healing, and likely, according to numerous theoretical and scientific works, at the core of religion itself.

Indeed, even before the advent of religion, such as we know it today, even before ancient religions, humankind encountered growing, essentially nonmotile, organisms in the deserts and the forests, in the tundras and in the tropics. Many of these were so-called "higher" plants; others were actually nonphotosynthetic fungi, i.e., mushrooms. Among these plants and mushrooms were some that, when imbibed by eating, drinking, smoking, snuffing, or rubbing on the body for transcutaneous absorption or inserted as suppositories into the rectum or vagina for absorption through the delicate mucosa of those places, produced psychodynamic effects whereby the imbiber felt himself or herself connected, exposed, part of, or humbled by the world of spirit, the world animated through the spark of the Divine. He or she felt continuity, gratitude, and humility before the infinite and holy essence that underlies the entirety of creation, or some reasonable approximation of that. That is why pundits and theoreticians have speculated that inebriation with such *entheogenic* plants and mushrooms may have facilitated the emergence of religion among women and men.

The anxiety of humankind may be founded in awareness of the impending demise of its individual members, who may likely lack the wherewithal to weather it. This is a monstrous and inconsolable reality for most, and the quest for inebriation to the point of connecting to the Beyond presents a kind of way out, or at least a kind of cognitive reconstruction matching this formidable task. In any event, modern research with phosphorylated 4-indole tryptamine, i.e., psilocybin, the predominant psychoactive compound in *Psilocybe* mushrooms, has been shown, from even a single inebriation with this compound lasting less than 12 h in a clinical context, to often trigger long-term beneficial clinical changes commensurate with alleviation of depression, loss of fear of death, and mollification of the terminal anxiety of advanced stage 4 patients with life-threatening cancer (Griffiths et al. 2016; Nichols 2016; Nutt 2016; Ross et al. 2016). It may just be, therefore, that premodern human sought these compounds for the relief they could provide in confronting the perennial human concern of whether or not there is life after death. In an extremely mysterious manner, psilocybin and possibly related entheogenic compounds such as other serotonergic hallucinogens like harmine, dimethyltryptamine, and lysergic acid amides seem to, if administered to the patient in a safe and secure therapeutic set and setting, allay this depression and associated anxiety related to the fear of death. Thus, this symptom avoidance, with fear at its core, may have been a contributor

55

to the human's early fascination with plants and mushrooms that could open a portal to the Beyond (Figures 3.1 and 3.2).

In his excellent overview (Grof 2006) of practical considerations in navigating one's own approaching posthumous journey, the eminent psychiatrist Stanislav Grof presented multiple experiential apprehensions of death, or simulated death through entheogenic compounds or consciousness-altering techniques without drugs, the work drew from his own extensive experience with psychedelic psychotherapy, and later nondrug methods such as hyperventilation with evocative music, i.e., holotropic therapy, moving toward wholeness. Grof's insight is based in no small degree on his earlier work in Maryland with cancer patients using LSD (Grof et al. 1973; Grof and Halifax 1977; Pahnke et al. 1970). Grof's mature narrative in *The Ultimate Journey* spans mythology, shamanism, and religious practices from many cultures to confirm that death is not the end, but a transition, and that humans may prepare for this great transition with a clear vision and an open mind through the use of these practices, and/or entheogenic plants and mushrooms or their derivatives, or synthetic approximations or possible improvements (Grof 2006). Throughout Grof's work (Grof 1975, 1980, 2006), the concept of four basic perinatal matrices (BPMs) based on the earlier psychoanalytic ideas of Otto Rank (1924, 1994) is presented. Rank emphasized the importance of the trauma of birth as a deep etiologic factor in anxiety experienced later on throughout life. According to Grof's representation of Rank's core concept, there are four stages of birth, giving rise to the psychological experience of the four BPMs, or in Grof's language, their associated condensed organizations of experience (COEXs).

In BPM 1, the fetus is still awash with the oceanic unity of the womb and its amniotic fluid. All is still at peace, other than anxiety transmitted from time to time to the child by the mother or external environment. In BPM 2, the smooth muscles of the womb begin their first contractions, and suddenly the tranquility of the amniotic sea is disturbed. The waves have begun, and the first anxiety can be perceived. In BPM 3,

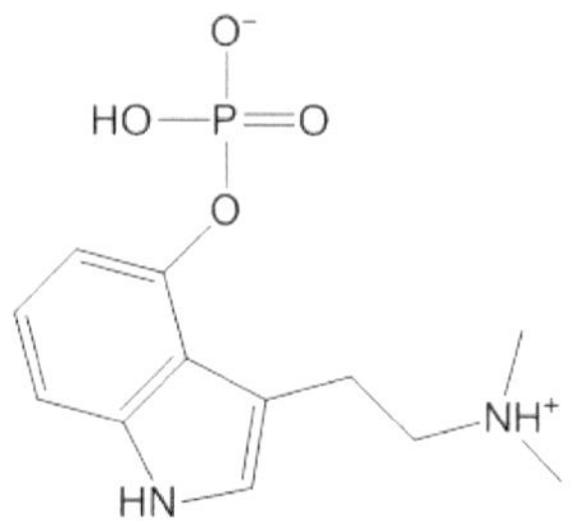

FIGURE 3.1   Psilocybin.

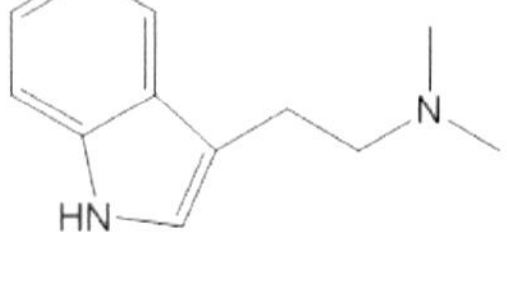

FIGURE 3.2   Dimethyltryptamine.

the contractions become stronger and closer together, and the experience for the fetus is one of tremendous pressure, turbulence, even violence, as it is recreated and experienced in Grof's subjects. In BPM 4, the gate finally opens and the child is born! There is enormous freedom and liberation, and in psychedelic recreation of this state, religious ecstasy and a new type of feeling of cosmic belonging in the Divine schema may occur. Also, to account for the "transpersonal" (i.e., beyond the personal) and archetypal experiences of BPM 4, Grof notes that the archetypal models cognized by Jung (1970) provide one with the possibility to travel further and with more accuracy and understanding in the transpersonal realms than the Rankian models by themselves allow.

Grof's work was conducted with LSD, at first in Czechoslovakia with low and frequent doses (50 µg), the psycholytic approach, and later in Maryland with high and infrequent doses (200 µg), the psychedelic approach, aimed at psychological breakthrough and spiritual rebirth. Although Grof later observed the same BPM progressions without LSD, but only with his holotropic breathwork, employing a combination of evocative music and hyperventilation (and later his models were clearly applicable to studies with psilocybin), it is not clear to what extent these models would help contextualize the psychopharmacological effects of harmal, which though in some ways similar to other classical serotonin-modulating hallucinogens like LSD and psilocybin, in other ways may vary.

One potential clue is the recent focus on the entheogenic brew, ayahuasca, which contains, from the vine *Banisteriopsis caapi*, harmala alkaloids such as are later described for *P. harmala*. So some overlap between the phenomenology of ayahuasca and inebriation with harmal might be reasonable to consider (Figure 3.3).

FIGURE 3.3 Preparing ayahuasca. (From Awkipuma via Wikimedia Commons https://commons.wikimedia.org/wiki/File:Ayahuasca_and_chacruna_cocinando.jpg, retrieved 3/7/2017.)

Ayahuasca is an aqueous decoction, which according to Luna (2011) contains either *Banisteriopsis caapi* with *Psychotria viridis* or *B. caapi* with *Diplopterys cabrerana*. It is the *B. caapi* that is pharmaceutically similar to *P. harmala* (Figure 3.4).

Ayahuasca is consumed in the modern world for entheogenic purposes, for example as a sacrament during religious services, which themselves likely emerged through appreciation of its sacrament. Considerable progress has been made by these groups in having their religions recognized and permission to employ the sacrament for religious reasons granted (Bullis 2008), though cultural taboos against the use of entheogenic substances, especially in the context of entheogen-centered religions, persist (Blainey 2015). Anyway, it is important to keep this context in mind when examining reports about clinical ameliorations following ayahuasca sessions... the sessions include not only the herbal drugs but also a religious ceremony. Clinical benefits to patients of both personal and intercessory prayer without ayahuasca are known.

Schenberg (2013) describes nine cases of patients with cancer who exhibited positive results following often only a single ayahuasca session. The group included regression of hepatic metastases from a colon carcinoma in one male patient, and a 50-year-old woman with ovarian cancer whose CEA 125 marker decreased from 4,000 to 600 following a single session with ayahuasca under both shamanic and medical supervision.

Aside from these intriguing reports from Schenberg, most of the reported putative medical benefits of ayahuasca are in the realm of psychiatry (Frecska et al. 2016). Dos Santos et al. (2016) reviewed documented anti-addictive, anxiolytic, and antidepressive properties of ayahuasca. One of the most striking characteristics of the research into the antidepressive action of ayahuasca is that, like in the case of psilocybin, a single session can produce deep and long lasting improvement (Sanches et al. 2016). Regular users adapt to the brew with less impairment in executive tasks (Barbosa et al. 2016; Bouso et al. 2013).

Ayahuasca enhances creative divergent thinking while decreasing conventional convergent thinking (Kuypers et al. 2016), increases mindfulness-related capacities (Soler et al. 2016), and treats addictions (Loizaga-Velder and Verres 2014; Tófoli and de Araujo 2016; Winkelman 2014). The unique combination of harmala alkaloids with serotonin-modulating properties and antioxidants in *Banisteriopsis caapi* supports its use for Parkinson's and Alzheimer's diseases, and related syndromes of neurodegeneration (Samoylenko et al. 2010; Wang et al. 2010). Ayahuasca may exert beneficial effects on human immunity (Dos Santos 2014) and at high doses is antidepressant in rat models (Pic-Taylor et al. 2015). The pharmacology of ayahuasca, traditionally centered on the harmala alkaloids (Freedland and Mansbach 1999), is expanding and in a dynamic state of research and development into understanding how ayahuasca achieves its effects: "...transient modified state of consciousness characterized by introspection, visions, enhanced emotions and recollection of personal memories... increases certain mindfulness facets related to acceptance and to the ability to take a detached view of one's own thoughts and emotions... enhancing self-acceptance and allowing safe exposure to emotional events" (Domínguez-Clavé et al. 2016).

We are very deeply indebted to Dr. David Stophlet Flattery and Professor Martin Schwartz of the University of California, Berkeley (Flattery and Schwartz 1989) for their fine, careful, and comprehensive thesis that *Peganum harmala*, and not the

(a)

(b)

FIGURE 3.4 *Banisteriopsis caapi*, used for making ayahuasca, is pharmaceutically similar to *Peganum harmala*. ([a] From Costa, P.P.P.R. via Wikimedia Commons, https:// commons.wikimedia.org/wiki/File:Banisteriopsis_caapi-CPPPR2.jpg; [b] 2/22/2014, Haiku, Maui, Hawaii. From Forest and Kim Starr, http://www.starrenvironmental.com/images /image/?q=25240510635.)

fungus *Amanita muscaria* (Wasson 1972; Wasson and Wasson 1957), was the actual "soma" of immortality in a tradition spanning both Iran and India. Flattery and Schwartz point out that according to the Avesta, the sacred writings of pre-Islamic Zoroastrianism, and a linguistic exploration of the word *haoma*, the word in Iran, or *soma*, the word in India, the substance had to be ground according to instructions in a mortar and pestle. Certainly, mushrooms, at least if they were fresh, would not have required grinding. Seeds, the active part of *P. harmala* in the summer and fall, and the roots, the most active part of *P. harmala* in the spring, would have required the traditional processing by mortar and pestle.

Flattery and Schwartz (1989) develop their argument that *P. harmala* was the divine *soma/haoma* of Indo-Iranian antiquity according to several criteria. These may be summarized as follows:

1. *Geographic.* Common plant over entire Indo-Iranian region. Easy accessibility.
2. *Pharmacologic.* Common harmala alkaloids between *P. harmala* and *Banisteriopsis caapi* vine used in South American Ayahuasca beverage—both esteemed for "...visually revealing a simultaneous, intangible spirit world interpreted as being a higher reality...," with the visions serving as impetus for core religious beliefs and institutions.
3. *Folk religions. Haoma* is the more formal name for *soma* (also *sauma*) restricted to religious contexts and preserved in the pre-Islamic *Avesta* of Zorastrianism. It is said to deliver healing, victory, salvation, and protection, originating in mountains, promote childbirth, employed as incense, and to deflect evil (apotropaic effect). It is the most widely used incense in Iran and the only one that contains psychoactive alkaloids.
4. *Zoroastrian rituals.* The imbibing of the *hoama* (putatively *P. harmala* extract) is associated with companion plants, namely *Ephedra* sp., *Punica granatum* (pomegranate), and *Ruta graveolens*. Because the rituals associated with *haoma* use have persisted throughout India as well as Iran—even when the inclusion of the psychoactive element has ceased as sometimes *Ephedra* or *Punica* are used as surrogate harmal—it is concluded that only *P. harmala* of the three herbs is capable of eliciting the strong psychoactive effect associated with the religious induction.

Concerning *Ephedra*, *Ruta*, and *Punica*, as noted, any of *Punica*, *Ephedra*, or *Ruta* may be called on to "stand in" for *Peganum*, when the wishes of the evolving culture toward modernization is to make the *haoma* ceremony less controversial, safer, and without substantial "built-in" influence owing to psychoactive drug plants such as *Peganum*. On the other hand, any of *Ephedra*, *Ruta*, and *Punica* may be not replacements for *Peganum*, but complements (Figures 3.5 and 3.6)

Herbs used in a formula are generally *complementary* to the principal herb, presumably in ways that are, even if presently unknown, physiologically and pharmacologically sensible. In the case of the three putatively complementary to harmal (note: we use the spelling "harmal," though Flattery and Schwartz employ "harmel"—both of us are referring to the same *Peganum harmala*), i.e., of *Ephedra*, *Ruta*, and *Punica*, only *Ephedra* and *Punica* are really putatively complementary. *Ruta* is often spoken of

FIGURE 3.5   *Ruta* spp., rue, and *Peganum harmala*, called "Syrian rue," were regularly treated as two species of the same genus in the ancient and medieval herbals. (6/4/2012, Red Butte Garden, Salt Lake City, Utah: by Andrey Zharkikh, https://www.flickr.com.)

in the same breath as *Peganum: Ruta* is known as "rue," *Peganum harmala* as "Syrian rue," "wild rue" or "mountain rue," though by no means should the latter imply that *P. harmala* restricts itself to a mountainous niche. Though both *Ruta* and *Peganum* are grossly similar and possibly mistakable—each also having highly odiferous leaves— the real explanation for *Ruta* is that it is a garden species introduced to the Zoroastrian liturgy much later than *Punica* or *Ephedra* and is more of a "counterfeit" *Peganum*, and one without relevant complementary psychoactive effects. In ancient times, there was no *Ruta* in the Middle East to be had; it was a much later introduction, probably arriving on board long-distance European vessels.

However, *Ephedra* was widely available and native, growing wild in Iran and environs. In those days and in those places, *Ephedra* tea was the house stimulant. Real Chinese tea and coffee were later introductions from afar, and not easy to be had. *Ephedra* was what was available for a general pick-me-up (Figure 3.7).

If *Peganum* were to be left out of the *haoma* to be drunk, the presence of *Ephedra* in a way is capable of holding the day. Today we could call the *Ephedra* an "active placebo," because although *Ephedra* does not produce recognizable entheogenic

(a)

(b)

FIGURE 3.6 Fruit of *Ruta graveolens,* a four- to five-lobed capsule that somewhat resembles the three-lobed fruit of *Peganum harmala.* ([a] 9/23/2004: by Corin Royal Drummond via Flickr https://www.flickr.com; [b] 3/11/2015, Jerusalem, Israel: by Helena Paavilainen.)

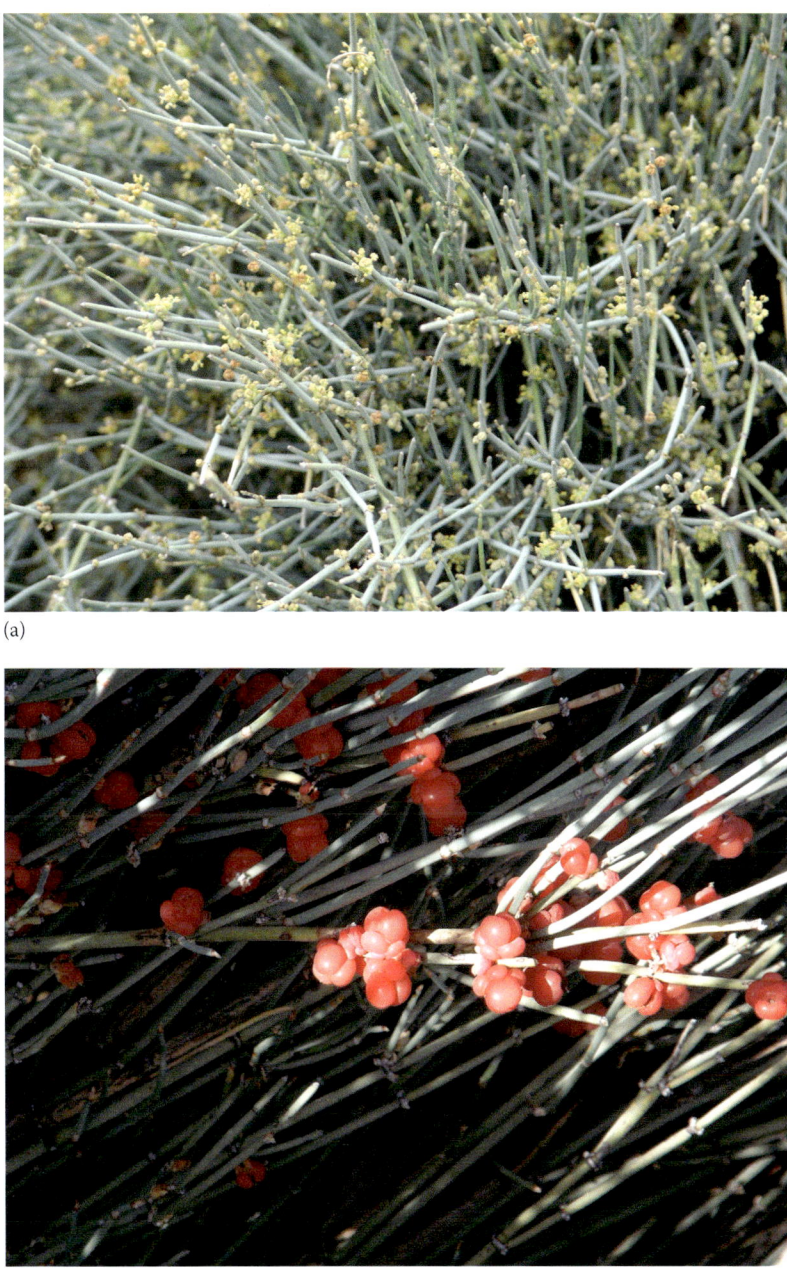

(a)

(b)

FIGURE 3.7 *Ephedra* spp. are known in ethnomedicine all over the world for their stimulant effects. (a) *Ephedra equisetina*, aerial parts. (4/24/2009, Denver Botanical Garden, Denver, Colorado: by Drew Avery, https://www.flickr.com/photos/33590535@N06/3495777730.) (b) *Ephedra* sp., fruit. (Lake Issyk-Kul, Kyrgyzstan: by Malcolm Manners, https://commons.wikimedia.org/wiki/File:Ephedra_sp._(4224427704).jpg.)

effects *à la* harmal, it is nonetheless stimulating, and the aspirant in those days, having the active *Peganum* left out of the sacrificial beverage, will at least "feel something," making it easier, perhaps, to get into a ceremonial or sacrificial mood.

Flattery and Schwartz, though, make an additional comment concerning *Ephedra,* namely that it is a true complement to *Peganum*—from the point of view of psychophar-macology, or perhaps, ethnopsychopharmacology. They base their intriguing sugges-tion in part on clinical studies of intravenous harmaline injection by legendary expert on entheogenic plants, Chilean psychiatrist Claudio Naranjo Cohen. Naranjo noted that harmaline produced essentially the same effect as harmine, but at less than half the dosage, and that the effect of harmaline could include lowering of blood pressure and somnolence (Naranjo 1967, 1973, 1975). Because *Ephedra* can cause blood pressure to rise, allay somnolence, and promote vigilance, its balancing effect on the *Peganum* prodrome seems sensible. Whether *Ephedra* could also in some way synergize with *Peganum* medically, say, as an anticancer agent, is a topic for further investigation.

Ephedra is not known as an entheogenic plant, though there are thousands of years of experience with it in China as *ma huang*, a hot and dry herb well-known for its benefit in treating asthma. Although ephedrine has long been considered as a kind of active component, there is improved safety in using complex extracts of *Ephedra* or even alkaloids extracted from the plant than synthetically obtained compounds. Unlike *Ruta*, it would be hard to mistake *Ephedra* for *Peganum* in the field, so it does not seem much of a replacement for *Peganum*. Yet as a complement it is most intriguing. In Israel, it is known as *sharvitan*, widely growing wild throughout the country. Following the report a few years ago, which appeared in an Israeli newspa-per, of a farmer whose goat (or sheep?) was apparently cured of her visible cancerous tumors, tea made from *Ephedra* stems has become very popular as a folk remedy for cancer among the Israeli public according to direct clinical observations of ESL. Recently, investigations on the anticancer potential of *sharvitan, ma huang/Ephedra* have begun and are showing great promise, clearly elucidating advantages of com-plex products over purified **ephedrine** alkaloids (Hyuga et al. 2016) and identifying non-ephedrine compounds, such as the flavonoid **herbacetin**, which may serve as an entourage compound contributing to an anticancer effect within the plant, and a useful marker for maintaining quality control within the *Ephedra* extract (Oshima et al. 2016). Therefore, the opportunity for designing and executing trials to investigate synergy between *Peganum* and *Ephedra*, particularly in the anticancer arena, beck-ons (Figures 3.8 and 3.9).

In the case of *Punica*, the connections to *Peganum* are more esoteric. First, it should be noted that pomegranate (*Punica granatum*) is a much-revered symbol in its own right. Its many seeds convey an image of fertility, for families-to-be and

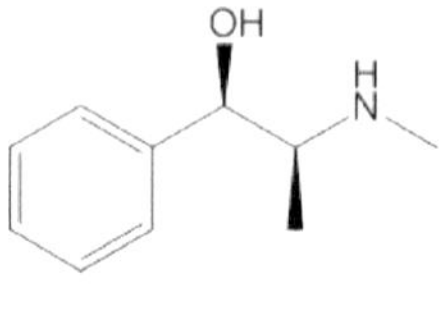

FIGURE 3.8   Ephedrine.

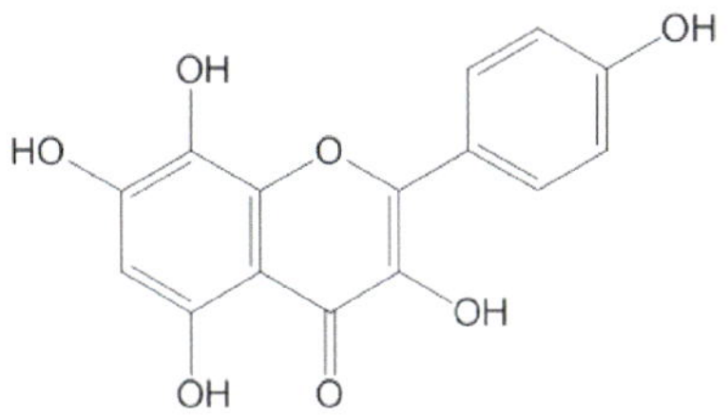

FIGURE 3.9 Herbacetin.

beyond. It is well-known as a symbol of medicine and was incorporated into the coats of arms of British medical societies for centuries—this is all reviewed beautifully by Langley (2000). In *Sefer Ha Rimon, The Book of the Pomegranate*, it is taught that, kabbalistically, pomegranate is betrothed to the Creator (Ben Shem-Tov 1988). According to Flattery and Schwartz, pomegranate was considered the mother of *Peganum*, *rimon* the mother of harmal. At one point it is written that if fresh *Peganum* juice cannot be obtained, then a couple of drops of pomegranate juice can suffice (Figure 3.10).

The imaginal associations begin with the multiplicity of seeds inside a *Peganum* capsule (Figure 3.11). The gross morphology of the *Peganum* fruit is similar to that of *Punica*. Further, the seeds within the pomegranate arils are of a shape reminiscent to the smaller seeds of *Peganum*—the seeds of both are very hard and have distinct surfaces like the facets of a jewel. Both the pomegranate seed and the harmal seed possess octadecatrienoic acids. Although the three double bonds in these 18 carbon fatty acids in *Peganum* are unconjugated, and those in *Punica* are conjugated (i.e., the double bonds alternate, in between which lies a single bond), their homology bespeaks just one more similarity between the seeds of these two

FIGURE 3.10 Opened pomegranate fruit. Note the seeds ensconced within the bright red arils. (10/11/2014: by Coniferconifer via Flickr https://www.flickr.com.)

FIGURE 3.11   Ripe seed capsules of *Peganum harmala*. (7/20/2015, Judaean Desert, by side of the Inn of the Good Samaritan, Israel: by Helena Paavilainen.)

(otherwise unrelated?) plants. The juice of the pomegranate arils is bright red, as is the ethanolic extract of harmal. By color, they appear surprisingly identical and could easily be misconstrued one for the other. The harmal fruits resemble miniature pomegranates. They clearly possess some of the spirit of their mother.

REFERENCES

Barbosa, P.C., R.J. Strassman, D.X. da Silveira, K. Areco, R. Hoy, J. Pommy, R. Thoma, and M. Bogenschutz. 2016. Psychological and neuropsychological assessment of regular hoasca users. *Compr Psychiatr* 71: 95–105.

Ben Shem-Tov, M. (Moses de Leon). 1988. *The Book of the Pomegranate: Moses de Leon's Sefer ha-Rimmon*. Ed. and trans. E.R. Wolfson. Atlanta, GA: Scholars Press.

Blainey, M.G. 2015. Forbidden therapies: Santo Daime, ayahuasca, and the prohibition of entheogens in Western society. *J Relig Health* 54(1): 287–302.

Bouso, J.C., J.M. Fábregas, R.M. Antonijoan, A. Rodríguez-Fornells, and J. Riba. 2013. Acute effects of ayahuasca on neuropsychological performance: Differences in executive function between experienced and occasional users. *Psychopharmacology (Berl)* 230(3): 415–24.

Bullis, R.K. 2008. The "vine of the soul" vs. the Controlled Substances Act: Implications of the hoasca case. *J Psychoactive Drugs* 40(2): 193–9.

Domínguez-Clavé, E., J. Soler, M. Elices, J.C. Pascual, E. Álvarez, M. de la Fuente Revenga, P. Friedlander, A. Feilding, and J. Riba. 2016. Ayahuasca: Pharmacology, neuroscience and therapeutic potential. *Brain Res Bull* 126(Pt 1): 89–101.

Dos Santos, R.G. 2014. Immunological effects of ayahuasca in humans. *J Psychoactive Drugs* 46(5): 383–8.

Dos Santos, R.G., F.L. Osório, J.A. Crippa, J. Riba, A.W. Zuardi, and J.E. Hallak. 2016. Antidepressive, anxiolytic, and anti-addictive effects of ayahuasca, psilocybin and lysergic acid diethylamide (LSD): A systematic review of clinical trials published in the last 25 years. *Ther Adv Psychopharmacol* 6(3): 193–213.

Flattery, D.S., and M. Schwartz. 1989. *Haoma and Harmaline: The Botanical Identity of the Indo-Iranian Sacred Hallucinogen "Soma" and Its Legacy in Religion, Language, and Middle-Eastern Folklore.* Berkeley: University of California Press.

Frecska, E., P. Bokor, G. Andrassy, and A. Kovacs. 2016. The potential use of ayahuasca in psychiatry. *Neuropsychopharmacol Hung* 18(2): 79–86.

Freedland, C.S., and R.S. Mansbach. 1999. Behavioral profile of constituents in ayahuasca, an Amazonian psychoactive plant mixture. *Drug Alcohol Depend* 54(3): 183–94.

Griffiths, R.R., M.W. Johnson, M.A. Carducci, A. Umbricht, W.A. Richards, B.D. Richards, M.P. Cosimano, and M.A. Klinedinst. 2016. Psilocybin produces substantial and sustained decreases in depression and anxiety in patients with life-threatening cancer: A randomized double-blind trial. *J Psychopharmacol* 30(12): 1181–97.

Grof, S. 1975. *Realms of the Human Unconscious: Observations from LSD Research.* New York: Viking Press.

Grof, S. 1980. *LSD Psychotherapy.* Alameda, CA: Hunter House.

Grof, S. 2006. *The Ultimate Journey: Consciousness and the Mystery of Death.* Santa Cruz, CA: Multidisciplinary Association for Psychedelic Studies (MAPS).

Grof, S., and J. Halifax. 1977. *The Human Encounter with Death.* New York: E.P. Dutton.

Grof, S., L.E. Goodman, W.A. Richards, and A.A. Kurland. 1973. LSD-assisted psychotherapy in patients with terminal cancer. *Int Pharmacopsychiatr* 8(3): 129–44.

Hyuga, S., M. Hyuga, N. Oshima, T. Maruyama, H. Kamakura, T. Yamashita, M. Yoshimura, Y. Amakura, T. Hakamatsuka, H. Odaguchi, Y. Goda, and T. Hanawa. 2016. Ephedrine alkaloids-free Ephedra Herb extract: A safer alternative to ephedra with comparable analgesic, anticancer, and anti-influenza activities. *J Nat Med* 70(3): 571–83.

Jung, C.G. 1970. Mysterium Coniunctionis: The components of the coniunctio. 1. The opposites. In: C. Jung, *Collected Works of C. G. Jung*, Vol. 14. 2nd ed. Princeton, NJ: Princeton University Press.

Kuypers, K.P., J. Riba, M. de la Fuente Revenga, S. Barker, E.L. Theunissen, and J.G. Ramaekers. 2016. Ayahuasca enhances creative divergent thinking while decreasing conventional convergent thinking. *Psychopharmacology (Berl)* 233(18): 3395–403.

Langley, P. 2000. Why a pomegranate? *BMJ* 321(7269): 1153–4.

Loizaga-Velder, A., and R. Verres. 2014. Therapeutic effects of ritual ayahuasca use in the treatment of substance dependence: Qualitative results. *J Psychoactive Drugs* 46(1): 63–72.

Luna, L.E. 2011. Indigenous and mestizo use of ayahuasca: An overview. In: R. Guimarães dos Santos, ed., *The Ethnopharmacology of Ayahuasca.* 1st ed. Trivandrum, India: Transworld Research Network, pp. 1–21 (cited by Schenberg 2013).

Naranjo, C. 1967. Psychotropic properties of the harmala alkaloids. In: E.H. Efron, ed., *Ethnopharmacologic Search for Psychoactive Drugs.* Washington, DC: US Government Printing Office.

Naranjo, C. 1973. Psychological aspects of the of the *yage* experience in an experimental setting. In: M.J. Harner, ed., *Hallucinogens and Shamanism.* Oxford: Oxford University Press.

Naranjo, C. 1975. Harmine and the collective unconscious. In: C. Naranjo, *The Healing Journey: New Approaches to Consciousness.* New York: Ballantine Books.

Nichols, D.E. 2016. Psychedelics. *Pharmacol Rev* 68(2): 264–355.

Nutt, D. 2016. Psilocybin for anxiety and depression in cancer care? Lessons from the past and prospects for the future. *J Psychopharmacol* 30(12): 1163–4.

Oshima, N., T. Yamashita, S. Hyuga, M. Hyuga, H. Kamakura, M. Yoshimura, T. Maruyama, T. Hakamatsuka, Y. Amakura, T. Hanawa, and Y. Goda. 2016. Efficiently prepared ephedrine alkaloids-free Ephedra Herb extract: A putative marker and antiproliferative effects. *J Nat Med* 70(3): 554–62.

Pahnke, W.N., A.A. Kurland, S. Unger, C. Savage, and S. Grof. 1970. The experimental use of psychedelic (LSD) psychotherapy. *JAMA* 212(11): 1856–63.

Pic-Taylor, A., L.G. da Motta, J.A. de Morais, W.M. Junior, F. Santos Ade, L.A. Campos, M.R. Mortari, M.V. von Zuben, and E.D. Caldas. 2015. Behavioural and neurotoxic effects of ayahuasca infusion (*Banisteriopsis caapi* and *Psychotria viridis*) in female Wistar rat. *Behav Processes* 118: 102–10.

Rank, O. 1994. *The Trauma of Birth*. Reprint of 1929 translation (original publication 1924). Mineola, NY: Dover Publications.

Ross, S., A. Bossis, J. Guss, G. Agin-Liebes, T. Malone, B. Cohen, S.E. Mennenga, A. Belser, K. Kalliontzi, J. Babb, Z. Su, P. Corby, and B.L. Schmidt. 2016. Rapid and sustained symptom reduction following psilocybin treatment for anxiety and depression in patients with life-threatening cancer: A randomized controlled trial. *J Psychopharmacol* 30(12): 1165–80.

Samoylenko, V., M.M. Rahman, B.L. Tekwani, L.M. Tripathi, Y.H. Wang, S.I. Khan, I.A. Khan, L.S. Miller, V.C. Joshi, and I. Muhammad. 2010. *Banisteriopsis caapi*, a unique combination of MAO inhibitory and antioxidative constituents for the activities relevant to neurodegenerative disorders and Parkinson's disease. *J Ethnopharmacol* 127(2): 357–67.

Sanches, R.F., F. de Lima Osório, R.G. Dos Santos, L.R. Macedo, J.P. Maia-de-Oliveira, L. Wichert-Ana, D.B. de Araujo, J. Riba, J.A. Crippa, and J.E. Hallak. 2016. Antidepressant effects of a single dose of ayahuasca in patients with recurrent depression: A SPECT study. *J Clin Psychopharmacol* 36(1): 77–81.

Schenberg, E.E. 2013. Ayahuasca and cancer treatment. *SAGE Open Med* 1: 2050312113508389.

Soler, J., M. Elices, A. Franquesa, S. Barker, P. Friedlander, A. Feilding, J.C. Pascual, and J. Riba. 2016. Exploring the therapeutic potential of Ayahuasca: Acute intake increases mindfulness-related capacities. *Psychopharmacology (Berl)* 233(5): 823–9.

Tófoli, L.F., and D.B. de Araujo. 2016. Treating addiction: Perspectives from EEG and imaging studies on psychedelics. *Int Rev Neurobiol* 129: 157–85.

Wang, Y.H., V. Samoylenko, B.L. Tekwani, I.A. Khan, L.S. Miller, N.D. Chaurasiya, M.M. Rahman, L.M. Tripathi, S.I. Khan, V.C. Joshi, F.T. Wigger, and I. Muhammad. 2010. Composition, standardization and chemical profiling of *Banisteriopsis caapi*, a plant for the treatment of neurodegenerative disorders relevant to Parkinson's disease. *J Ethnopharmacol* 128(3): 662–71.

Wasson, R.G. 1972. *Soma: Divine Mushroom of Immortality*. New York: Harcourt Brace Jovanovich.

Wasson, V.P., and R.G. Wasson. 1957. *Mushrooms, Russia and History*. New York: Pantheon Books.

Winkelman, M. 2014. Psychedelics as medicines for substance abuse rehabilitation: Evaluating treatments with LSD, Peyote, Ibogaine and Ayahuasca. *Curr Drug Abuse Rev* 7(2): 101–16.

4 Overview of Employment of Harmal in Traditional Iranian Medicine

Mina Cheraghi Niroumand and coworkers in Tehran (Cheraghi Niroumand et al. 2015) systematically reviewed the literature on major medical databases pertinent to *Peganum harmala*, *P. harmala*, and "Syrian rue" concerning traditional Persian, or traditional Iranian, medicine (TIM), and presented their comprehensive findings within the context of the TIM doctrine of four elements and temperament found in ancient Persian medical texts that they surveyed (Aghili 2009; Tonkaboni 2007; both as reported by Cheraghi Niroumand et al. 2015). Unlike in modern allopathic medicine, which takes no real interest in the individual constitution of the patient when prescribing medicines, the TIM prescriber views each patient within the context of the "quadratic theory of temperament." Accordingly, the four elements are air, which is hot and moist; water, which is cold and wet; fire, which is hot and dry; and soil, which is cold and dry. The seeds of harmal, also known in Persian as *esphand*, are considered to be in the realm of fire, i.e., hot and dry, and therefore, great care must be exercised in prescribing harmal to patients having a "fire" (hot and dry) constitution. In fact, it may be considered generally contraindicated to prescribe harmal/esphand seeds to patients of this constitutional type. "Fire" constitution patients, according to TIM, will be high energy, get hot more readily than normal in hot conditions, get warm more readily than normal in cold conditions, are more intolerant than normal of hunger, and are more prone than normal to sore eyes, headaches, a sudden hot feeling in a normal condition, and arterial hypertension. The TIM concept is such that hot, dry (fire) persons are more susceptible than their peers to the toxicity of harmal, a constitutionally hot and dry plant coincidentally arising in and demanding hot and dry ecological niches in nature. According to theses authors, toxic reactions to the harmal seed may be ameliorated or eliminated by coprescribing juice of sour fruits or oxymel, the latter a mixture of honey and vinegar, or similarly, prescribing harmal in a blend with other herbal materials to balance the effect and mediate the "fire."

The authors review the known therapeutic and toxic effects of esphand/harmal in TIM (Figure 4.1). Harmal is a reliable abortifacient and emmenagogue, and so its use in pregnant women is contraindicated absolutely. Beneficial effects of *P. harmala* seeds according to TIM include (1) analgesic (for the management of sciatica, joint pain, coxalgia, chronic headache, and toothache), (2) disinfectant (antibacterial and antifungal), (3) anthelmintic, (4) insect repellant, (5) carminative, (6) benefiting colic disorder (which they relate to known modern concepts of calcium channels blockade), (7) antispasmodic, (8) antiepileptic, (9) antidepressant, (10) improving visual

FIGURE 4.1   Ripe *Peganum*. (7/20/2015, Judaean Desert, by side of the Inn of the Good Samaritan, Israel: by Helena Paavilainen.)

acuity, (11) aphrodisiac, (12) phlegmatic, (13) purgative, and (14) benefiting kidney stone. Antithrombotic effects, according to the authors, are presumptively related to known anti-inflammatory properties of harmal (via inhibition of NFkappaB and TNFalpha). Benefits for dropsy, jaundice, psychosis, and memory loss are also noted, but as yet unstudied by modern pharmacognosy.

Aqueous extract of the seeds is said to enable blood purification, while a mixture of the ground seeds with flax seeds and honey purportedly benefits dyspnea. Aqueous decoction has been utilized for numbness, liver, and lung diseases. The poultice from the seeds is applied externally against paralysis, numbness, joint pain, coxalgia, and back pain. The incense of the seeds is used against toothache and for repelling mosquitos. The most popular traditional use of harmal is as a disinfectant, the smoke for which is also obtained by direct heat to the seeds. The disinfection may also be used to ward off the "evil eye" (Cheraghi Niroumand et al. 2015), just as the strong tobacco smoke of another harmine-containing plant, *Nicotinia* sp., is used by Peruvian shamans to ward off hostile attacks from their rivals (Luna 1984a,b).

Cheraghi Niroumand et al. (2015) also report on the dosage of harmal seed in TIM as being between 4 and 9 g. This is in contrast to other reports, such as the recommended dose given by Kapoor (2000) for Ayurvedic prescribing as between 3 and 6 g a day. According to Cheraghi Niroumand, however, the discrepancy between their finding of dosage in TIM and doses that produce either mild stimulation (25–50 mg) or hallucinations (300–750 mg) may be due to the presence of additional materials (e.g., other herbs, honey, etc), which they refer to as "correctives" in the TIM harmal medicament.

Toxic reactions to harmal recognized by TIM include dizziness, nausea, vomiting, and headache. Neurosensorial symptoms, visual hallucination, slight elevation of body temperature, cardiovascular disorder (bradycardia and low blood pressure), ataxia, diffuse tremors, psychomotor agitation, and headache may all occur (Cheraghi Niroumand et al. 2015), while high doses may lead to liver degeneration, paralysis, spongiform alterations in the central nervous system, euphoria, hypothermia, convulsions, and bradycardia (Frison et al 2008; Lamchouri et al. 2002; Mahmoudian et al. 2002—cited by Cheraghi Niroumand et al. 2015), which are suggested to derive largely from the known monoamine oxidase (MAO) inhibition property of harmal. Cheraghi Niroumand et al. further note that the DNA intercalating property of harmine may lead to genotoxic effects, a further reason for the general contraindication of harmal during pregnancy, though therapeutic doses are considered to be safe in rodents.

External uses of harmal are also not, and should not be neglected. Peganum Oil is a feature of traditional Iranian pharmacy prepared according to traditional methods (Abolhassanzadeh et al. 2015a). Recently, a clinical trial (double blind) was conducted with this oil against an olive oil placebo. Patients with osteoarthritis rubbed their oil into their affected knee, two drops, three times per day (TID). After 4 weeks, the group receiving the active oil had a significant improvement in their pain and discomfort according to the McMaster arthritis questionnaire and other instruments (Abolhassanzadeh et al. 2015b). The work was conducted in laboratories under the supervision of Professor Abdolali Mohaghegzadeh at the Department of Traditional Pharmacy, Faculty of Pharmacy, Shiraz University of Medical Sciences, Shiraz, Iran.

REFERENCES

Abolhassanzadeh, Z., E. Aflaki, G. Yousefi, and A. Mohagheghzadeh. 2015a. Peganum Oil in Iranian traditional pharmacy. *Pharm Hist (Lond)* 45(2): 34–6.

Abolhassanzadeh, Z., E. Aflaki, G. Yousefi, and A. Mohagheghzadeh. 2015b. Randomized clinical trial of peganum oil for knee osteoarthritis. *J Evid Based Complement Altern Med* 20(2): 126–31.

Aghili, M.H. 2009. *Makhzan-al-Advia* [in Persian]. *Tehran: Tehran University of Medical Sciences* 9: 328.

Cheraghi Niroumand, M., M.H. Farzaei, and G. Amin 2015. Medicinal properties of *Peganum harmala* L. in traditional Iranian medicine and modern phytotherapy: A review. *J Tradit Chin Med* 35(1): 104–9.

Frison, G., D. Favretto, F. Zancanaro, G. Fazzin, and S.D. Ferrara. 2008. A case of beta-carboline alkaloid intoxication following ingestion of *Peganum harmala* seed extract. *Forensic Sci Int* 179: e37–43.

Kapoor, L.D. 2000. *Handbook of Ayurvedic Medicinal Plants*. Herbal Reference Library. Boca Raton, FL: CRC Press.

Lamchouri, F., A. Settaf, Y. Cherrah, M. El Hamidi, N. Tligui, B. Lyoussi, and M. Hassar. 2002. Experimental toxicity of *Peganum harmala* seeds. *Ann Pharm Fr* 60(2): 123–9.

Luna, L.E. 1984a. The healing practices of a Peruvian shaman. *J Ethnopharmacol* 11(2): 123–33.

Luna, L.E. 1984b. The concept of plants as teachers among four mestizo shamans of Iquitos, northeastern Peru. *J Ethnopharmacol* 11(2): 135–56.

Mahmoudian, M., H. Jalilpour, and P. Salehian. 2002. Toxicity of *Peganum harmala*: Review and a Case Report. *Iran J Pharmacol Therapeut* 1: 1–4. http://ijpt.iums.ac.ir.

Tonkaboni, M.M. 2007. *Tohfeh al-Momenin* [Persian]. *Tehran: Shahid Beheshti University of Medical Sciences Publication*, 150.

5 Harmal as Anticancer Agent in Traditional Chinese and Iranian Medicine

Sobhani et al. (2002) explained that *Peganum harmala* is the most important and common medicinal herb used in the treatment of cancer by contemporary Iranian folk healers, with at least some of these field practitioners having had some considerable degree of success in treating neoplastic disease by this method. Sobhani et al. (2002) note, as have other authors (Jahaniani et al. 2005; Talari et al. 2014), that the harmal in TIM is typically used in concert with another Iranian medicinal herb, *Dracocephalum kotschyi* Boiss.

Harmal has enjoyed substantial ethnographic employment as an anticancer medicament in China, where *P. harmala* established itself after migrating from the Middle Eastern lands of its likely origin. A Chinese name for *P. harmala* sometimes contains the ideogram for camel, possibly because it was on the backs of camels from afar that the plant undoubtedly made its way to China along the Silk Road. Harmal is also known in Chinese as "camel artemisia." The toxicity of the plant—to decrease respiratory rate and blood pressure and produce convulsions in high doses—is recognized, and its principal harmala alkaloids are noted to be "more hallucinogenic than mescaline."

Many preclinical studies on *P. harmala* of course have been conducted in China, and those few reported in this volume represent only a fraction of what has been learned regarding *P. harmala* in China itself. Among the reports of such studies, a few points are worthy of mention.

- Harmal seed extract inhibits not only topoisomerase Type 1, but also topoisomerase Type 2, and the complex extract was more effective in this inhibition than pure harmine or harmaline (Wang et al. 2008).
- Harmal seed extract promotes apoptosis in many human cancer cell lines, including pancreatic.

What is of most interest is the clinical usage of harmal for cancer treatment.

There is one such report with the title: "*Peganum harmala*, musk, Laladeng, sea buckthorn—anticancer effect." It was published online on October 13, 2007 and was a collaborative effort between authors from Logan University of Chiropractic in Chesterfield, Missouri, and from **Tongji** and **Zhejiang** Universities of traditional

Chinese medicine in China. These workers report successes regarding topical employment of harmal preparations for

- Skin cancer
- Cervical cancer
- Lip cancer
- Anal cancer

Obviously, all these cancers share in common their ready accessibility for the placement of suitable external preparations, even when used intravaginally for cervical cancer or intrarectally for anal cancer.

A preparation from Xinjiang containing a total alkaloid extract of *P. harmala* seed, used for tumors of esophagus and stomach in 21 cases, was reported. A powder of the seeds (possibly diluted) 4 g po bid or tid was reportedly employed in eight cases of gastric cancer. Improved appetite or alleviated visible symptoms were noted in six cases. Patients were treated for 2–9 months, using total alkaloid tablets, 10 mg/tablet, or injectable 5 mg/ml, with total 30 mg for intramuscular infectious or intravenous infusions. Complete remission was described in 19%, partial remission in 66.7% of cases, effective in some way 85.7%, and two patients had a >5 year survival. A reference given is the *Chinese Journal of Thoracic and Cardiovascular Surgery* [7,1 (1991): 37–38].

Awareness of the anticancer effects of *P. harmala* in China increased with preclinical reports in the 1980s. Synonyms included camel artemisia, bitter vegetables, smelly grass, and stinky peony. Mention is made of its occurrence in Avicenna's "Medical Code." In Uighur medicine (Adrasman), mention is made of a flat, bitter, toxic medicine that strengthens tendons, helps *yang* to warm *yin*, and eliminates viscous body fluids and cold dampness. It is bitter, astringent, and warm; moves through the meridians; and expels dampness and pain.

An anticancer preparation associated with the Xinjiang Uygur Autonomous Region Uyghur Hospital consisted of *P. harmala*, cardamom, and five other herbs in "honey cream." It was noted that 100 cases were treated there over 20 years with an effective rate of 77.2%. A harmal oral emulsion and also capsules were reported to be well-tolerated and efficacious against cancers of stomach, liver, breast, and lung.

REFERENCES

Jahaniani, F., S.A. Ebrahimi, N. Rahbar-Roshandel, and M. Mahmoudian. 2005. Xanthomicrol is the main cytotoxic component of *Dracocephalum kotschyii* and a potential anticancer agent. *Phytochemistry* 66(13): 1581–92.

Sobhani, A.M., S.A. Ebrahimi, and M. Mahmoudian. 2002. An *in vitro* evaluation of human DNA topoisomerase I inhibition by *Peganum harmala* L. seeds extract and its betacarboline alkaloids. *J Pharm Pharm Sci* 5(1): 19–23.

Talari, M., E. Seydi, A. Salimi, Z. Mohsenifar, M. Kamalinejad, and J. Pourahmad. 2014. *Dracocephalum*: Novel anticancer plant acting on liver cancer cell mitochondria. *Biomed Res Int* 2014: 892170.

Wang C.H., X.M. Cheng, and Z.Y. Liu. 2008. Inhibitory effects of Peganum harmala seed extract and β-carbobenzidine alkaloid on DNA topoisomerase II activity [J]. *Chin J Clin Pharmacol* 24(5): 422–5.

6 Phytochemistry of Harmal

ALKALOIDS

Alkaloids are organic compounds, i.e., compounds based on carbon, which also contain at least one nitrogen atom. Alkaloids are frequently highly active physiologically in animals, and may cause death, nervous system intoxication, paralysis, and other effects related to poisoning. Catching these effects and riding their waves in a controlled manner is the art of pharmacotherapeutics and, to a not insignificant degree, the art of medicine.

The alkaloids in *P. harmala* seeds are of two known types, both of which are highly physiologically active in animals and also in other plants of different genera that *P. harmala* contacts. Accordingly, all the alkaloids of *P. harmala*, in spite of or because of their specific toxicities, possess marked medicinal potential for treatment of humankind. Thus, *P. harmala* alkaloids are of either the β-carboline class of indole alkaloids, represented by harmine, or quinazoline alkaloids, exemplified by peganine.

The β-carboline indole alkaloids within *P. harmala* seeds are derived from **tryptamine**, a derivative of the indolic amino acid **tryptophan**. The parent skeleton is **indole** itself, the key plant growth hormone. *P. harmala* β-carboline indole alkaloids are also commonly known as "harmala alkaloids" after the plant species in question (Figures 6.1 through 6.3). Harmine is further found in other plants besides *P. harmala*, including the South American "vine of the soul," *Banisteriopsis caapi*, the "engine" of the hallucinogenic brew, ayahuasca (Shanon 2010), tobacco, *Nicotinia tabacum*, and passion flowers (*Passiflora incarnata*).

One author emphasized the ubiquity of harmine across many plant species in 100% of insect-pollinated species studied, including lemon balm (*Melissa officinalis*), common rue (*Ruta graveolens*), meadow rue (*Thalictrum aquilegifolium*), hydrangea (*Hydrangea arborescens*), spirea (*Spirea japonica*), forget-me-not (*Myosotis scorpioides*), and blue star grass (*Sisyrinchium augustifolium*). Wind-pollinated plants such as sugar maple (*Acer saccharum*), white velvet (*Tradescantia sillamontana*), another plant also known as meadow rue (*Thalictrum ichangense*), and rhoeo (*Rhoeo spathacea*) did not contain harmine. She observed that harmine and its harmala alkaloid congeners harmol, harmalol, and harmaline fluoresce in the visible range of 300–700 nm, which coincidentally overlays bees' visual range of 300–600 nm, speculating that these compounds, in spite of their insecticidal properties, may also serve as important insect attractants in species dependent on insect pollination (Harrington 2012).

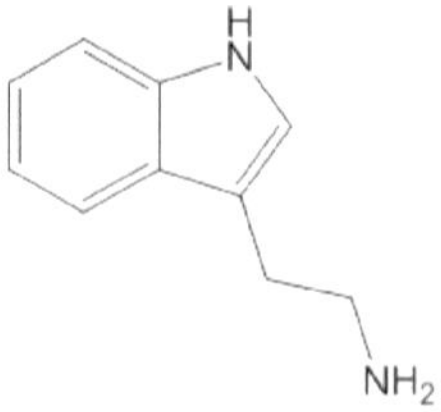

FIGURE 6.1   Tryptamine.

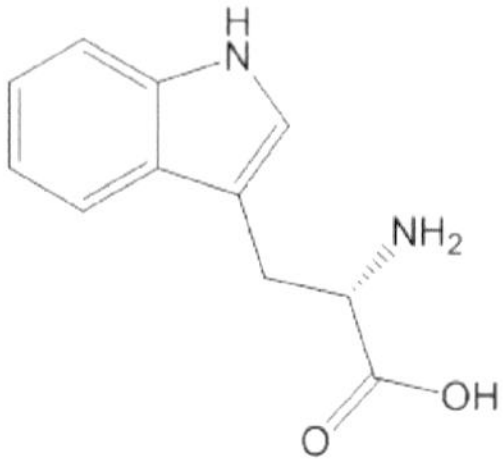

FIGURE 6.2   Tryptophan.

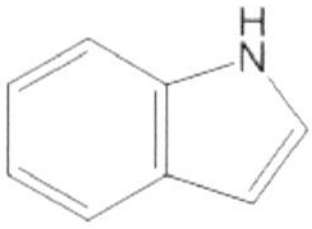

FIGURE 6.3   Indole.

Related β-carbolines are produced endogenously by man, e.g., harmane, puta-tively an atypical neurotransmitter (Abu Ghazaleh et al. 2015). The β-carboline, **10-methoxyharmalan** (10-Me-HML, also known as 6-methoxyharmalan), cyclized in the pineal gland from another tryptamine derivative, the cousin of **serotonin**, **melatonin**, could account for psychotic phenomena during mental illness (McIsaac et al. 1961), or for ecstatic states associated with "rising *kundalini*" during yogic ecstasy (Lansky 1979) (Figures 6.4 through 6.6).

The second great class of *Peganum* alkaloids are the quinazoline alkaloids, deriv-ative of the parent compound **quinazoline**, a fusion product of benzene and pyrimi-dine structures. Quinazoline alkaloids represent a vigorous area of drug exploration, with derivatives being developed as oncologics (Ravez et al. 2015), neurologics (Kaczmarek et al. 2015), and cardiologics (Patanè 2015). Potential of quinazolines

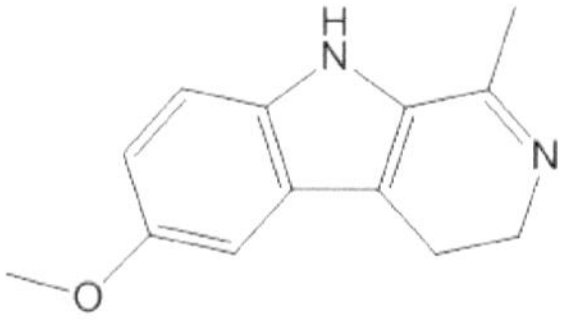

FIGURE 6.4   6-Methoxyharmalan/10-methoxyharmalan.

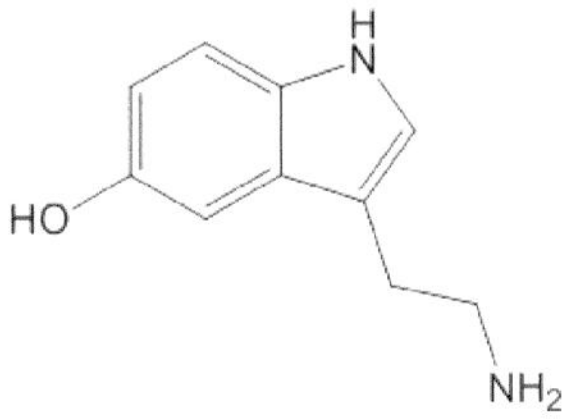

FIGURE 6.5   Serotonin.

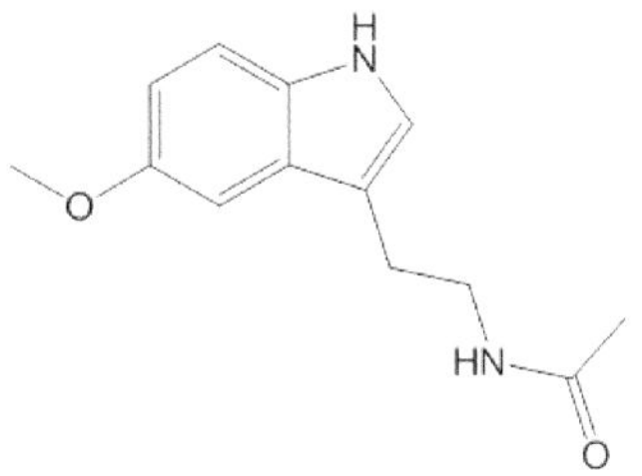

FIGURE 6.6   Melatonin.

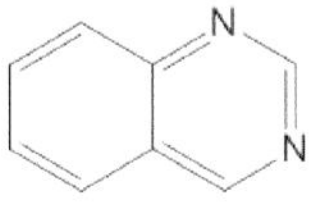

FIGURE 6.7   Quinazoline.

includes anti-HIV, anticancer, antifungal, antibacterial, antimutagenic, anticoccidial, anticonvulsant, anti-inflammatory, antidepressant, antimalarial, antioxidant, antileukemic, and antileishmanial applications (Asif 2014) (Figure 6.7).

SPECIFIC ALKALOIDS

Harmala Alkaloids

Harmine

In the July 2015 issue of *Nature Immunology*, Fehervari (2015) speaks of "harmine-ization" of the immune system, whereas only harmine of 3,000 compounds tested by NIH and Harvard inhibited the tyrosine-phosphorylation-regulated kinase DYRK1A, thus enhancing differentiation of T-regulatory T cells (T_{reg}), while attenuating differentiation of Th17 (a type of helper T cell itself with pro-inflammatory potential) "without affecting known pathways of T_{reg}/Th17 differentiation." Thus harmine did also reduce inflammation in the authors' multiple experimental models of systemic autoimmunity and mucosal inflammation (Khor et al. 2015). Pursuit of more powerful and selective synthetic harmine analogues for use as DYRK1A inhibitors is vigorous (Rüben et al. 2015) (Figure 6.8).

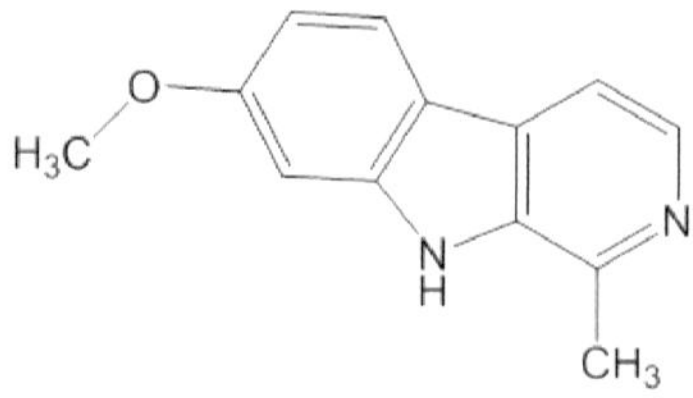

FIGURE 6.8   Harmine (7-MeO-1-Me-9H-pyrido[3,4-b]-indole).

The ability of harmine to intercalate between DNA strands due to its biophysical properties (Smythies and Antun 1969) led one of us (ESL) to speculate that such powerful binding to DNA by both harmine and the stronger hallucinogen lysergic acid diethylamide (LSD) could underlie experiential "unfolding" of archetypal memories during exposure to these agents (Lansky 1975) (Figure 6.9). Indeed, the active ingredient of the liana *Banisteriopsis caapi*, the "vine of the soul" of the Amazonian brew Ayahauasca, was initially named telepathine. Later, its structure was proved identical to harmine, the alkaloid known from *P. harmala*.

Binding of harmine or harmaline to DNA is accompanied by fluorescence, though only harmine may intercalate between DNA strands (Duportail and Lami 1975). Chicken blood cell smears, *Trypanosoma cruzi epimastigotes*, Ehrlich ascites tumor cells, or mouse spleen treated with harmine and the other harmala alkaloids exhibited a "strong bluish white fluorescence from condensed chromatin and basophilic cytoplasm under ultraviolet excitation" (Molero et al. 1995).

Harmine is a classic *fluorophore* or *fluorochrome*, meaning it is a fluorescent compound, one whose aromatic structures, when excited by light, are able to re-emit light at characteristic frequencies usually different from the frequency of the light that caused the excitation. In combination with lasers, fluorophores are given to multiple technological uses, including flow cytometry and spectrophotometry, and harmine is sometimes the molecule of choice in high technology measurements. The high affinity of harmine for nucleic acids lends to its efficacy as a fluorescent tag in the quantification and identification of macromolecules.

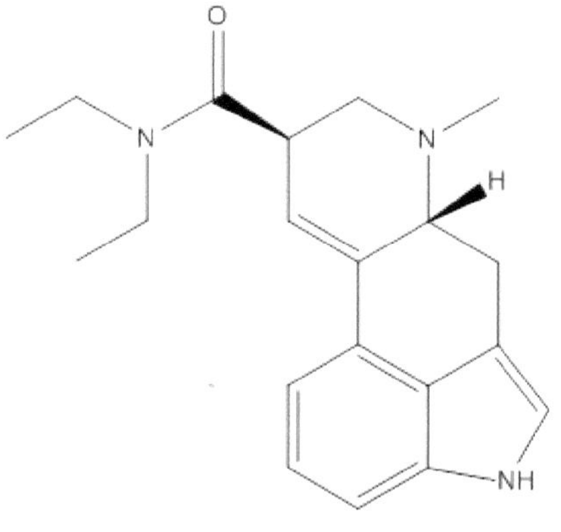

FIGURE 6.9   LSD—lysergic acid diethylamide.

This high affinity for nucleic acids also helps explain harmine's crippling cytotoxicity in diverse biological contexts, for example in tobacco, as an insecticide, and leads to the search for even more effective and specific insecticides or related cytotoxins through synthetic modification of harmine as lead compound (Zeng et al. 2010). Widely distributed throughout mammals, marine invertebrates, and insects, harmine was originally isolated from the seeds of *P. harmala* in 1847 with its characteristic core indole and pyrimidine ring, and antimicrobial, antifungal, antitumor, cytotoxic, antiplasmodial, antioxidant, antimutagenic, antigenotoxic, and entheogenic/psychedelic/hallucinogenic/psychotomimetic properties. Harmine further acts on gamma-aminobutyric acid (GABA) type A and monoamine oxidase (MAO) A or B receptors, enhances insulin sensitivity, is vasorelaxant, and prevents bone loss by suppressing osteoclastogenesis, suggesting a wide array of potential medical as well as biotechnological applications (Patel et al. 2012).

Harmine is an effective *in vitro* inhibitor of angiogenesis, the *sine qua non* of cancer metastasis, since *de novo* blood vessel formation allows new satellite tumors (metastases) sufficient blood supply. At 10 mg/kg intraperitoneal injection (i.p.) into C57BL/6 mice, harmine significantly decreased tumor-directed capillary formation, serum pro-angiogenic vascular endothelial growth factor (VEGF), nitric oxide (NO), and pro-inflammatory cytokines, nuclear factor (NF)-κB, CREB, and ATF-2, while *enhancing* antitumor interleukin-2 (IL-2) and tissue inhibitor metalloprotease (TIMP) actions (Hamsa and Kuttan 2010). Activation of tumor antigen protein P-53 contributed to the anti-angiogenic effect of harmine (Dai et al. 2012).

Harmine is a potent, reversible monoamine oxidase inhibitor (MAOI), selective for, and with high affinity to, monoamine oxidase A (i.e., with affinity for MAO-A, but not for MAO-B). Selectivity to MAO-A and harmine's reversible binding as its substrate make harmine a promising lead MAOI with potential as an antidepressant or in the treatment of neurodegenerative diseases like Parkinson's and Alzheimer's. Furthermore, radioactively labelled carbon isotope [11C] harmine is a radioligand for MAO-A, which, when displaced by an MAO-A inhibitor with greater affinity for the core of the MAO-A complex, aids in assessing the therapeutic potential of newer, emerging MAOIs for both antidepressant and anti-neurodegenerative employment (Chiuccariello et al. 2015). Moreover, in addition to the MAO-A inhibition, harmine promotes **dopamine** efflux from the brain's nucleus accumbens shell, which further accentuates its heuristic efficacy for treating Parkinson's disease (Brierley and Davidson 2013). This dopamine efflux also suggests benefit for reversing toxicity and addiction to cocaine, as is borne out from success in treating addictions with Ayahuasca. Additionally, harmine mollifies **cocaine** toxicity through its inhibition of glutamatergic pathways (Owaisat et al. 2012) and likely helps reduce recidivism to cocaine or alcohol through MAO inhibition, effects at 5-HT(2A), and, via imidazoline receptors, inhibition of dual-specificity tyrosine-phosphorylation regulated kinase 1A (DYRK1A) and the dopamine transporter (Brierley and Davidson 2012) (Figures 6.10 through 6.12).

The action of harmine against cancers both *in vitro* and *in vivo* has been extensively studied. For example, harmine, as well as its congener, harmaline (see below), significantly reduced the induction of the carcinogen-activating enzyme Cyp1a1 in murine liver and lung caused by the standard carcinogen 2,3,7,8-tetrachlorodibenzo-p-dioxin

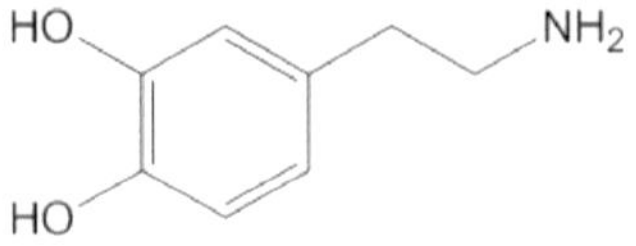

FIGURE 6.10 Dopamine.

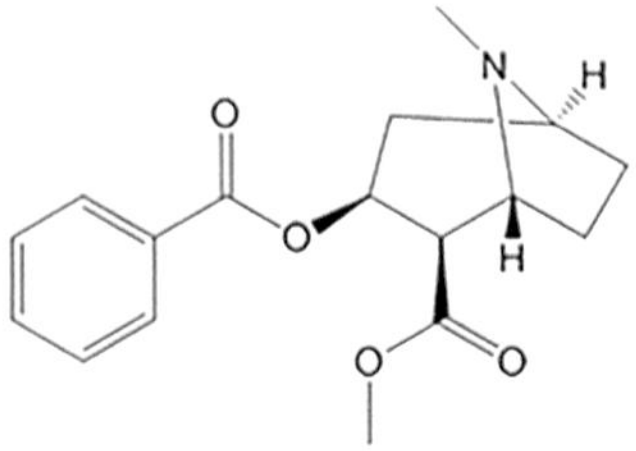

FIGURE 6.11 Cocaine.

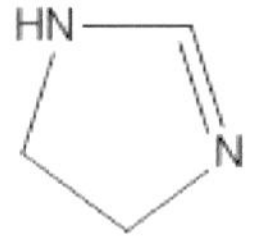

FIGURE 6.12 Imidazoline.

(TCDD), suggesting anticarcinogen action and further directions for basic as well as clinical research (El Gendy and El-Kadi 2013). In a study of the effects of four different *P. harmala* seed alkaloids, harmalicidine, harmine, peganine (vasicine), and vasicinone, against four different carcinogen-induced cancer cell lines, harmine was shown to have the most potent growth inhibitory effect (Lamchouri et al. 2013) (Figure 6.13).

A newly published study at the time of this writing (June 2016) from the Zewail City of Science and Technology, Giza, Egypt, and the University of Sheffield, UK, found harmine a model natural compound that could attenuate **doxyrubicininduced cardiotoxicity** by significantly reducing the therapeutic dose of doxyrubicin required, owing to synergy between the two compounds as anticancer cytotoxics (Atteya et al. 2016). Atteya's study reinforces the work published a year prior by Sun et al. (2015), which proved harmine synergistic with **paclitaxel** in inhibiting *in vitro* invasion and migration of human gastric carcinoma cells of two different

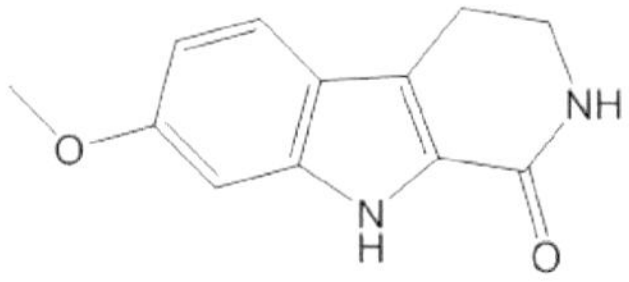

FIGURE 6.13 Harmalicidine. (From Lamchouri, F. et al., *Pak J Pharm Sci*, 26(4): 699–706, 2013.)

human gastric cancer cell lines (SGC-7901 and MKN-45) through downregulation of expression of cyclooxygenase (COX-2) and matrix metalloproteinase (MMP-9), well-known enzyme targets of anti-inflammatory drugs (Sun et al. 2015). In both of these examples, the combination of harmine with either paclitaxel or doxyrubicin proved more lethal than either harmine or the other cytotoxic alone (Figure 6.14).

Synthetic derivatives of harmine are as much as four times more cytotoxic as harmine against HGC-27 human gastric carcinoma and HT-29 human colon cancer cells (Shankaraiah et al. 2016). Harmine is one of very few natural compounds known to inhibit breast cancer resistance protein (BCRP), a key mediator of breast cancer resistance to chemotherapeutic drugs (Ma and Wink 2010), and similarly, newly synthesized harmine analogues at the University of Bonn, Germany, have been shown to be as potent as another known inhibitor of BCRP, the mycotoxin fumitremorgin C (FTC), Ko143, isolated from the fungus *Aspergillus fumigatus* (Spindler et al. 2016) (Figure 6.15), newly synthesized harmine analog at China's Shihezi University, in cooperation with the Xinjiang Huashidan Drug-Discovery Co., Ltd., Urumchi, was shown to have strong antiproliferative activity in human lung, stomach, ovary, and prostate cancer cell lines through multiple anti-angiogenic mechanisms related to vascular epithelial growth factor (VEGF) activity and its receptor (VEGFR) (Ma et al. 2016). Another synthetic harmine derivative proved effective against B cell lymphoma through caspase activation (Gao et al. 2014).

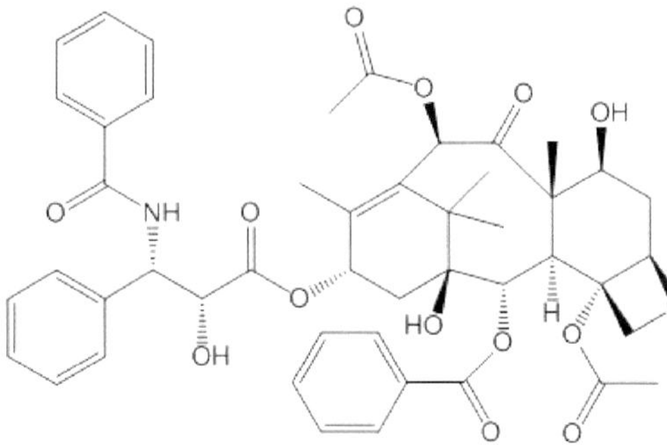

FIGURE 6.14 Paclitaxel.

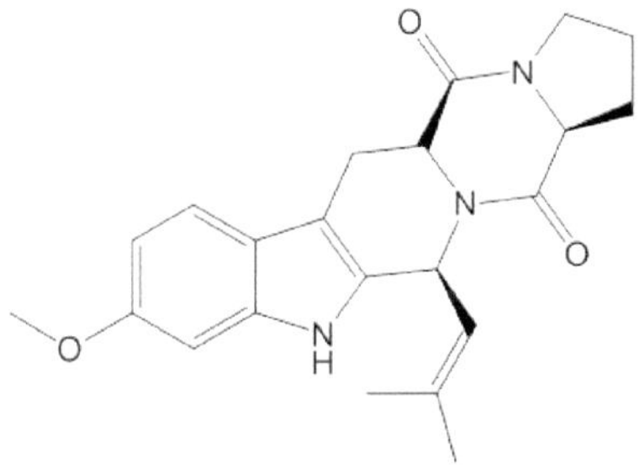

FIGURE 6.15 Fumitremorgin C (FTC), Ko143. (From Spindler, A. et al., *J Med Chem*, 6/2017 [Epub ahead of print].)

A different group at University of Namur, Belgium, created an analog of harmine with augmented potency and complexed it with maltodextrin to achieve superior drug solubility (Meinguet et al. 2015a) in a portfolio of harmine analogs for antiproliferative purposes in experimental oncology (Meinguet et al. 2015b). Harmine was the reference compound for all of the aforementioned studies (Bruel et al. 2014).

Precise targeting for an antitumor drug may be desirable, and modifications in the harmine structure have been achieved synthetically in order to increase the anticancer potency of harmine while simultaneously reducing its "neurotoxicity." This was the objective reported to be successfully accomplished through a collaboration between the departments of Bioengineering from Nanjing University (China) and the University of Texas, Arlington (Li et al. 2015).

An approach to solubilizing the hydrophobic harmine is the creation of polysaccharide micelles that are biodegradable aimed to insinuate harmine into liver cancer tumors (Bei et al. 2013, 2014). Chitosan-coated liposomes have been developed for solubilization of harmine (Chen et al. 2016), as have stabilized harmine nanocrystals derived from harmine nanosuspensions (Yue et al. 2015) (Table 6.1).

TABLE 6.1
Synthetic Harmine Analogs

Description or Names of Compounds	Advantages	Reference
JKA97, a benzylidene analog of harmine	Dose-dependent inhibition of growth, a c proliferation of MCF-7 (p53 wild-type), MCF-7 (p53 knockdown), and MDA-MB-468 (p53 mutant) breast cancer cells, G1 arrest, apoptosis, suppressed MCF7 and MDA-MB-468 xenografts, regulated expression of G1 regulators p21, p27, cyclinE, cylinD1, activated p21 transcription, independent of p53, with little effect on p21	Yang X et al. 2012
Several novel 1,2,3,4-tetrahydro-β-carboline, β-carboline, and 1-substituted-β-carboline derivatives bearing a substituted carbohydrazide group at C-3 Especially N'-benzylidene-1-phenyl-β-carboline-3-carbohydrazide (C(25) H(18)N(4)O, m.w. 390.4) (c2) and N'-(4-trifluoromethyl-benzylidene)-1-phenyl-β-carboline-3-carbohydrazide (C(26)H(17)N(4)OF(3), m.w. 458) (d2)	Good inhibitory activity against dicotyledonous and monocotyledonous weeds, with EC(50) values of 4.83 and 14.25 μM, respectively	Weng et al. 2012
Harmine derivatives designed with silicon docking based on crystal structure of haspin kinase	Increased haspin kinase inhibitory potency, less activity towards DYRK2	Cuny et al. 2012

Yet, in spite of the advances in synthetic harmine analogs and novel delivery aggregates, simple harmine (or harmine hydrochloride) continues to provoke interest as an anticancer drug. Harmine caused mitochondrial apoptosis in human gastric carcinoma and hepatocellular cancer cells, but not in LO-2 immortalized liver cells via upregulated expression of p21, activation of Myt1, inhibition of cdc2 by phospho-cdc2, and triggering G2 phase arrest, in both MGC-803 human gastric cancer cells and SMMC-7721 human hepatocellular carcinoma cells. Harmine caused as well mitochondria-related cell apoptosis via caspase-8/Bid, inhibiting the ERK/Bad pathway (Zhang et al. 2016).

Tongji Hospital in Shanghai and collaborators found harmine a potent inhibitor of homologous recombination repair mechanisms in human hepatoma cells, resulting in profound cytotoxicity and inhibition of proliferation. Although harmine did not disrupt nonhomologous end joining (NHEJ), a parallel mechanism of cancer cell DNA repair to homologous recombination repair, it synergized with agents that did disrupt NHEJ (Zhang et al. 2015).

As alluded to above, Khor et al. (2015), a large group from Harvard University and the National Institutes of Health, Bethesda, Maryland, exhibited the efficacy of harmine as an inhibitor of dual specificity tyrosine-phosphorylation-regulated kinase DYRK1A, which resulted in attenuation of mucosal inflammation and unwanted autoimmunity. Harmine application further resulted in a balancing between the Th17 and T regulatory (T reg) lymphocytes, immunologically specialized cells necessary for achieving "immunological homeostasis" and modulation of inflammation. Such modulation of DYRK1A has particular relevance for the treatment of hematological and brain malignancies, as well as neurodegenerative diseases (Abbassi et al. 2015). Inhibition of DYRK1A also destabilized epithelial growth factor receptor, a favorable event in reducing growth of EGFR-dependent glioblastoma multiforme (Pozo et al. 2013), while harmine also inhibited Akt phosphorylation and reduced the pool of glioblastoma stem-like cells (Liu et al. 2013). In the quest for new modulators of DYRK1A, harmine has been the reference compound (Bruel et al. 2014) and considered a first in its class of DYRK1A inhibitors, though strongly suppressing MAO-A (Cozza et al. 2013). The cysteine aspartyl protease caspase 9, a critical component of the intrinsic apoptotic pathway, was a substrate of DYRK1A that, when activated, resulted in inhibition of the apoptosis. This inhibition was potently blocked by harmine, partially explaining its mechanism for inducing apoptosis (Seifert et al. 2008). Harmine-induced inhibition of the pleiotropic DYRK1A also underlay beneficial effects against cancer and neurodegeneration (Abbassi et al. 2015), while stimulating beneficial β-cell proliferation in the pancreas (Wang P. et al. 2015).

Paradoxically, however, DYRK1A overexpression resulted in overexpression of master antioxidant switch nuclear factor (erythroid-derived 2)-like 2, a transcription factor (NRF2) (Noll et al. 2012), which is also upregulated by certain beneficial natural products, including pomegranate emulsion (PE), a complex fully derived from the fermented aqueous phase and the lipid phase of *Punica granatum*, which also helped prevent hepatocellular carcinoma via Nrf2 *in vivo* (Bishayee et al. 2011). In short, therefore, DYRK1A is linked to Nrf2, though the possible interactions between harmine and PE remain to be elucidated. Harmine also induced apoptosis in HepG2 human hepatocellular carcinoma cells (HCC) *in vitro* (Cao et al. 2011). Additional research is indicated to clarify the relationship between the mechanisms of harmine and PE in retarding HCC growth.

Harmine promoted apoptosis and inhibited proliferation in human gastric cancer cells (Song et al. 2006) through downregulation of COX-2 (Zhang et al. 2014), and inhibited proliferation and telomerase in MCF-7 breast cancer cells, inducing an accelerated senescence phenotype by overexpressing elements of the p53/p21 pathway (Zhao and Wink 2013). A structure–activity assessment for harmine analogs has reported according to both neurotoxicity and antitumor criteria, namely, formate substitution at R3 of the tricyclic skeleton attenuated neurotoxicity, while short alkyl or aryl substitution at R9 enhanced antitumor potency (Chen et al. 2005). The biodistribution and radiation dosimetry for 11C-harmine was established in baboons, exploiting its MAO-A binding proclivity, and found to be especially prominent in lung (Murthy et al. 2007). The contribution of P-450 isozymes to the O-demethylation of harmine was delineated (Yu et al. 2003). Design and synthesis of new C3-tethered 1,2,3-triazolo-analogs in India were more cytotoxic to gastric and colon cancer cells compared to harmine (Shankaraiah et al. 2016). Structure–activity assessments indicated that the 7-methoxy structural moiety was responsible for the neurotoxic effect and that substituents in positions 2 and 9 modulated antitumor actions (Cao et al. 2013).

An important mechanism contributing to the anticancer effect of harmine derives from its DNA intercalating ability, preventing enzyme topoisomerase 1 from catalyzing the unwinding of tumor DNA required for cancer cell propagation (Sobhani et al. 2002).

Harmine may benefit disorders of bone and cartilage, including osteoporosis, bone fracture, and osteoarthritis. This proceeds from regenerative and protective effects on these tissues through regulation of the proliferation, differentiation, and metabolism of osteoclasts, osteoblasts, and chondrocytes (Hu and Xie 2016; Saeki and Egusa 2014). Harmine meaningfully upregulated connective tissue growth factor (i.e., CTGF or CCN2) in chondrocytes, a pheonomeon which also elicited chondroprotective (Hara et al. 2013) and skeletogenic effects (Arnott et al. 2011).

Harmine was a drug lead for treating myotonic dystrophy type 1, a disabling neuromuscular disease (Herrendorff et al. 2016). A series of newly synthesized harmine derivatives possessed antilipoxygenase and anti-acetylcholinesterase activities, in addition to cancer cytotoxic actions (Filali et al. 2016). Another series of synthetic harmine derivatives possessed up to 100 times the anticancer action of harmine, but without inhibition of DYRK1A (Frédérick et al. 2012). **Inhibition of DYRK1A by harmine had the possibly unique effect of stimulating human β-cell (i.e., cells of the Islets of Langerhans) proliferation** (Wang P. et al. 2015), a highly desirable outcome for both type-1 and type-2 diabetes mellitus patients (Dirice et al. 2016).

Radioactive harmine has been used in many imaging studies especially where monoamine oxidase A was concerned, for example, in investigation of major depressive illness (Chiuccariello et al. 2014), antisocial personality disorder (Kolla et al. 2015, 2016b), and borderline personality disorder (Kolla et al. 2016a). Radioactive harmine was also used to show the dynamic nature of MAO-A V(T), total MAO-A volume, an index of MAO-A density, in response to substrate availability. That is,

MAO-A V(T) was decreased (mean: 14% ± 9%) following tryptophan depletion in prefrontal cortex ($P < 0.031$), and elevated (mean: 17% ± 11%) in striatum following carbidopa–levodopa administration ($P < 0.007$) (Sacher et al. 2012). An automated synthesis of the radioactive isotope required for all of these investigations, 11-C harmine, was established (Philippe et al. 2015).

Intraperitoneal injections of harmine on each of 5 days prior to **traumatic brain injury** significantly reduced cerebral edema and improved functional recovery in rats through anti-inflammatory and anti-excitotoxic mechanisms (Zhong Z. et al. 2015). Chen et al. (2015), from Nanjing University Medical School in China, showed *in vitro* that harmine potently inhibited human simplex viruses (HSV), types 1 and 2, through blockage of NFkappaB and other anti-inflammatory and antioxidant mechanisms. Rüben's group at Aachen University, Germany, synthesized more new harmine analogs with extra potency in inhibiting the pleiotropic protein kinases DYRK1A but without effects on MAOI-A (Rüben et al. 2015). Through its inhibition of DYRK1A, harmine **inhibited progression from myocardial infarction to heart failure** *in vivo* (He J. et al. 2015). Harmine is a potent inhibitor of acetylcholinesterase, and 2 weeks treatment by oral gavage was sufficient to rescue mice from cognitive impairment caused by a single i.p. injection of scopolamine, while long-term dosing with harmine over 10 weeks exerts a mild cognitive impairment in a murine model of **Alzheimer's** disease (He D. et al. 2015). In rats subject to chronic mild stress procedure for 40 days, citrate synthase activity (an index of oxidative energy metabolism) in the prefrontal cortex was significantly increased, but 7 days of harmine treatment following the procedure in the same rats reversed this effect (Abelaira et al. 2013). Old rats pretreated with either low or high dose of harmine exhibited superior performance compared to controls in the Delayed Match to Sample (asymmetrical 3 choice water maze) Test, an instrument for assessing spatial working and recent memory (Mennenga et al. 2015). Harmine exerted neuroprotective effects in a rat model of amyotropic lateral sclerosis (ALS) and attenuated *in vivo* damage from global cerebral ischemia, if given after the acute event, concurrent with glutamate transporter-1 activation and attenuation of astrocyte activation (Sun et al. 2014). Dr. Sorin Hostiuc of the Carol Davila University of Medicine and Pharmacy in Bucharest, Romania, described three cases of catatonic schizophrenia successfully treated by Dr. Petre Tomescu in the 1920s (Hostiuc et al. 2014). In the study, published only in Romanian and later suppressed by the Romanian government, Tomescu studied the effects of 0.3–0.4 mg injection of harmine in three different patients diagnosed with catatonic schizophrenia. **After single injections, muscular rigidity disappeared after an hour, and positive effects persisted for 3 days**. In a case where the treatment was continued for 25 days, improvements in affect (seeking a romantic liaison with a nurse), appetite (asking for favorite foods), and modest weight gain were also noted. In a 1929 study, also published only in Romanian, clinical improvement of Parkinson's disease (Beringer 1929) was observed.

In Japan, Kondoh et al. (2014) showed harmine to lengthen circadian period of the mammalian molecular clock in the suprachiasmatic nucleus, a finding also known from lithium and associated with mood stabilization (Noguchi et al. 2016). The 11C isotope of harmine was employed to visualize total brain MAO-A volume using positron emission tomography (PET) in postpartum women as an index of their situational depression and postpartum crying (Sacher et al. 2015), which is similarly elevated during the perimenopause (Rekkas et al. 2014).

The flourochromic and nucleic acid binding capacities of harmine (and 10-methylharmine) combine to render harmine an important photosensitizing and phototoxic molecule (Vignoni et al. 2014). Specifically, once harmine intercalates between DNA strands, it becomes active and deadly to the cells containing that DNA, thus illuminating another dimension in harmine's potential antitumor action (Hazen et al. 2002). Although the mechanism of the phototoxic effect is as yet unknown, at neutral pH most of the DNA damage is generated from the protonated form of the excited harmine or related 9-methylated congeners (Vignoni et al. 2013). Further, the ground and excited state of harmine in various micellar assemblies, such as the micelles and advanced nanodelivery systems discussed above, were differentially modulated (Paul et al. 2012), with DNA damage in such instances primarily mediated by direct photoreaction of the protonated harmine after nonintercalative electrostatic binding (Gonzalez et al. 2012). To some degree, the photoreaction is affected by the matrix in which the harmine is dissolved (Becker et al. 2005), and harmine is more effective in its photosensitizing than conventional DNA fluorescent dyes used for such purpose (Gutiérrez-Gonzálvez et al. 1988). Complexed with silver nanoparticles, harmine acts as the proton donor, with the silver nanoparticle as proton acceptor (Amjadi and Farzampour 2014).

In a library screen of compounds against malaria, harmine proved to be a potent inhibitor of *Plasmodium falciparum* heat shock protein 90 (PfHsp90) ATP-binding domain and to synergize with chloroquine and artemisinin *in vitro* and in the *Plasmodium berghei* mouse model (Shahinas et al. 2012). Harmine and harmol were moderately potent inhibitors (the best two) out of a 140,000 compound screening library as inhibitors of haspin, a serine/threonine kinase that phosphorylates Thr-3 of histone H3 in mitosis, and a potential cancer therapeutic target (Cuny et al. 2012).

Harmine was a potent destroyer of the insect pest *Plodia interpunctella*. When fed to its larvae, harmine induced strong reduction of larvae weight, cannibalism between larvae, and significant mortality. Harmine also caused development disruption, i.e., in both delay and reduction of pupation and adult emergence. With spectrophotometric assays, it was shown that harmine ingestion resulted in marked decrease in this insect's protein, glycogen and lipids, and activity of the digestive enzyme α-amylase (Bouayad et al. 2012).

Harmine, and its metabolite, harmol, inhibited *in vitro*, in murine and human hepatoma cells, the induction of carcinogen-activating enzyme P-450 1A1, aka, CYP 1A1, induced by compounds of the dioxin class, in a concentration-dependent manner for a significant post-translational period through a ubiquitin-proteasomal pathway (El Gendy et al. 2012b). Similar data regarding the related CYP 2D6 and harmine demand awareness of a likely substrate dependency (VandenBrink et al. 2012).

Harmine was over three times as potent as **imipramine** in decreasing creatine kinase activity, an index of mitochondrial energy metabolism, in rat brain following periods of acute and chronic dosing (Réus et al. 2012). Harmine as a base and salt in water and in mixture of dimethyl sulfoxide (DMSO) and water had the highest inhibition activities against acetyl cholineasterase (ACHE) using eserine as reference substance among β-carbolines studied, though harmalol as a salt in water and harmine as a base and salt in a mixture of water and DMSO were the best suppressors of butyrylcholinesterase (BUCHE) (Krsková et al. 2011) (Figures 6.16 and 6.17).

Dos Santos and Hallak (2016) reviewed the basic literature on harmine with reference to its effects on memory and in the limbic system's hippocampus. They sought to evaluate the possible support of the preclinical literature on neuroprotection and memory enhancement for the reports emanating from the harmine-containing Ayahuasca community of improved neuropsychological functioning. They found in their literature search two studies of hippocampal neurons, and nine *in vivo* regarding the favorable effects of harmine on memory. Factors supporting neuroprotection included reduced excitotoxicity, inflammation, and oxidative stress, and increased brain-derived neurotrophic factor (BDNF). Memory learning enhancement *in vivo* was attributed to monoamine oxidase or acetylcholinesterase inhibition, upregulation of glutamate transporters, decreases in reactive oxygen species, increases in neurotrophic factors, and anti-inflammatory mechanisms.

Back to benchtop oncology, Liu et al. (2016) studied the effect of harmine on SW620 human colon carcinoma cells and observed powerful promotion of caspases 3 and 9 apoptosis through mediated Akt and ERK signaling. At 5.13 µg/ml, harmine decreased the proportion of cells in the G0–G1 phase of cell cycle and increased the proportion of those in S and G2-M, with decreased expression of cyclin D1 and increased expression of cyclin A, E2 and B1, CDK1/cdc2, Myt-1, and p-cdc2 (Tyr15) consistent with the changes in cell cycle. Blue fluorescent Hoechst staining revealed nuclear fragmentation, chromosomal condensation, and cell shrinkage in the colon cancer cytology.

Gu et al. (2015) identified harmine in *P. nigellastrum* while screening plants for possible inhibition of dipeptidyl peptidase IV (DPP-4). DPP-4 is a protein coded by

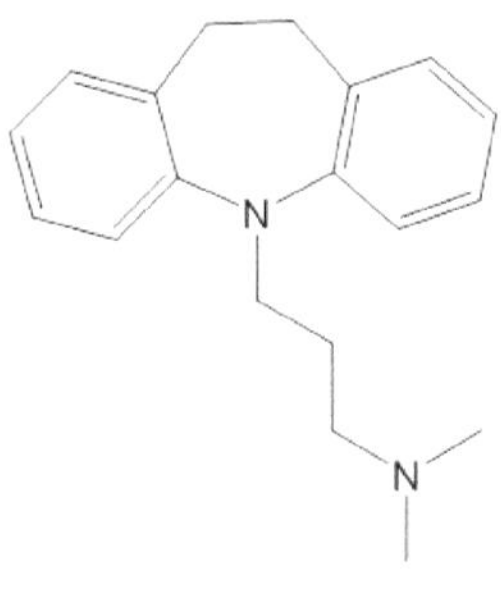

FIGURE 6.16   Imipramine.

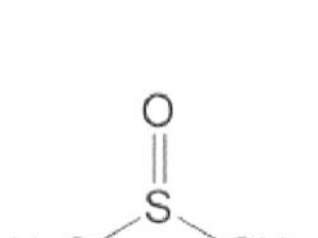

FIGURE 6.17   Dimethylsulfoxide (DMSO).

the DPP-4 gene that plays a key role in glucose metabolism. DPP-4 inhibition results in lowering of blood sugar, and this mechanism is responsible for its use in treatment of diabetes mellitus. DPP-4 inhibitors are the second most prescribed hypoglycemic drugs after insulin, with a world market share of over $7 billion (Cahn et al. 2016). Gu et al. found harmine hydrochloride to be a potent DPP-4 inhibitor, with a 32.4% inhibition at 10 mM compared to 54.8% at 50 µM for diprotin A, a synthetic antidiabetic pharmaceutical.

Harmol

Harmol has been noted in the leaves of *Passiflora incarnata* (Avula et al. 2012), though not in appreciable quantity as in *P. harmala* seeds (Figure 6.18). Harmol does occur in *P. harmala*'s roots (1.4% w/w) (Herraiz et al. 2010) and Zheng et al. (2009) showed it in the seeds of *Peganum nigellastrum*, and also showed harmol to inhibit MAO-A. Their finding of harmol being an MAO-A inhibitor was corroborated (Maschauer et al. 2015) and elaborated with semisynthetic harmol derivatives (i.e., analogs) (Schieferstein et al. 2015). Similarly, radioactively labeled harmol was used to detect MAO-A in rat *in situ* under PET (Cumming et al. 2015).

Abe and Kokuba (2013), from Tokyo Medical University, revealed that harmol induced both apoptosis (programmed cell suicide) and autophagy (cell eats itself) through suppressing expression of the survivin protein in human glioma cells. Nevertheless, Abe's group had earlier shown that harmol induced autophagy, but not apoptosis, in A-549 human non-small cell lung cancer cells, though autophagy preceded cell death through a non-caspase dependent mechanism (Abe et al. 2011). However, in the human adenosquamous carcinoma H-596 lung cancer cells, apoptosis did occur in response to harmol through a caspase-dependent mechanism independent of Fas/Fas ligand interaction (Abe and Yamada 2009).

Harmol hydrochloride was shown *in vitro* by Jones et al. (2009) at the University of California, San Francisco, to noncompetitively inhibit androgen receptor binding of other ligands, a finding of relevance to the treatment of prostate cancer, benign prostatic hypertrophy, and hirsutism. Harmol moderately inhibited collagen-induced platelet aggregation *in vitro*, though less potently than harmine or harmane (Im et al. 2009), suggesting potential for all three compounds as possible lead antithrombotic drugs. Along with the other harmala alkaloids, harmol exhibited antigenotoxic and antimutagenic effects in *Saccharomyces cerevisiae* strains proficient and deficient in antioxidant defenses (Moura et al. 2007) and improved performance on an object recognition test when injected intraperitoneally into mice prior to testing (Moura et al. 2006).

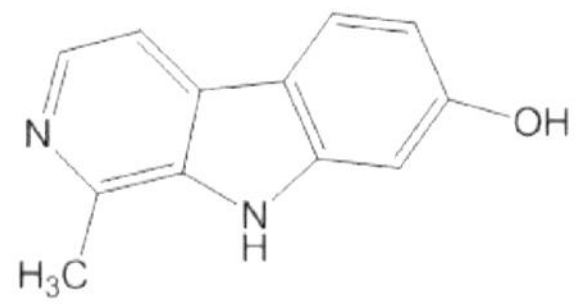

FIGURE 6.18 Harmol (1-methyl-2,9-dihydropyrido[3,4-b]indol-7-one).

Harmalol

Along with harmol, harmine, and harmaline, harmalol helped reverse chloroquine resistance in *Plasmodium falciparum*, the causative protozoan in malaria (Ibraheem et al. 2015) (Figure 6.19). Harmalol was utilized in several recent studies to assess binding affinity and kinetics to RNAs of differing motifs, for example harmalol bound more securely to double strands such as poly (C)–poly (G) compared to clover-leaf tRNA. Such elucidation was seen as preliminary to designing new antiviral drugs (Bhattacharjee et al. 2016). Similarly, harmalol bound strongly with hetero GC polymer via intercalation while resisting overlap to the DNA base pairs inside the intercalation cavity, and showed maximum cytotoxicity on human hepatoma cells HepG2 with IC50 value of 14 μM (Sarkar et al. 2014). A theory promoting the alteration of the structure of harmalol between protonated and deprotonated forms of the molecule over a range of pHs was presented (Sarkar and Bhadra 2014), though the chemical calculations leading to its principal supposition were challenged (Alomar et al. 2014). Harmalol was shown to bind well to human serum albumin (HSA) due, to a major degree, to hydrogen bonding and Van der Waal's forces (Hemmateenejad et al. 2012). And, like harmol and harmine as well as harmaline, harmalol, a metabolite of harmaline (see the Harmaline section below), was able to protect against dioxin-mediated carcinogenesis-related effects in murine hepatoma cells *in vitro* through inhibition of the carcinogen-activating enzyme CYP1A1 (El Gendy et al. 2012a). Similarly, both harmalol and its parent, harmaline, likely helped prevent skin cancer by stimulating **melanin** synthesis. This was associated with decreased carcinogenesis, through an upregulated oxidation of melanin's precursor, the amino acid **tyrosine**, increased production of tyrosinase, and tyrosinase-related proteins 1 and 2 (TRP-1 and TRP-2), and time-dependent phosphorylation of enzyme p38 mitogen activated protein kinase (MAPK), an enzyme responsive to stress stimuli, in a significant concentration-dependent manner (Park et al. 2010). Nevertheless Yamazaki and Kawano (2010) showed harmol and related compounds to be tyrosinase inhibitors (Figure 6.20).

Affinity of harmalol to RNA through *in vitro* assays was on the order of harmine > harmaline > harmane > harmalol > **tryptoline** (tetrahydro-β-carboline) (Nafisi et al. 2010b) but for DNA binding is on the order of harmine > harmalol > harmaline > harmane > tryptoline (Nafisi et al. 2010a) (Figure 6.21).

Harmalol was an effective protector against mitochondrial dysfunction leading to Parkinson's disease, more so than was **deprenyl** (Han et al. 2005), a putative well-known drug used for treating this disease (Tetrud and Langstan 1987). In that dopaminergic system emboldens opiate withdrawal symptoms (Diaz et al. 2005),

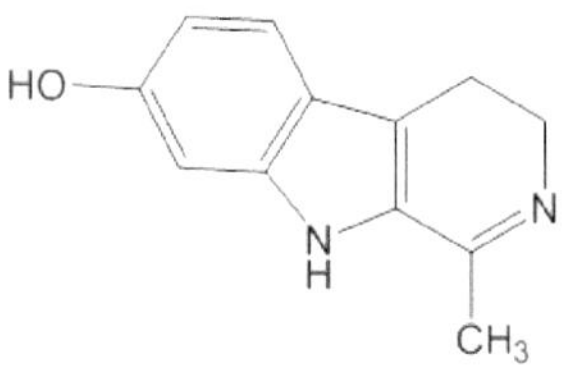

FIGURE 6.19 Harmalol (1-Methyl-4,9-dihydro-3*H*-β-carbolin-7-ol).

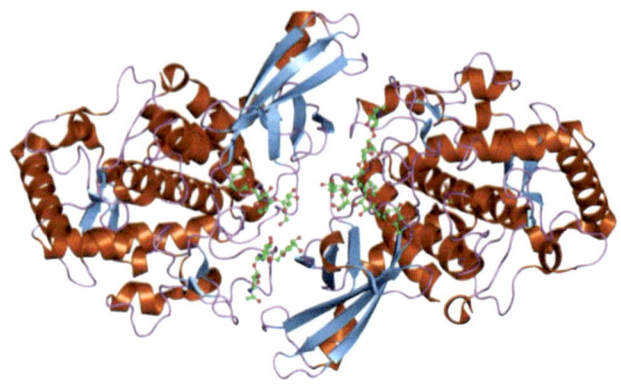

FIGURE 6.20   Tyrosinase. (From Jawahar Swaminathan and MSD staff at the European Bioinformatics Institute, http://www.ebi.ac.uk/pdbe-srv/view/images/entry/1js8600.png, displayed on http://www.ebi.ac.uk/pdbe-srv/view/entry/1js8/summary, Public Domain, https://commons.wikimedia.org/w/index.php?curid=5936046, retrieved 3/2/2017.)

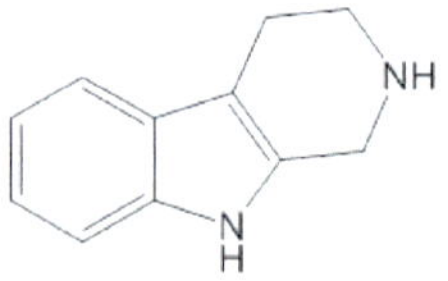

FIGURE 6.21   Tryptoline (tetrahydro β carboline).

suppression of DOPA synthesis via harmalol is important, viz., harmalol blocked the effects of brain **3,4-dihydroxyphenylalanine (DOPA)** synthesis in morphine-dependent rats in (K(iH) nM) according to potency of **noreleagnine** (12) > **norharman** (20) > harmalol (82) > harmaline (177) >> harmine (630) > harman (700) >> FG-7142 (>100,000), indicating a high affinity of these compounds to **imadazoline** receptors and potential of harmalol for blocking **opiate withdrawal** behavioral effects (Miralles et al. 2005). In brain mitochondria, harmalol and harmaline, like antioxidant enzymes superoxide dismutase/SOD and catalase, decreased the alteration of mitochondrial swelling and membrane potential induced by 200 µM dopamine or 100 µM 6-hydroxydopamine (Kim et al. 2001). Of three harmala alkaloids tested, the order of their potency in causing vasorelaxation in isolated rat thoracic aorta was harmine > harmaline > harmalol. Only with harmalol was the effect not attenuated by a nitric oxide inhibitor (Shi et al. 2001) (Figures 6.22 through 6.26).

Harmalol, along with harmaline and harmine, reduced the oxidative damage to dopamine-containing neurons caused by the Parkinson's disease–inducing proneurotoxin, **MPTP (1-methyl-4-phenyl-1,2,3,6-tetrahydropyridine)**, namely 500 µM MPP(+)-induced inhibition of electron flow and membrane potential formation and 100 µM dopamine-induced thiol oxidation and carbonyl formation in mitochondria (Lee et al. 2000) (Figure 6.27).

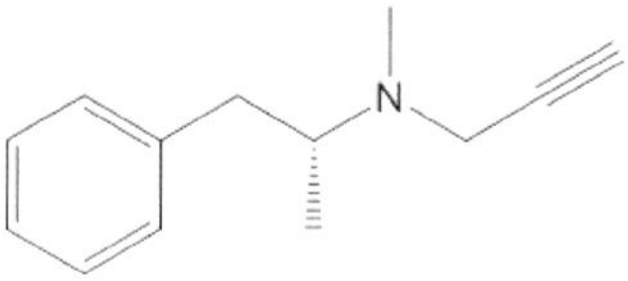

FIGURE 6.22 Deprenyl.

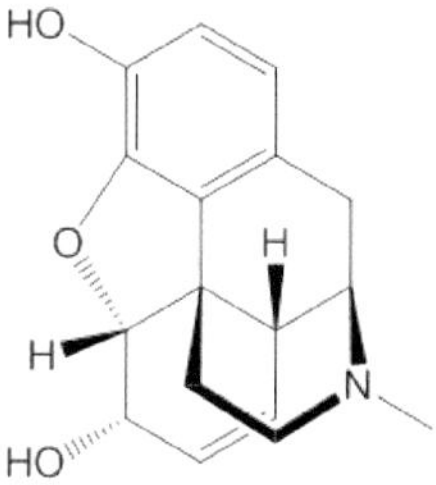

FIGURE 6.23 Morphine.

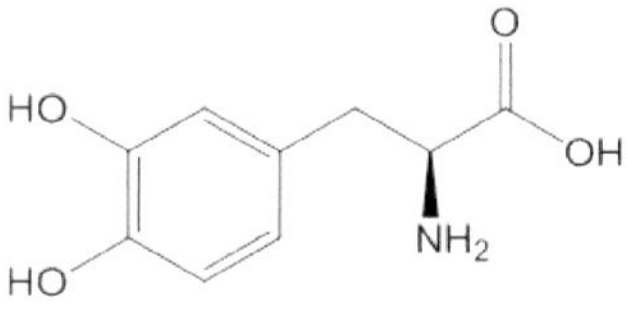

FIGURE 6.24 L-3,4-dihydroxyphenylalanine (DOPA).

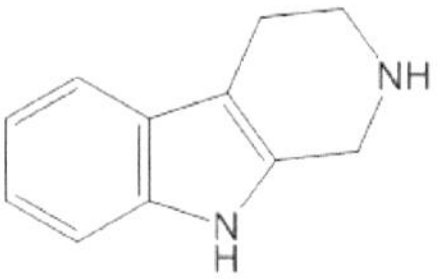

FIGURE 6.25 Noreleagnine.

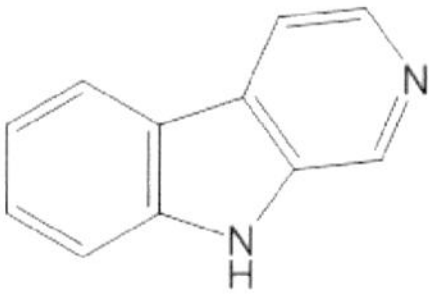

FIGURE 6.26 Norharman.

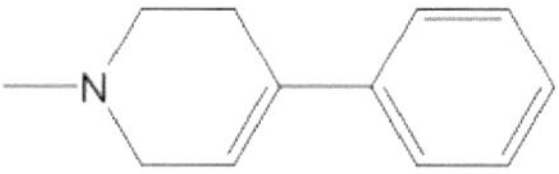

FIGURE 6.27 MPTP (1-methyl-4-phenyl-1,2,3,6-tetrahydropyridine).

Harmaline

Among all the harmala alkaloids in the seeds of *P. harmala*, Khan et al. (2013) considered harmaline to be the most medicinal owing to its established antileishmanial, antimicrobial, antiplatelet, antiplasmodial, antitumoral, hypothermic, and vasorelaxant properties (Figure 6.28). Nonetheless, harmaline is reliable for producing experimental tremors *in vivo*, and is specifically, along with the other harmala alkaloids, implicated in the pathogenesis of clinical essential tremors (Laviță et al. 2016). The Harmaline Model of Essential Tremor is based on the effect of harmaline on cerebellar activity: inducing the inferior olive (IO) to burst fire rhythmically, and recruiting rhythmic activity in Purkinje cells (PCs) and deep cerebellar nuclei (DCN) (Handforth 2016). Because of the homology between the acute neurotoxicity of harmaline and idiopathic essential tremor (ET) in patients, creating tremor in rodents by dosing with harmaline has become a model for the testing of various potential interventions for human ET (Handforth 2012; Miwa 2007), an extremely widespread health problem for which standard current treatments may include "gamma knife" ablation of parts of the brain with a highly focused beam of ionizing radiation (Colino 2015).

Harmaline has been supposed to exert its potential tremorogenic effect in rodents at least in part through inhibition of serotonin and other tryptamine receptor binding (Airaksinen et al. 1987), and natural products have been shown to reduce tremor in the harmaline essential tremor model. Among these are both aqueous and ethanolic extracts of saffron, i.e., *Crocus sativa*, and low doses of its principal component, **safranal** (Amin et al. 2015), and **berberine**, the yellow alkaloid characteristic of *Hydrastis canadensis* (golden seal) and *Coptis chinensis* (*huang lian* or "golden thread") roots (Vaziri et al. 2015) (Figures 6.29 and 6.30).

Agonism of cannabinoid receptors with a synthetic cannabinoid known as WIN blocked tremors in the harmaline model in rats and partially reversed associated passive avoidance memory loss (Abbassian et al. 2016). Along with ibogaine, in addicted

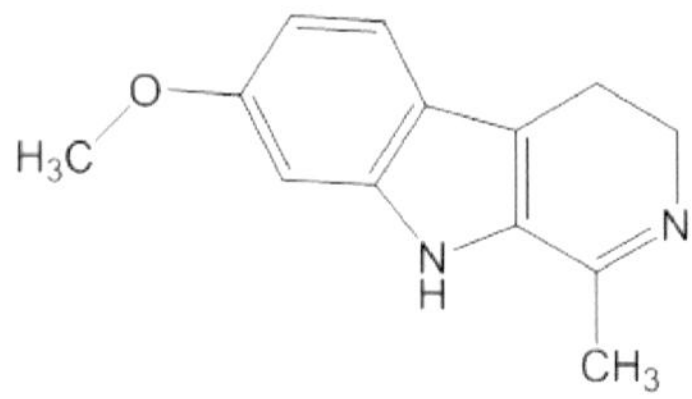

FIGURE 6.28   Harmaline (7-methoxy-1-methyl-4,9-dihydro-3*H*-pyrido[3,4-*b*]indole).

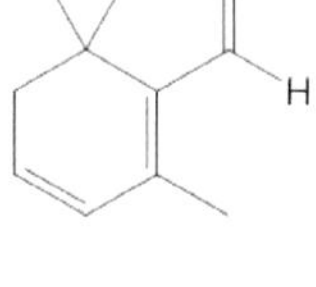

FIGURE 6.29   Safranal.

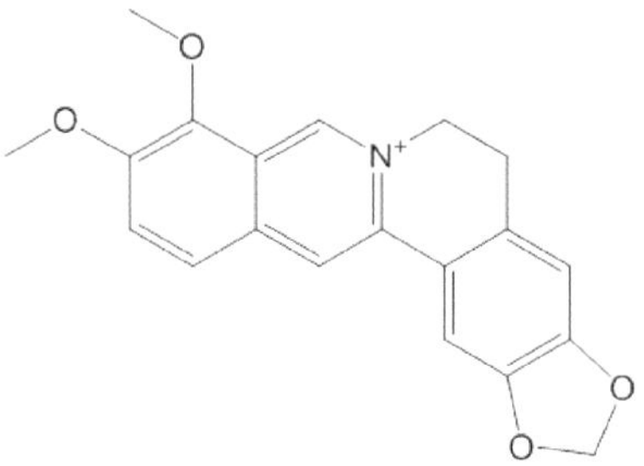

FIGURE 6.30 Berberine.

rats, harmaline dose-dependently (2.5–80 mg/kg) decreased morphine and cocaine intake in the hour after treatment, but unlike ibogaine, harmaline did not decrease morphine or cocaine intake on the day after dosing (Glick et al. 1994). In rats trained with ibogaine, in one study harmaline was able to completely substitute with ibogaine as a discriminative stimulus, but in another only **10-methoxyharmalan** was able to completely substitute (Helsley et al. 1998). As potential modulators of immune function, harmaline and ibogaine showed comparable inhibitor effects, but only at very high doses (House et al. 1995). Both drugs given intravenously induced tremor in photosensitive baboons at similar doses (Brailowsky et al. 1975) (Figure 6.31).

Excitatory amino acids **kainic acid** and **quisqualic acid** had differing effects on harmaline-induced tremors. Kainic acid, which increased the tremorgenic effect of the synthetic tremorgenic drug tremorine, suppressed the tremorgenic effect of harmaline, but quisqualic acid depressed the tremorgenic effects of both **tremorine** and harmaline dose-dependently (Shinozaki et al. 1987) (Figures 6.32 through 6.34).

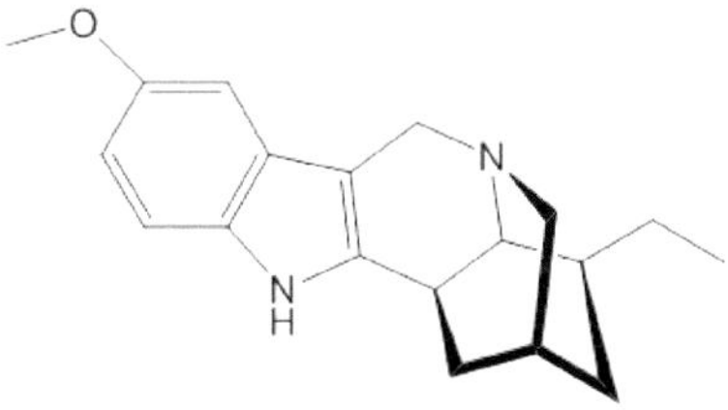

FIGURE 6.31 Ibogaine.

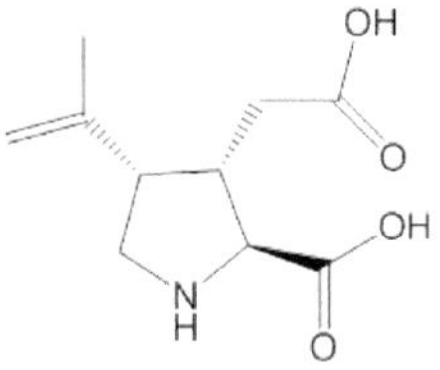

FIGURE 6.32 Kainic acid.

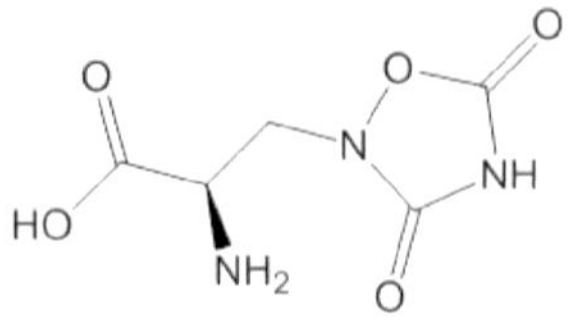

FIGURE 6.33 Quisqualic acid.

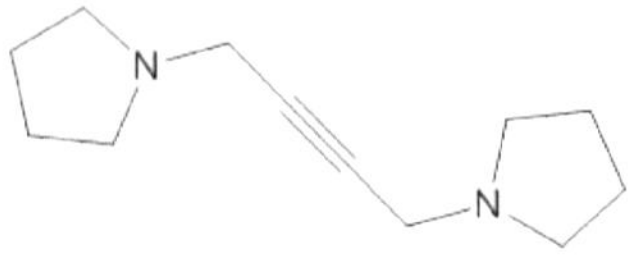

FIGURE 6.34 Tremorine.

Tetrahydroharmine

At 0.1% w/w, tetrahydroharmine is the fourth most copious harmala alkaloid in *P. harmala* seeds, compared to harmaline (5.6%), harmine (4.3%), and harmalol (0.6%), and along with these other compounds, is an inhibitor of MAO (Herraiz et al. 2010) (Figure 6.35). In addition to being found in *P. harmala* seeds, tetrahydroharmine is found in human urine following acute *P. harmala* seed intoxication (Frison et al. 2008). Tetrahydroharmine is also important as a circulating compound following Ayahuasca inebriation, correlated with neurophysiological phenomena such as reduced power in the alpha band (8–13 Hz) after 50 min from ingestion of the brew and increased slow- and fast-gamma power (30–50 and 50–100 Hz, respectively) between 75 and 125 min. "Alpha power reductions were mostly located at left parieto-occipital cortex, slow-gamma power increase was observed at left centro-parieto-occipital, left fronto-temporal and right frontal cortices while fast-gamma increases were significant at left centro-parieto-occipital, left fronto-temporal, right frontal and right parieto-occipital cortices" (Schenberg et al. 2015). New N-acyl derivatives of tetrahydroharmine have been developed for their spasmolytic actions with special attention to their heuristic application for Parkinson's disease (Begum et al. 2006), while a novel naturally occurring tetrahydroharmine derivative, **2-aldehyde tetrahydroharmine**, was recently discovered in *P. harmala* seeds and shown to possess potent inhibitory action against enzymes acetylcholinesterase and butyrylcholinesterase with IC50 values of 12.35 ± 0.24 and 5.51 ± 0.33 μM, respectively (Yang et al. 2015) (Figure 6.36).

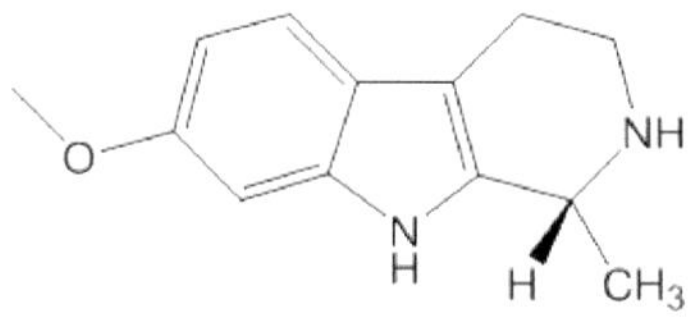

FIGURE 6.35 Tetrahydroharmine ((1R)-7-methoxy-1-methyl-2,3,4,9-tetrahydro-1H-pyrido[3,4-b]indole).

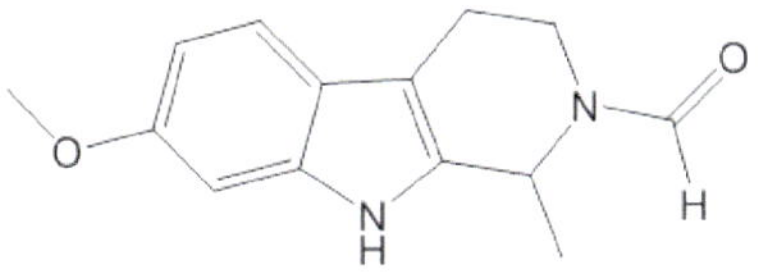

FIGURE 6.36   2-Aldehyde tetrahydroharmine. (From Yang, Y. et al., *J Ethnopharmacol*, 168: 279–86, 2015.)

Harmane

Unlike the more classical harmala alkaloids such as harmine, tetrahydroharmine, harmaline, harmalol, and harmol—usually listed as constituents of *P. harmala* or of the Ayahuasca liana, *Banisteriopsis caapi*—harmane seems rarely mentioned in these contexts (Figure 6.37). However, the presence of harmane in *P. harmala* has been recently confirmed (Tascón et al. 2016)—it is probably a major constituent (Bensalem et al. 2014), and so, its presence should be taken into consideration when considering the pharmacology of the whole plant, its extracts, and its total alkaloid fractions. Harmane is also among the β-carbolines found in the flowering tops of another plant better known as a carrier of harmine, i.e., the passion flower, *Passiflora incarnata* (Avula et al. 2012), and is among the MAO-A inhibitors in tobacco smoke (Truman et al. 2017) (Figure 6.38).

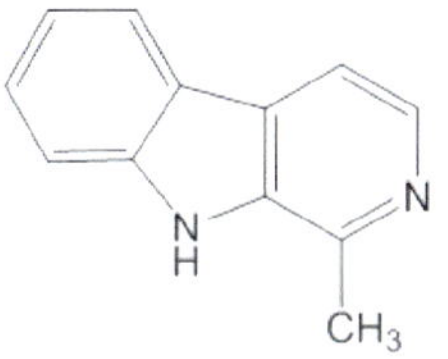

FIGURE 6.37   Harmane (1-methyl-9H-pyrido[3,4-b]indole).

FIGURE 6.38   *Passiflora incarnata* contains several harmala alkaloids, among others, harmane, harmine and harmol (Avula et al. 2012). (Courtesy of Flannery, J., Richmond County, North Carolina, USA via Wikimedia Commons.)

In a study of six harmala alkaloids (harmine, harmane, harmalol, harmol, harmaline, and norharmane), harmane and harmine had the highest activity as potential inhibitors of collagen-induced platelet aggregation, and they are thus of greatest potential interest as drugs for atherothrombotic diseases (Im et al. 2009). It is postulated further that harmane induces epithelial cells to release nitric oxide, thus contributing to the vasorelaxant and hypotensive effects of harmane and of *harmal* extracts in general (Shi et al. 2000).

Harmane may also contribute significantly to the photosensitizing effect of *harmal* on cancer cells, at least in part due to its binding affinity for RNA and DNA (Paul and Guchhait 2011a,b) and electrostatic interaction between its cationic charge and the surface charge of lipids (Paul and Guchhait 2012). Harmane is also unfortunately sometimes implicated as a procarcinogenic compound, owing to factors such as elevated serum levels in patients with essential tremors and cancer, especially as a comutagen (Louis et al. 2008) and also as a promoter of Parkinson's disease and addiction, as well as cancer (Pfau and Skog 2004). Others have confirmed elevated levels of harmane in patients with Parkinson's disease (Kuhn et al. 1995a,b, 1996), though harmane ameliorated experimental depression in the mouse forced swimming test (Farzin and Mansouri 2006). Harmane is also produced endogenously and has even been postulated as an "atypical" neurotransmitter in *Homo sapiens*—a putative imidazoline binding site ligand (Abu Ghazaleh et al. 2015).

Harmane was looking better and better, except for one issue. It seems that just as harmaline was roped into the role of poster child for causing tremors, complete with a harmaline tremor scale named in its honor, harmane is associated with something called harmane-induced amnesia (Nasehi et al. 2010, 2012, 2013a,b). The phenomenon was measurable in mice according to a battery of parameters and elegant pharmacological interventions, and involved modulation of multiple receptor systems by harmane, including benzodiazepine, serotoninergic, dopaminergic, and nicotinic cholinergic ones.

Novel Indoles

Several novel indolic compounds have been recently identified in *P. harmala* seeds that had not been previously observed. So far, the work in this field has been mainly accomplished by two independent research groups.

Wang et al. (2014) described a dimeric β-carboline alkaloid characterized by a unique 3,9-diazatetracyclo[6.5.2.0(1,9).0(3,8)]pentadec-2-one scaffold, which they called **Peganumine A**. Its structure, including absolute configuration, was revealed via spectroscopy, x-ray crystallography, electron capture detection, and circular dichromism exciton chirality. The compound showed moderate cytotoxic activity against MCF-7 (breast), PC-3 (prostate), and HepG2 (liver) human cancer cells with selective effects on HL-60 (leukemia) cells with IC50 of 5.8 μM (Figure 6.39).

Wang C. et al. (2015) employed antileukemic bioactivity guided fractionation to elucidate 2 new indole alkaloids and 10 known ones. Among the known alkaloids was a compound, **harmalacidine**, which had the highest cytotoxicity against human histiocytic lymphoma U-937 cells with IC50 value of 3.1 ± 0.2 μmol/L. Its mechanism was targeting the mitochondrial and protein tyrosine kinase signaling pathways (PTKs-Ras/Raf/ERK). The new alkaloids were also cytotoxic and identified

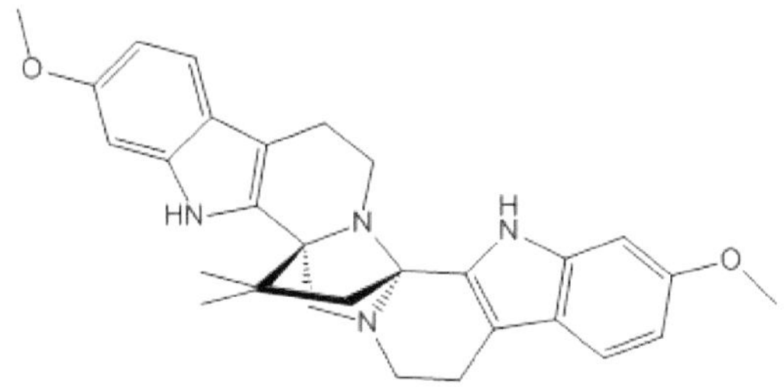

FIGURE 6.39 Peganumine A.

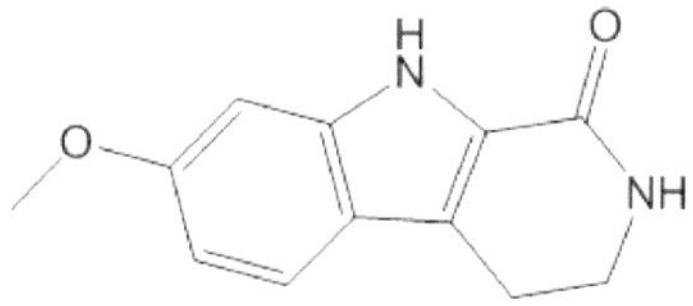

FIGURE 6.40 Harmalacidine.

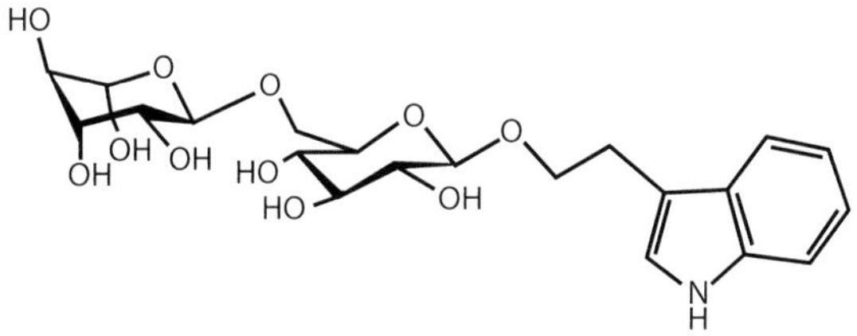

FIGURE 6.41 2-(Indol-3-yl)ethyl-α-L-rhamnopyranosyl-(1 → 6)-β-D-glucopyranoside.

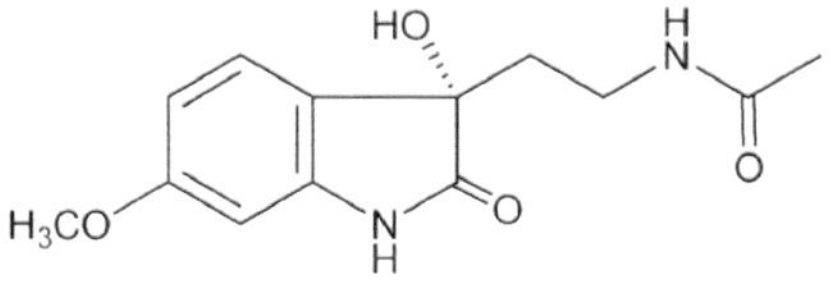

FIGURE 6.42 3-Hydroxy-3-(N-acetyl-2-aminoethyl)-6-methoxyindol-2-one.

as **2-(indol-3-yl) ethyl-α-L-rhamnopyranosyl-(1 → 6)-β-D-glucopyranoside** and **3-hydroxy-3-(N-acetyl-2-aminoethyl)-6-methoxyindol-2-one** (Figures 6.40 through 6.42).

Additional Notes on Harmala Alkaloids

Aarons et al. (1977) noted common cardiovascular actions among harmine, harmaline, and harmalol, resulting in lower heart rate, increased pulse pressure, increased peak aortic flow, and increased myocardial contractile force in intact normotensive anesthetized dogs, while harmine reduced systemic arterial pressure and total peripheral vascular resistance.

Nafisi et al. (2010a) studied binding of harmala alkaloids to yeast RNA. The observed binding affinities were according to harmine > harmaline > harmane > harmalol > tryptoline. Jiménez et al. (2008) investigated the safety of one full aromatic β-carboline alkaloid (harmine) and one dihydro-β-carboline alkaloid (harmaline) with two *in vitro* human cell assays: the cytochalasin-B blocked micronucleus (CBMN) assay and the viability/colony formation assay with four different human cultured nontransformed (CCD18Lu) and transformed (HeLa, C33A, and SW480) cells. Harmine was shown to be the least toxic with the nontransformed cells and most toxic to the tumor cells overall relative to harmaline.

Quinazoline Alkaloids

Quinazoline

Quinazoline and its derivatives are important drug leads (Figure 6.43). One interesting application is in the prevention and even reversal of damage to pancreatic β cells. These are the cells in the anatomic Islets of Langerhans that are responsible for insulin metabolism. Dysfunction of these cells is correlated with Type 2 and especially Type 1 diabetes. Duan et al. (2016) showed the **β cell protective and reparative effects of quinazoline** and its newer synthetic congeners in this context.

Another group (Sompalle et al. 2016) synthesized a series of quinazoline alkaloids and tested them for their antioxidant and antifungal properties *in vitro*. Many had potent free radical scavenging power, and all showed antifungal activity superior to a conventional antifungal drug (i.e., **fluconazole**) against the test organisms *Aspergillus flavus* and *A. niger* (Figure 6.44).

Potential for quinazoline and quinazolinone derivatives as anticancer agents has been noted. One relevant mechanism for exerting anticancer effect of quinazolinone derivatives is on epithelial growth factor receptor (EGFR), discussed by Jafari et al. (2016) in their excellent review on the antimicrobial and cancer cytotoxic effects of the quinazoline family of compounds. EGFR is a member of a group of related

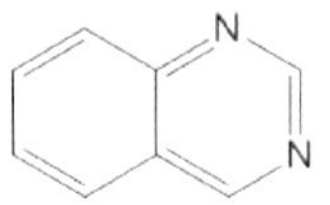

FIGURE 6.43 Quinazoline (1,3-diazanaphthalene, benzopyrimidine, phenmiazine, benzo-1,3-diazine).

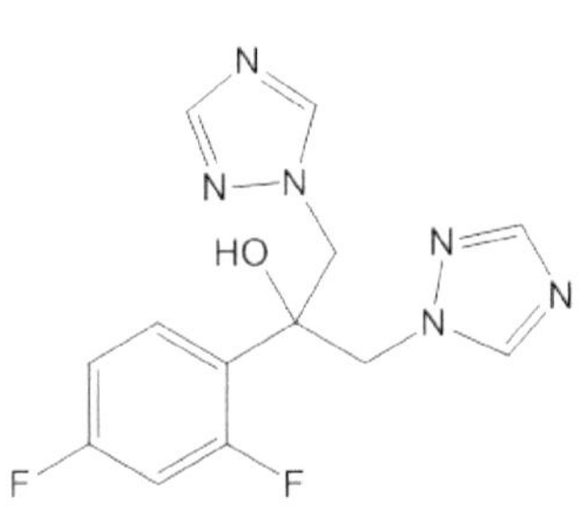

FIGURE 6.44 Fluconazole.

proteins, the others being Her2/neu, Her3, and Her4, that are associated with upregulating cancer growth. The search for new, more effective, and safer EGFR inhibitors is a vigorous frontier of anticancer research, with a substantial part focused on quinazoline and congeners (Asadollahi-Baboli 2016; Bai et al. 2016; Hou et al. 2016; Wang Z. et al. 2015, 2016).

Many of these derivatives of quinazoline are already being manufactured as bonafide drugs, have passed clinical trials and approvals, and are being prescribed (Brower and Robins 2016).

Although other herb-derived natural products, such as those derived from *Scutellaria baicalensis*, have also demonstrated anti-EGFR activity (He et al. 2012; Ye et al. 2009), other medicinal herbs used as integrative cancer therapy may exert paradoxical EGFR upregulation. There is indeed concern that **tetrahydrocannabinol** (THC), the main psychoactive cannabinoid in *Cannabis* sp., may upregulate EGFR (Hart et al. 2004; Idrizi et al. 2016) leading to cancer progression, not apoptosis, though recent evidence suggests that ligands for the CB2 cannabinoid receptor downregulate EGFR (Elbaz et al. 2016), or alternatively, that CBD (**cannabidiol**), an only minimally psychoactive cannabinoid, downregulates EGFR (Elbaz et al. 2015). Quinazolines with their notable anti-EGFR effect may help balance drugs from other herbal products that promote EGFR, and thus, as they do in conventional oncology, render synergism with chemotoxic agents (Howe et al. 2016; Lee et al. 2011; Leto et al. 2015; Shingu et al. 2016; Yang and Tam 2016). Synergy between EGFR inhibitors and other natural drugs suggests ongoing and intriguing research opportunities (Leeman-Neill et al. 2010; Liu et al. 2014; Ren et al. 2012; Suárez-Arroyo et al. 2016; Tarozzi et al. 2016; Tian et al. 2015; Wu et al. 2013; Xu et al. 2016). Possible novel applications for EGFR inhibitors include chordoma, a rare, highly resistant neoplasm of the vestigial notochord (Scheipl et al. 2016) (Figures 6.45 through 6.47).

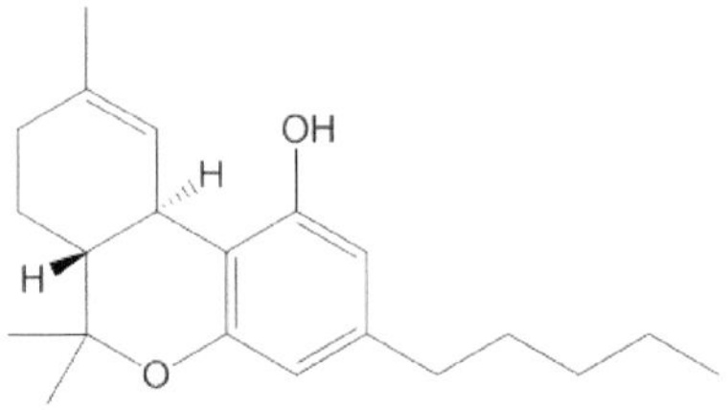

FIGURE 6.45 Tetrahydrocannabinol (THC).

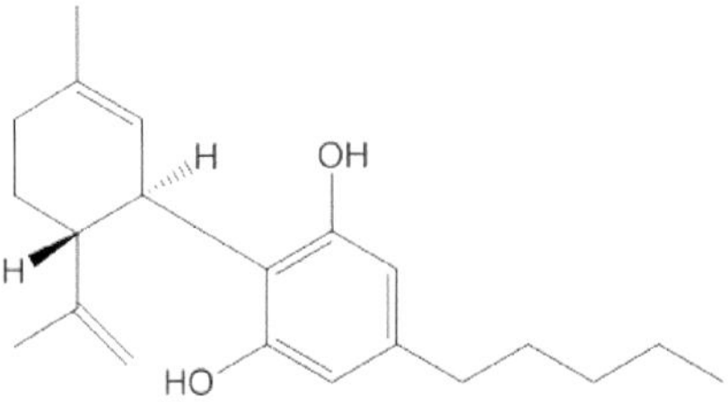

FIGURE 6.46 Cannabidiol.

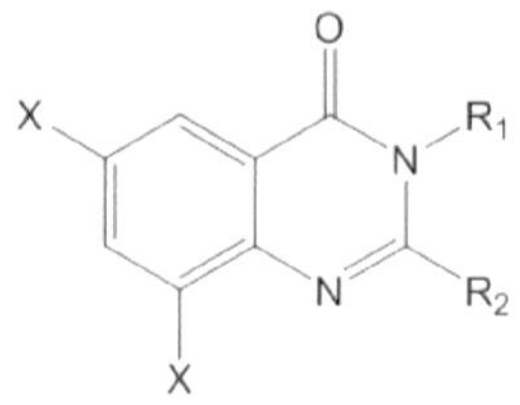

FIGURE 6.47   Quinazolinone basic structure. (From Jafari, E. et al., *Res Pharm Sci*, 11(1): 1–14, 2016.)

Vasicine (Peganine)

The name *vasicine* derives from the common names *vasa* or *vasaka*, which refer to *Justicia adhatoda*, or by an alternative name, *Adhatoda vasica*, a medicinal plant native to Asia and employed in India, Sri Lanka, Malaysia, and China that contains over 1% vasicine in its dried material (Figures 6.48 and 6.49). *Vasaka* or *ana-toda* is well-known to Ayurveda, Siddha, and Unani medicines with a reputation for improving respiratory function, though recent pharmacognostic investigations showing oxytocic or abortifacient actions of the plant have called into question its safety for use in treatment (Claeson et al. 2000), presumably owing at least in part to this vasicine and its congener, vasicinone (see below). Nevertheless, the plant has been a subject of several clinical trials, which have demonstrated its efficacy as an antitussive (Barth et al. 2015). In mice and guinea pigs, *P. harmala*–derived vasicine and deoxyvasicine, and vasicinone which was synthetically derived from vasicine, showed significant antitussive activity at the highest dose studied (45 mg/kg), which was comparable to **codeine**, and demonstrated bronchodilating and expectorant properties comparable to **aminophylline** (Liu et al. 2015c) (Figures 6.50 and 6.51).

Vasicine is, of course, also found in *P. harmala* seeds, where it may more commonly be called by its *Peganum* name, i.e., peganine. Most recently, a new method of microwave-assisted extraction was applied to *P. harmala* seeds and was found to be faster with higher yields than previous methods for isolating vasicine. Further, vasicine, as well as harmala alkaloids found in the same seeds, were individually efficacious as ascarides (killing arachnids, especially ticks and mites) (Shang et al. 2016).

Wang C.H. et al. (2015), employing spectroscopic analysis including IR, HR-ESI-MS, 1D and 2D NMR, and specific rotation as well as comparison of their data with those in the literature concerning their methanolic extract of *P. harmala* seeds, found vasicine and previously noted **(R)-vasicinone-1-O-β-d-glucopyranoside**, vasicinone, and deoxyvasicinone, in addition to the novel

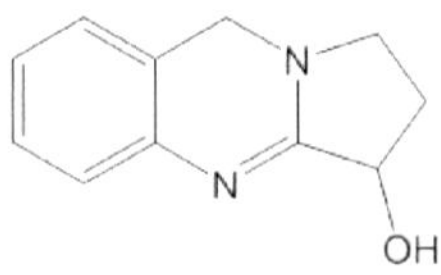

FIGURE 6.48   Vasicine (1,2,3,9-Tetrahydropyrrolo[2,1-*b*]quinazolin-3-ol).

FIGURE 6.49   Both *Peganum harmala* and *Justicia adhatoda* contain vasicine. (The Hong Kong Zoological and Botanical Gardens, Hong Kong: by Daderot [Public domain], via Wikimedia Commons.)

(S)-vasicinone-1-O-β-d-glucopyranoside, and showed vasicine and their novel compound active against proliferation of MCG-803 human gastric cancer cells according to the MTT assay (Figures 6.52 and 6.53). Vachnadze et al. (2015) found peganine and d,l peganine to slow the heartbeat consistent with a putative anticholinesterase action, and to comprise at certain stages of the plant's life as much as 5% of its total alkaloids. Vasicine (i.e., peganine) isolated from *Adhatoda vasica*, as well as vasicine acetate prepared from the vasicine, possessed good cytotoxic activity against human A-549 lung adenocarcinoma cells *in vitro*, and exhibited antibacterial and antioxidant activities (Duraipandiyan et al. 2015).

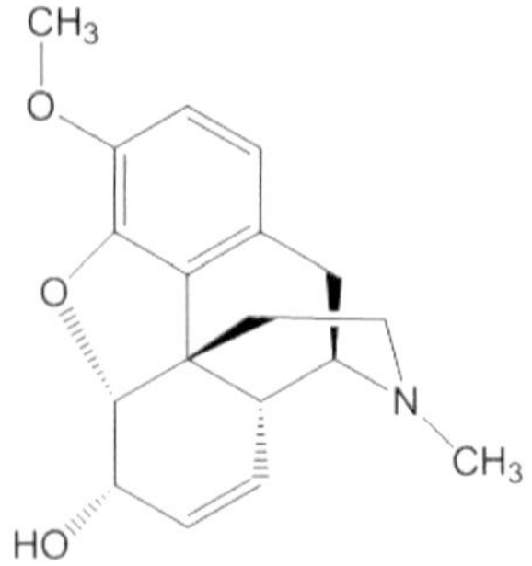

FIGURE 6.50 Codeine.

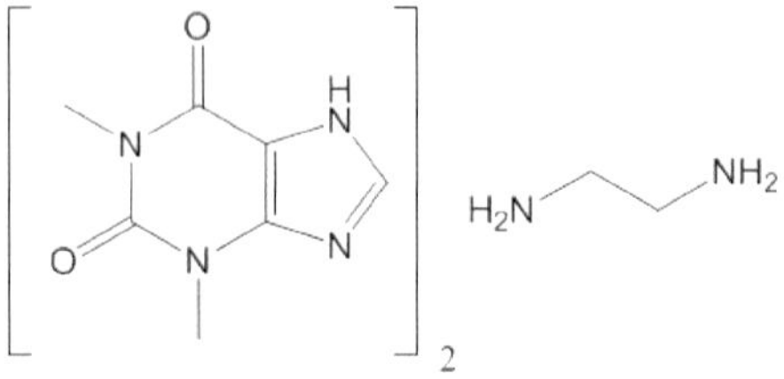

FIGURE 6.51 Aminophylline.

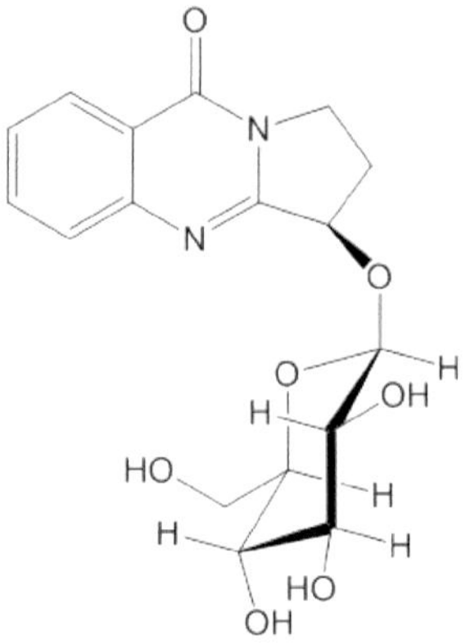

FIGURE 6.52 (R)-vasicinone-1-O-β-d-glucopyranoside. (From Wang, C.H. et al., *J Asian Nat Prod Res*, 17(5): 595–600, 2015.)

Vasicine found service as a biological catalyst (Sharma et al. 2014), for example in the synthesis, through intramolecular arylation, of pharmacologically active **phenanthridinones** and **dihydrophenanthridines** (Sharma et al. 2016), and it was a promising modern pharmaceutical lead for an antitubercular drug (Chaliha et al. 2016; Grange and Snell 1996; Ignacimuthu and Shanmugam 2010). A bioinformatic approach (Jha et al. 2012) may aid in the guest for an antitubercular drug against strains of the mycobacterium resistant to existing therapies (Figures 6.54 and 6.55).

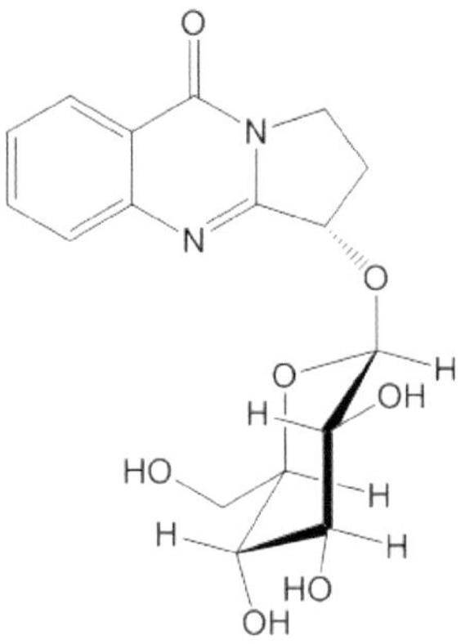

FIGURE 6.53 (S)-vasicinone-1-O-β-d-glucopyranoside. (From Wang, C.H. et al., *J Asian Nat Prod Res*, 17(5): 595–600, 2015.)

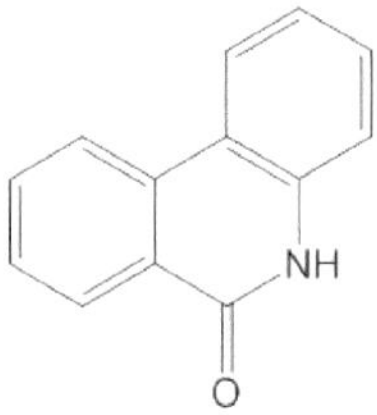

FIGURE 6.54 Phenanthridinones (basic structure).

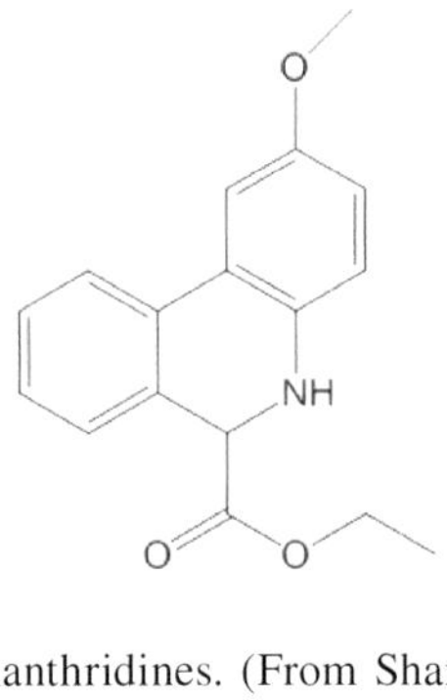

FIGURE 6.55 5,6-Dihydrophenanthridines. (From Sharma, S. et al., *Org Biomol Chem*, 14(36): 8536–44, 2016.)

Against allergic asthma, vasicine or its synthetically modified analogues reduced inflammatory cell infiltration of the airways in mice (Rayees et al. 2015). Vasicine was investigated as a putative drug against Alzheimer's disease, owing to its potent anticholinesterase and antibutylesterase activities (Liu et al. 2015b; Yang Y. et al. 2015). These workers also studied the drug's metabolism and showed it to be inactivated in the kidney *in vivo*.

Peganine possesses potent gastroprotective action against ulcers in rat. The drug reduced free acidity (33.38%) and total acidity (38.09%), while upregulating mucin secretion by 67.91%. It also significantly inhibited H(+) K(+)-ATPase activity *in vitro*, with IC50 of 73.47 μg/ml vs. 30.24 μg/ml for **omeprazole**, illustrating an antisecretory component of its gastroprotection (Singh et al. 2013) (Figure 6.56). In addition to its antibacterial and antioxidant capacities, vasicine inhibited the proteolytic enzyme trypsin, an additional potentially desirable effect for individuals with pancreatitis (Shahwar et al. 2012). Vasicine has been standardized in a lyophilized aqueous extract of *Adhatoda vasica*—a small part of the vasicine in the extract contained in gelatin capsules was converted to vasicinone over time (Vyas et al. 2011). Vasicine is the basis of Ambroxol, a widely used "secretolytic" agent used in treating bronchitis, asthma, and rheumatism through common anti-IgE, that is, anti-allergenic, mechanisms (Gibbs 2009). Peganine hydrochloride was a potent antileishmanial agent *in vitro* and *in vivo* (hamsters) (Khaliq et al. 2009), likely related to an apoptosis-promoting mechanism (Misra et al. 2008).

Peganine further inhibited alpha-glucosidase, more potently than 40 other traditional Chinese herbs, with sucrose as substrate. Inhibition of this enzyme, which catalyzes catabolism of sucrose and related carbohydrates, is a key mechanism for a drug directed at lowering blood sugar especially in type 2 (adult onset) diabetes mellitus (Gao et al. 2008).

Peganine is conducive to synthetic modifications aimed at building a library of bioactive compounds (Shevyakov et al. 2006) or for creating new compounds with superior biological activities, for example, improved bronchodilating (Johri and Zutshi 2002; Racle et al. 1976; Zabeer et al. 2006) and/or broncho-mucolytic capacity (Bruce and Kumar 1968). As opposed to its roots and fruits, shoots, flowers, and shoot cultures of *P. harmala* contain more peganine than harmala alkaloids, suggesting that the harmala alkaloids are more important to *harmal's* defense systems, with the quinazolines such as peganine serving non-defense functions (Zayed and Wink 2005).

Clinically, peganine may be combined with herbal "correctives" such as the Ayurvedic "Trikatu" (1:1:1 ratio of dried fruits of *Piper nigrum* and *Piper longum*, and dried rhizomes of *Zingiber officinale*) to improve bioavailability (Lala et al. 2004). Most likely, these "acrids" achieve their influence to improve bioavailability "either by promoting rapid absorption from the gastrointestinal tract, or by protecting the drug from being metabolised/oxidised in its first passage through the liver after being absorbed, or by a combination of these two mechanisms" (Atal et al. 1981).

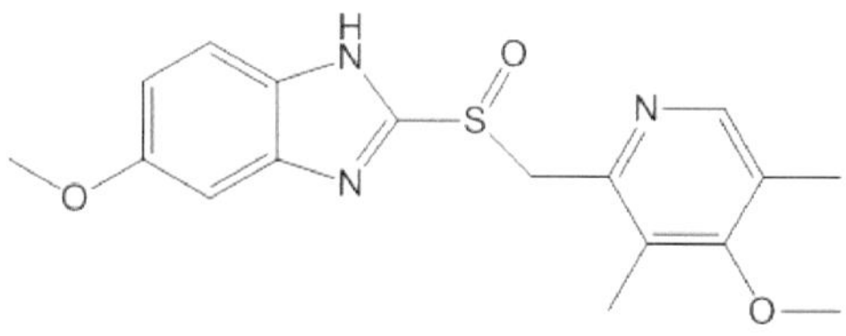

FIGURE 6.56   Omeprazole.

Pharmacokinetic studies of a single dose of intravenous peganine in six Indian volunteers revealed a peak serum level after 15 min and a bronchodilatory effect similar to theophylline lasting 4–5 h (Amla et al. 1987). Toxicology studies in rats and monkeys were completed (Pahwa et al. 1987).

Peganine was shown to be thrombopoietic (Atal et al. 1982). As noted above, its reliable oxytocic action (Gupta et al. 1977, 1978) is a potential contraindication in some cases, though it may also benefit, facilitate, and hasten delayed childbirth (Gautam and Sharma 1982; Lal and Sharma 1981; Zutshi et al. 1980).

In addition to being found in *Peganum* and *Adhatoda vasica*, peganine is also known in the goat rue, *Galega officinalis*, which may account in part for that plant's reputation for toxicity (Schreiber et al. 1962), in *Anisotes trisulcus* (El-Shanawany et al. 2014), a traditional Yemeni medicinal plant with hepatoprotective properties (Fleurentin et al. 1986; Lanhers et al. 1986), and, as its 7-hydroxy form, in *Linaria vulgaris* (Hua et al. 2002) (=*L. acutiloba*), called in English with various names, i.e., "toadflax," brideweed, bridewort, butter and eggs, butter haycocks, bread and butter, bunny haycocks, bunny mouths, calf's snout, Continental weed, dead men's bones, devil's flax, devil's flower, doggies, dragon bushes, eggs and bacon, eggs and butter, false flax, flaxweed, fluellen (but see *Kickxia*), gallweed, gallwort, impudent lawyer, Jacob's ladder, lion's mouth, monkey flower, North American ramsted, rabbit flower, rancid, ransted, snapdragon, wild flax, wild snapdragon, wild tobacco—actually an evolutionary cousin of *Nicotiana* (Kovacova et al. 2014), yellow rod or yellow toadflax (Mabey 1996), a European folk remedy and notorious invasive species.

The biosynthesis of peganine *in situ*, i.e., in *Adhatoda vasica*, was elaborated (Groeger and Mothes 1960; Johne et al. 1968), and facile laboratory syntheses of peganine, deoxyvasicine, and deoxyvasicinone described (Richers et al. 2013). In an *in vivo* study that measured inflammation using carrageenan and/or Complete Freund's Adjuvant (CFA)–induced rat paw edema and antimicrobial activity with the microdilution method, vasicine, 20 mg/kg injected 6 h after carrageenan, was the most potent anti-inflammatory (59.51%), after vasicinone (63.94%) at 10 mg/kg 4 *days* after CFA. Peganine potently inhibited *Escherichia coli* at 20 μg/ml and *Candida albicans* > 55 μg/ml (Singh and Sharma 2013).

Desoxypeganine

In desoxypeganine (also, deoxypeganine, desoxyvasicine, deoxyvasicine), the hydroxyl group (-OH) is lost, leaving the streamlined form of vasicine/peganine, which may have important physiological advantages, perhaps in potency or bio-availability, over the parent compound (Figure 6.57). For example, desoxypeganine isolated from *P. harmala* seeds was 10 times as potent an anticholinesterase as peganine hydrochloride *in vivo* (Tuliaganov et al. 1986). According to the authors, desoxypeganine was being utilized in Russia at that time for the treatment of "lesions of

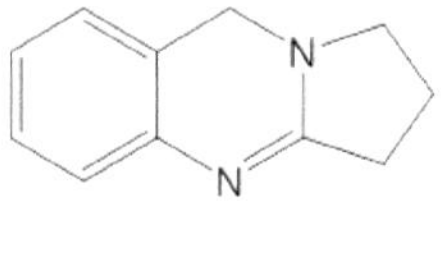

FIGURE 6.57   Desoxypeganine (1,2,3,9-tetrahydropyrrolo[2,1-b]quinazoline).

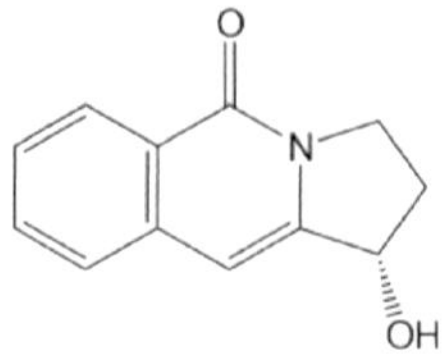

FIGURE 6.58   Vasicinone ((3S)-3-hydroxy-2,3-dihydro-1H-pyrrolo[2,1-b]quinazolin-9-one).

the peripheral nervous system." More recently, the selective inhibition of monoamine oxidase, type A (MAO-I), by desoxypeganine was noted in addition to its anticholinesterase property, and the compound had been the subject of Phase I clinical trials for putative application as an agent to decrease craving for alcohol, i.e., as an anti-addictive drug (Algorta et al. 2008, 2009; Doetkotte et al. 2005; Mucke 2005). However, in spite of this clinical potential in addiction treatment or Alzheimer's disease, its use may be limited by the facile conversion of desoxypeganine to the inactive metabolite pegeneone *in situ* (Illgner and Matusch 2005).

Desoxypeganine derivatives were developed as inhibitors of NEDD8-activating enzyme (NAE), which regulates specific degradation of proteins controlled by cullin-RING ubiquitin E3 ligases. The inhibition was believed, based on molecular modeling, to be related to blocking of ATP binding domains, and the derivatives were considered lead molecules for the development of more effective NAE inhibitors (Zhong H.J. et al. 2012, 2015).

Vasicinone

Vasicinone possessed hepatoprotective (Sarkar et al. 2014), antiasthmatic (Nilani et al. 2009), and anticancer (Wang C.H. et al. 2015) properties (Figure 6.58). Its presence was confirmed in *P. harmala* (Pulpati et al. 2008) and *P. multisectum* (Liu 2011) (Figure 6.59).

PROTEINS

Like all seeds in nature, the seeds of *P. harmala* contain proteins. Many different proteins of course naturally occur, and in ways yet to be understood undoubtedly contribute to the pharmacological effect of whole extracts of the seeds, particularly of the aqueous extracts generally utilized in clinical therapy. Four of these proteins were investigated and particular pharmacological activities noted. These four proteins and their activities are discussed in the next paragraphs.

Two of the proteins were isolated by Amel M. Soliman and her group at the Zoology Department, Faculty of Science, Cairo University, Egypt. Although the approximate molecular weights of the two compounds were established, further characterization lay in the future. The first compound, with a MW of about 15,000 or 15 kilodalton (15 KD) is referred to by its weight: 15KD. Soliman et al. (2011) showed the 15 KD protein to significantly improve oxidation status in the liver following assault with the laboratory carcinogen, carbon tetrachloride. Similarly, Soliman et al. (2013) showed a larger protein they isolated from *P. harmala* seeds, which at 132 KD (known simply

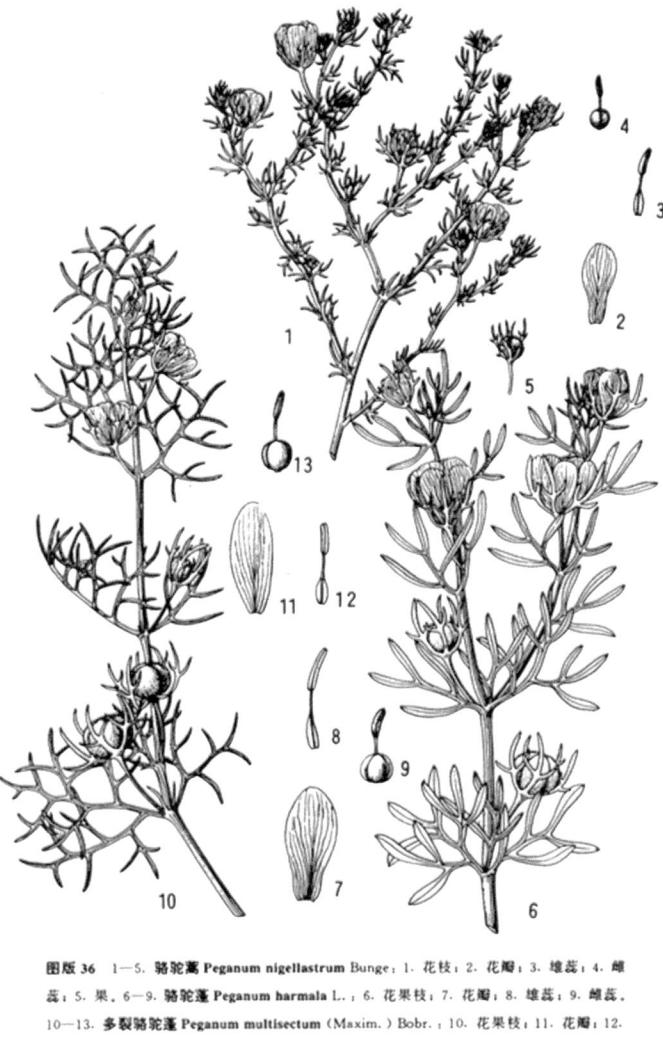

图版 36 1—5. 骆驼蒿 Peganum nigellastrum Bunge: 1. 花枝; 2. 花瓣; 3. 雄蕊; 4. 雌蕊; 5. 果。6—9. 骆驼蓬 Peganum harmala L.; 6. 花果枝; 7. 花瓣; 8. 雄蕊; 9. 雌蕊。10—13. 多裂骆驼蓬 Peganum multisectum (Maxim.) Bobr.; 10. 花果枝; 11. 花瓣; 12. 雄蕊; 13. 雌蕊。(陶明琴绘)

FIGURE 6.59 Line drawings of *Peganum nigellastrum*, *P. harmala*, and *P. multisectum*. (From Wu, Z.Y. et al., *Flora of China. Vol. 11, Oxalidaceae through Aceraceae*. Science Press, Beijing, and Missouri Botanical Garden Press, St. Louis, 2008.)

as the "132 KD protein") also possesses strong antioxidant activity in rat hepatocytes *in vivo* following exposure of the animal to carbon tetrachloride.

Shi et al. (2013) elucidated the structure of nonspecific lipid docking proteins (ns-LDPs) found in *P. harmala* seeds that may belong to ns-LTP1 family of ns-LDPs. Overall, their work shed much light on the lipid–protein dynamics involved in lipid docking and is a great advance toward the eventual characterization of these proteins' three-dimensional structures.

Ma et al. (2013) found a protein in *P. harmala* seeds weighing about 16 KD. It also had lipid binding activity, and further, suppressed growth of the pathogenic fungi *Alternaria alternata*, *Penicillium digitatum*, *Rhizopus stolonifer*, and *Magnaporthe grisea*, and retarded proliferation of human esophageal carcinoma cells (Eca-109), cervical carcinoma (HeLa), gastric carcinoma (MGC-7), and melanoma (B16) cells with IC(50)s of 0.7, 2.74, 3.13, and 1.47 µM, respectively. Moreover, their 16 KD protein, which they called PHP (for *P. harmala* protein), significantly inhibited HIV-1 reverse transcriptase (RT) with an IC(50) of 1.26 µM. In short, the proteins in *P. harmala* are also apparently quite active pharmacologically as antioxidants, antipest, and antineoplastic compounds. There is good reason to consider keeping these proteins in the mix and in mind.

POLYSACCHARIDES

Polysaccharides belonging to the class of compounds known as pectin or pectins were isolated from *P. harmala* and several other plants from the Mongolian desert (Golovchenko et al. 2012). The presence of these compounds in the diet inhibited absorption of ovalbumin from mouse gut *in vivo*. In general, pectins may be a source for molecules with cytotoxic activity, as was recently shown for citrus pectin (Leclere et al. 2016). Pectins may be a source of immunomodulating activity (Gong et al. 2015; Hamedi et al. 2015; Kerperien et al. 2014) in addition to their well-known salubrious effect on bowel motility.

POLYPHENOLS

Khadhr et al. (2016) in Tunisia noted polyphenols in general as components of the oil of local *P. harmala* seed, and as a contributor, along with γ-tocopherol and linoleic acid, to the oil's antioxidant and anti-inflammatory potency. The polyphenols in the oil may also have contributed to analgesic effects following topical applications, as well as anti-inflammatory effecs at 20% typically 5 h after carrageenan application, comparable to 1% of the nonsteroidal anti-inflammatory drug diclofenac.

LIPIDS

Moussa and Almaghrabi (2016) studied the fatty acid content of *P. harmala* plant using gas chromatography–mass spectroscopy. They found primarily saturated fatty acids, namely tetradecanoic, pentadecanoic, tridecanoic, hexadecanoic, heptadecanoic, and octadecanoic acids, with the derivatives 12-methyl tetradecanoic, 5,9,13-trimethyl tetradecanoic, and 2-methyl octadecanoic acids (Figures 6.60 through 6.68). Four unsaturated fatty acids were (E)-9-dodecenoic, (Z)-9-hexadecenoic, (Z,Z)-9,12-octadecadienoic, and (Z,Z,Z)-9,12,15-octadecatrienoic acids (Figures 6.69 through 6.72). The latter were the most abundant of the unsaturated fatty acids: 14.79%, followed by (Z,Z)-9,12-octadecadienoic at 10.61%. Also in this oil eight nonfatty acid compounds were characterized: eight nonfatty acid compounds **1-octadecene**, **6,10,14-trimethyl-2-pentadecanone**, **(E)-15-heptadecenal**, **oxacyclohexadecan-2-one**, **1,2,2,6,8-pentamethyl-7-oxabicyclo[4.3.1]dec-8-en-10-one**, **hexadecane-1,2-diol**, **n-heneicosane**, and **eicosan-3-ol** (Figures 6.73 through 6.79).

FIGURE 6.60 Tetradecanoic acid.

FIGURE 6.61 Pentadecanoic acid.

FIGURE 6.62 Tridecanoic acid.

FIGURE 6.63 Hexadecanoic acid.

FIGURE 6.64 Heptadecanoic acid.

FIGURE 6.65 Octadecanoic acid.

FIGURE 6.66 12-Methyl tetradecanoic acid.

FIGURE 6.67 5,9,13-Trimethyl tetradecanoic acid.

FIGURE 6.68   2-Methyl octadecanoic acid.

FIGURE 6.69   (E)-9-Dodecenoic acid.

FIGURE 6.70   (Z)-9-Hexadecenoic acid.

FIGURE 6.71   (Z,Z)-9,12-Octadecadienoic acid.

FIGURE 6.72   (Z,Z,Z)-9,12,15-Octadecatrienoic acid.

FIGURE 6.73   6,10,14-Trimethyl-2-pentadecanone.

FIGURE 6.74   (E)-15-Heptadecenal.

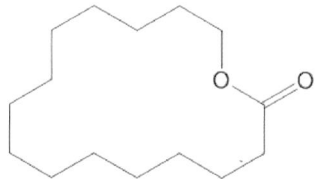

FIGURE 6.75   Oxacyclohexadecan-2 one.

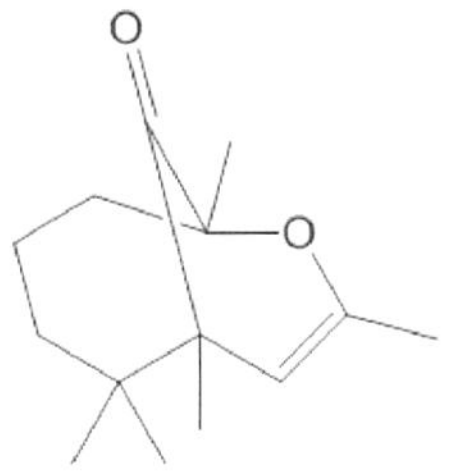

FIGURE 6.76 1,2,2,6,8-Pentamethyl-7-oxabicyclo[4.3.1]dec-8-en-10-one. (From Moussa, T.A., and O.A., Almaghrabi, *Saudi J Biol Sci*, 23(3): 397–403, 2016.)

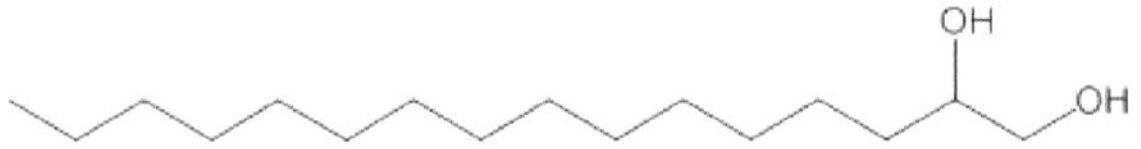

FIGURE 6.77 Hexadecane-1,2-diol.

FIGURE 6.78 n-Heneicosane.

FIGURE 6.79 Eicosan-3-ol.

VOLATILES

Apostolico et al. (2016) studied with gas chromatography/mass spectrophotometry the volatile essential oil fraction of the seed oil of *P. harmala* (Pègano). The botanical samples were obtained from regions of Algeria, Egypt, Libya, Morocco, and Tunisia. A total of 105 compounds were identified, mainly monoterpenes and oxygenated sesquiterpenes, with **eugenol** the most prominent (Figure 6.80). The complete essential oil fraction, as well as its individual components, were bactericidal to *Staphylococcus aureus*, *Bacillus cereus*, *Escherichia coli*, and *Pseudomonas aeruginosa*, and phytotoxic to the initial shoots and radicles of *Raphanus sativus*, *Lepidium sativum*, and *Ruta graveolens*.

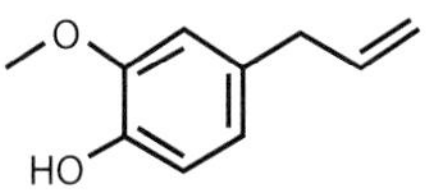

FIGURE 6.80 Eugenol.

OTHERS

Using anticancer bioactivity guided fractionation, Wang C. et al. (2016) discovered two new compounds in *harmal* seed: a terpenoid, **3α-acetoxy-27-hydroxyolean-12-en-28-oic acid methyl ester** (**OA**), and a phenolic glycoside, **N-acetyl-9-syringinoside**. Of these, the first, OA, showed exceptional potency *in vitro* against human non-small cell lung cancer cells (NSCLC), obstructing epidermal growth factor receptor (EGFR) and thus inactivating EGFR-driven antiapoptotic pathway (Figures 6.81 and 6.82).

ELEMENTS

Employing methods including microwave digestion of plants, flame atomic absorption, emission spectrometry, and electrothermal atomic absorption spectrometry, elements such as Fe, Cd, Cu, Mn, Pb, Cr, Ni, and Na were quantified in *P. harmala* (Küçükbay and Kuyumcu 2014; Ghani et al. 2014). *P. harmala* seeds were also found to contain 75% carbohydrate (Ghani et al. 2014) (Figure 6.83).

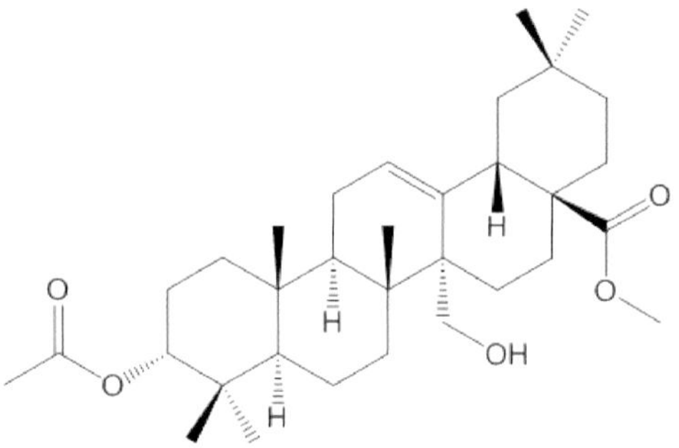

FIGURE 6.81   3α-Acetoxy-27-hydroxyolean-12-en-28-oic acid methyl ester. (From Wang, C., *Chem Biodivers*, 13(7): 961–8, 2016.)

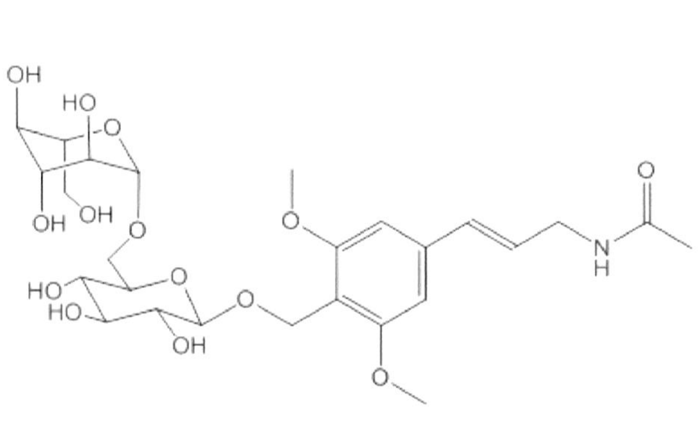

FIGURE 6.82   N-acetyl-9-syringinoside.

FIGURE 6.83 *Peganum harmala* seed pods ripen in Israel in the middle of the summer. (7/20/2015, Judaean Desert, by side of the Inn of the Good Samaritan, Israel: by Helena Paavilainen.)

REFERENCES

Aarons, D.H., G.V. Rossi, and R.F. Orzechowski. 1977. Cardiovascular actions of three harmala alkaloids: Harmine, harmaline, and harmalol. *J Pharm Sci* 66: 1244–8.

Abbassi, R, T.G. Johns, M. Kassiou, and L. Munoz. 2015. DYRK1A in neurodegeneration and cancer: Molecular basis and clinical implications. *Pharmacol Ther* 151: 87–98.

Abbassian, H., P. Esmaeili, M. Tahamtan, I. Aghaei, Z. Vaziri, V. Sheibani, B.J. Whalley, and M. Shabani. 2016. Cannabinoid receptor agonism suppresses tremor, cognition disturbances and anxiety-like behaviors in a rat model of essential tremor. *Physiol Behav* 164(Pt A): 314–320.

Abe, A., and H. Kokuba. 2013. Harmol induces autophagy and subsequent apoptosis in U251MG human glioma cells through the downregulation of survivin. *Oncol Rep* 29(4): 1333–42.

Abe, A., and H. Yamada. 2009. Harmol induces apoptosis by caspase-8 activation independently of Fas/Fas ligand interaction in human lung carcinoma H596 cells. *Anticancer Drugs* 20(5): 373–81.

Abe, A., H. Yamada, S. Moriya, and K. Miyazawa. 2011. The β-carboline alkaloid harmol induces cell death via autophagy but not apoptosis in human non-small cell lung cancer A549 cells. *Biol Pharm Bull* 34(8): 1264–72.

Abelaira, H.M., G.Z. Réus, G. Scaini, E.L. Streck, J.A. Crippa, and J. Quevedo. 2013. β-Carboline harmine reverses the effects induced by stress on behaviour and citrate synthase activity in the rat prefrontal cortex. *Acta Neuropsychiatr* 25(6): 328–33.

Abu Ghazaleh, H., M.D. Lalies, D.J. Nutt, and A.L. Hudson. 2015. Harmane: An atypical neurotransmitter? *Neurosci Lett* 590: 1–5.

Airaksinen, M.M., A. Lecklin, V. Saano, L. Tuomisto, and J. Gynther. 1987. Tremorigenic effect and inhibition of tryptamine and serotonin receptor binding by beta-carbolines. *Pharmacol Toxicol* 60(1): 5–8.

Algorta, J., M.A. Pena, A. Alvarez, C. Maraschiello, D. Maruhn, M. Windisch, and H.A. Mucke. 2009. Randomized, crossover, single-blind, placebo-controlled, human pharmacology clinical trial with desoxypeganine, a new cholinesterase and selective MAO-A inhibitor: Multiple-dose pharmacokinetics. *Int J Clin Pharmacol Ther* 47(7): 483–90.

Algorta, J., M.A. Pena, C. Maraschiello, A. Alvarez-González, D. Maruhn, M. Windisch, and H.A. Mucke. 2008. Phase I clinical trial with desoxypeganine, a new cholinesterase and selective MAO-A inhibitor: Tolerance and pharmacokinetics study of escalating single oral doses. *Methods Find Exp Clin Pharmacol* 30(2): 141–7.

Alomar, M.L., M.M. Gonzalez, R. Erra-Balsells, and F.M. Cabrerizo. 2014. Comment on "Binding of alkaloid harmalol to DNA: Photophysical and calorimetric approach." *J Photochem Photobiol B* 136: 26–8.

Amin, B., M. Malekzadeh, M.R. Heidari, and H. Hosseinzadeh. 2015. Effect of *Crocus sativus* extracts and its active constituent safranal on the harmaline-induced tremor in mice. *Iran J Basic Med Sci* 18(5): 449–58.

Amjadi, M., and L. Farzampour. 2014. Selective turn-on fluorescence assay of 6-thioguanine by using harmine-modified silver nanoparticles. *Luminescence* 29(6): 689–94.

Amla, V., G. Bano, R.K. Johri, U. Zutshi, and C.K. Atal. 1987. Pharmacokinetics of vasicine in healthy Indian volunteers. *Zhongguo Yao Li Xue Bao* 8(2): 190–2.

Apostolico, I., L. Aliberti, L. Caputo, V. De Feo, F. Fratianni, F. Nazzaro, L.F. Souza, and M. Khadhr. 2016. Chemical composition, antibacterial and phytotoxic activities of *Peganum harmala* seed essential oils from five different localities in northern Africa. *Molecules* 21(9): E1235.

Arnott, J.A., A.G. Lambi, C. Mundy, H. Hendesi, R.A. Pixley, T.A. Owen, F.F. Safadi, and S.N. Popoff. 2011. The role of connective tissue growth factor (CTGF/CCN2) in skeletogenesis. *Crit Rev Eukaryot Gene Expr* 21(1): 43–69.

Asadollahi-Baboli, M. 2016. In silico evaluation, molecular docking and QSAR analysis of quinazoline-based EGFR-T790M inhibitors. *Mol Divers* 20(3): 729–39.

Asif, M. 2014. Chemical characteristics, synthetic methods, and biological potential of quinazoline and quinazolinone derivatives. *Int J Med Chem* 2014: 395637.

Atal, C.K., M.L. Sharma, A. Khajuria, A. Kaul, and R.K. Arya. 1982. Thrombopoietic activity of vasicine hydrochloride. *Indian J Exp Biol* 20(9): 704–9.

Atal, C.K., U. Zutshi, and P.G. Rao. 1981. Scientific evidence on the role of Ayurvedic herbals on bioavailability of drugs. *J Ethnopharmacol* 4(2): 229–32.

Atteya, R., M.E. Ashour, E.E. Ibrahim, M.A. Farag, and S.F. El-Khamisy. 2016. Chemical screening identifies the β-carboline alkaloid harmine to be synergistically lethal with doxorubicin. *Mech Ageing Dev* Jun 6 [Epub ahead of print].

Avula, B., Y.H. Wang, C.S. Rumalla, T.J. Smillie, and I.A. Khan. 2012. Simultaneous determination of alkaloids and flavonoids from aerial parts of *Passiflora* species and dietary supplements using UPLC-UV-MS and HPTLC. *Nat Prod Commun* 7(9): 1177–80.

Bai, X.Y., X.C. Zhang, S.Q. Yang, S.J. An, Z.H. Chen, J. Su, Z. Xie, L.Y. Gou, and Y.L. Wu. 2016. Blockade of hedgehog signaling synergistically increases sensitivity to epidermal growth factor receptor tyrosine kinase inhibitors in non-small-cell lung cancer cell lines. *PLoS One* 11(3): e0149370.

Barth, A., A. Hovhannisyan, K. Jamalyan, and M. Narimanyan. 2015. Antitussive effect of a fixed combination of *Justicia adhatoda*, *Echinacea purpurea* and *Eleutherococcus senticosus* extracts in patients with acute upper respiratory tract infection: A comparative, randomized, double-blind, placebo-controlled study. *Phytomedicine* 22(13): 1195–200.

Becker, R.S., L.F. Ferreira, F. Elisei, I. Machado, and L. Latterini. 2005. Comprehensive photochemistry and photophysics of land- and marine-based beta-carbolines employing time-resolved emission and flash transient spectroscopy. *Photochem Photobio* 81(5): 1195–204.

Begum, S., S. Imran Hassa, B.S. Siddiqui, R. Ifzal, S. Perwaiz, T. Kiran, F. Shaheen, M.N. Ghayur, and A.H. Gilani. 2006. Preparation, structure and spasmolytic activities of some derivatives of harmine series of alkaloids. *Nat Prod Res* 20(3): 213–27.

Bei, Y.Y., X.F. Zhou, B.G. You, Z.Q. Yuan, W.L. Chen, P. Xia, Y. Liu, Y. Jin, X.J. Hu, Q.L. Zhu, C.G. Zhang, X.N. Zhang, and L. Zhang. 2013. Application of the central composite design to optimize the preparation of novel micelles of harmine. *Int J Nanomedicine* 8: 1795–808.

Bei, Y.Y., Z.Q. Yuan, L. Zhang, X.F. Zhou, W.L. Chen, P. Xia, Y. Liu, B.G. You, X.J, Hu, Q.L., Zhu, Zhang, C.G., X.N. Zhang, and Y. Jin. 2014. Novel self-assembled micelles based on palmitoyl-trimethyl-chitosan for efficient delivery of harmine to liver cancer. *Expert Opin Drug Deliv* 11(6): 843–54.

Bensalem, S., J. Soubhye, I. Aldib, L. Bournine, A.T. Nguyen, M. Vanhaeverbeek, A. Rousseau, K.Z. Boudjeltia, A. Sarakbi, J.M. Kauffmann, J. Nève, M. Prévost, C. Stévigny, F. Maiza-Benabdesselam, F. Bedjou, P. Van Antwerpen, and P. Duez. 2014. Inhibition of myeloperoxidase activity by the alkaloids of *Peganum harmala* L. (*Zygophyllaceae*). *J Ethnopharmacol* 154(2): 361–9.

Beringer, K. 1929. Zur Banisterin- und Harminfrage. *Nervenarzt* 2: 548–9 (cited by Hostiuc 2014).

Bhattacharjee, P., S. Sarkar, P. Pandya, and K. Bhadra. 2016. Targeting different RNA motifs by beta carboline alkaloid, harmalol: A comparative photophysical, calorimetric, and molecular docking approach. *J Biomol Struct Dyn* 6: 1–19.

Bishayee, A., D. Bhatia, R.J. Thoppil, A.S. Darvesh, E. Nevo, and E.P. Lansky. 2011. Pomegranate-mediated chemoprevention of experimental hepatocarcinogenesis involves Nrf2-regulated antioxidant mechanisms. *Carcinogenesis* 32(6): 888–96.

Bouayad, N., K. Rharrabe, M. Lamhamdi, N.G. Nourouti, and F. Sayah. 2012. Dietary effects of harmine, a β-carboline alkaloid, on development, energy reserves and α-amylase activity of *Plodia interpunctella* Hübner (*Lepidoptera: Pyralidae*). *Saudi J Biol Sci* 19(1): 73–80.

Brailowsky, S., S. Walter, G. Vuillon-Cacciuttolo, and T. Serbanescu. 1975. Indole alkaloids induction of tremors: Effect on photosensible epilepsy in Papiopapio. *C R Seances Soc Biol Fil* 169(5): 1190–3.

Brierley, D.I., and C. Davidson. 2012. Developments in harmine pharmacology—Implications for ayahuasca use and drug-dependence treatment. *Prog Neuropsychopharmacol Biol Psychiatry* 39(2): 263–72.

Brierley, D.I., and C. Davidson. 2013. Harmine augments electrically evoked dopamine efflux in the nucleus accumbens shell. *J Psychopharmacol* 27(1): 98–108.

Brower, J.V., and H.I. Robins. 2016. Erlotinib for the treatment of brain metastases in non-small cell lung cancer. *Expert Opin Pharmacother* 17(7): 1013–21.

Bruce, R.A., and V. Kumar. 1968. The effect of a derivative of vasicine on bronchial mucus. *Br J Clin Pract* 22(7): 289–92.

Bruel, A., R. Bénéteau, M. Chabanne, O. Lozach, R. Le Guevel, M. Ravache, H. Bénédetti, L. Meijer, C. Logé, and J.M. Robert. 2014. Synthesis of new pyridazino[4,5-b]indol-4-ones and pyridazin-3(2H)-one analogs as DYRK1A inhibitors. *Bioorg Med Chem Lett* 24(21): 5037–40.

Cahn, A., S. Cernea, and I. Raz. 2016. An update on DPP-4 inhibitors in the management of type 2 diabetes. *Expert Opin Emerg Drugs* 21(4): 409–19.

Cao, M.R., Q. Li, Z.L. Liu, H.H. Liu, W. Wang, X.L. Liao, Y.L. Pan, and J.W. Jiang. 2011. Harmine induces apoptosis in HepG2 cells via mitochondrial signaling pathway. *Hepatobiliary Pancreat Dis Int* 10(6): 599–604.

Cao, R., W. Fan, L. Guo, Q. Ma, G. Zhang, J. Li, X. Chen, Z. Ren, and L. Qiu. 2013. Synthesis and structure–activity relationships of harmine derivatives as potential antitumor agents. *Eur J Med Chem* 60: 135–43.

Chaliha, A.K., D. Gogoi, P. Chetia, D. Sarma, and A.K. Buragohain. 2016. An in silico approach for identification of potential anti-mycobacterial targets of vasicine and related chemical compounds. *Comb Chem High Throughput Screen* 19(1): 14–24.

Chen, D., A. Su, Y. Fu, X. Wang, X. Lv, W. Xu, S. Xu, H. Wang, and Z. Wu. 2015. Harmine blocks herpes simplex virus infection through downregulating cellular NF-κB and MAPK pathways induced by oxidative stress. *Antiviral Res* 123: 27–38.

Chen, Q., R. Chao, H. Chen, X. Hou, H. Yan, S. Zhou, W. Peng, and A. Xu. 2005. Antitumor and neurotoxic effects of novel harmine derivatives and structure–activity relationship analysis. *Int J Cancer* 114(5): 675–82.

Chen, W.L., Z.Q. Yuan, Y. Liu, S.D. Yang, C.G. Zhang, J.Z. Li, W.J. Zhu, F. Li, X.F. Zhou, Y.M. Lin, and X.N. Zhang. 2016. Liposomes coated with N-trimethyl chitosan to improve the absorption of harmine in vivo and *in vitro*. *Int J Nanomedicine* 11: 325–36.

Chiuccariello, L., R.G. Cooke, L. Miler, R.D. Levitan, G.B. Baker, S.J. Kish, N.J. Kolla, P.M. Rusjan, S. Houle, A.A. Wilson, and J.H. Meyer. 2015. Monoamine oxidase-A occupancy by moclobemide and phenelzine: Implications for the development of mono-amine oxidase inhibitors. *Int J Neuropsychopharmacol* Aug 27 [Epub ahead of print].

Chiuccariello, L., S. Houle, L. Miler, R.G. Cooke, P.M. Rusjan, G. Rajkowska, R.D. Levitan, S.J. Kish, N.J. Kolla, X. Ou, A.A. Wilson, and J.H. Meyer. 2014. Elevated monoamine oxidase a binding during major depressive episodes is associated with greater severity and reversed neurovegetative symptoms. *Neuropsychopharmacology* 39(4): 973–80.

Claeson, U.P., T. Malmfors, G. Wikman, and J.G. Bruhn. 2000. *Adhatoda vasica*: A critical review of ethnopharmacological and toxicological data. *J Ethnopharmacol* 72(1–2): 1–20.

Colino, S. 2015. The truth about essential tremor: It's not just a case of nerves. *US News and World Report* Nov 11.

Cozza, G., S. Sarno, M. Ruzzene, C. Girardi, A. Orzeszko, Z. Kazimierczuk, G. Zagotto, E. Bonaiuto, M.L. Di Paolo, and L.A. Pinna. 2013. Exploiting the repertoire of CK2 inhibitors to target DYRK and PIM kinases. *Biochim Biophys Acta* 1834(7): 1402–9.

Cumming, P., D. Skaper, T. Kuwert, S. Maschauer, and O. Prante. 2015. Detection of monoamine oxidase a in brain of living rats with [18F]fluoroethyl-harmol PET. *Synapse* 69(1): 57–9.

Cuny, G.D., N.P. Ulyanova, D. Patnaik, J.F. Liu, X. Lin, K. Auerbach, S.S. Ray. 2012. Structure–activity relationship study of beta-carboline derivatives as haspin kinase inhibitors. *Bioorg Med Chem Lett* 22(5): 2015–9.

Dai, F., Y. Chen, Y. Song, L. Huang, D. Zhai, Y. Dong, L. Lai, T. Zhang, D. Li, X. Pang, M. Liu, and Z. Yi. 2012. A natural small molecule harmine inhibits angiogenesis and suppresses tumour growth through activation of p53 in endothelial cells. *PLoS One* 7(12): e52162.

Diaz, S.L., A.K. Kemmling, M.C. Rubio, and G.N. Balerio. 2005. Morphine withdrawal syndrome: Involvement of the dopaminergic system in prepubertal male and female mice. *Pharmacol Biochem Behav* 82(4): 601–7.

Dirice, E., D. Walpita, A. Vetere, B.C. Meier, S. Kahraman, J. Hu, V. Dančík, S.M. Burns, T.J. Gilbert, D.E. Olson, P.A. Clemons, R.N. Kulkarni, and B.K. Wagner. 2016. Inhibition of DYRK1A stimulates human β-cell proliferation. *Diabetes* 65(6): 1660–71.

Doetkotte, R., K. Opitz, K. Kiianmaa, and H. Winterhoff. 2005. Reduction of voluntary ethanol consumption in alcohol-preferring Alko alcohol (AA) rats by desoxypeganine and galanthamine. *Eur J Pharmacol* 522(1–3): 72–7.

Dos Santos, R.G., and J.E. Hallak. 2016. Effects of the natural β-carboline alkaloid harmine, a main constituent of ayahuasca, in memory and in the hippocampus: A systematic literature review of preclinical studies. *J Psychoactive Drugs* Dec 5: 1–10 [Epub ahead of print].

Duan, H., J.W. Lee, S.W. Moon, D. Arora, Y. Li, H.Y. Lim, Wang. 2016. Discovery, synthesis, and evaluation of 2,4-diaminoquinazolines as a novel class of pancreatic β-cell-protective agents against endoplasmic reticulum (ER) stress. *J Med Chem* 59(17): 7783–800.

Duportail, G., and H. Lami. 1975. Studies of the interaction of the fluorophores harmine and harmaline with calf thymus DNA. *Biochim Biophys Acta* 402(1): 20–30.

Duraipandiyan, V., N.A. Al-Dhabi, C. Balachandran, S. Ignacimuthu, C. Sankar, and K. Balakrishna. 2015. Antimicrobial, antioxidant, and cytotoxic properties of vasicine acetate synthesized from vasicine isolated from *Adhatoda vasica* L. *Biomed Res Int* 727304.

El Gendy, M.A., and A.O. El-Kadi. 2013. Harmine and harmaline downregulate TCDD-induced Cyp1a1 in the livers and lungs of C57BL/6 mice. *Biomed Res Int* 258095.

El Gendy, M.A., A.A. Soshilov, M.S. Denison, A.O. El-Kadi. 2012a. Harmaline and harmalol inhibit the carcinogen-activating enzyme CYP1A1 via transcriptional and posttranslational mechanisms. *Food Chem Toxicol* 50(2): 353–62.

El Gendy, M.A., A.A. Soshilov, M.S. Denison, and A.O. El-Kadi. 2012b. Transcriptional and posttranslational inhibition of dioxin-mediated induction of CYP1A1 by harmine and harmol. *Toxicol Lett* 208(1): 51–61.

Elbaz, M., D. Ahirwar, J. Ravi, M.W. Nasser, and R.K. Ganju. 2016. Novel role of cannabinoid receptor 2 in inhibiting EGF/EGFR and IGF-I/IGF-IR pathways in breast cancer. *Oncotarget* May 17 [Epub ahead of print].

Elbaz, M., M.W. Nasser, J. Ravi, N.A. Wani, D.K. Ahirwar, H. Zhao, S. Oghumu, A.R. Satoskar, K. Shilo, W.E. Carson 3rd, and R.K. Ganju. 2015. Modulation of the tumor microenvironment and inhibition of EGF/EGFR pathway: Novel anti-tumor mechanisms of cannabidiol in breast cancer. *Mol Oncol* 9(4): 906–19.

El-Shanawany, M.A., H.M. Sayed, S.R. Ibrahim, and M.A. Fayed. 2014. New nitrogenous compounds from *Anisotes trisulcus*. *Z Naturforsch C* 69(5–6): 209–18.

Farzin, D., and N. Mansouri. 2006. Antidepressant-like effect of harmane and other beta-carbolines in the mouse forced swim test. *Eur Neuropsychopharmacol* 16(5): 324–8.

Fehervari, Z. 2015. Harmine-izing immunity. *Nat Immunol* 16(8): 801.

Filali, I., M.A. Belkacem, A. Ben Nejma, J.P. Souchard, H. Ben Jannet, and J. Bouajila. 2016. Synthesis, cytotoxic, anti-lipoxygenase and anti-acetylcholinesterase capacities of novel derivatives from harmine. *J Enzyme Inhib Med Chem* 31(sup1): 23–33.

Fleurentin, J., C. Hoefler, A. Lexa, F. Mortier, and J.M. Pelt. 1986. Hepatoprotective properties of *Crepis rueppellii* and *Anisotes trisulcus*: Two traditional medicinal plants of Yemen. *J Ethnopharmacol* 16(1): 105–11.

Frédérick, R., C. Bruyère, C. Vancraeynest, J. Reniers, C. Meinguet, L. Pochet, A. Backlund, B. Masereel, R. Kiss, and J. Wouters. 2012. Novel trisubstituted harmine derivatives with original *in vitro* anticancer activity. *J Med Chem* 55(14): 6489–501.

Frison, G., D. Favretto, F. Zancanaro, G. Fazzin, and S.D. Ferrara. 2008. A case of beta-carboline alkaloid intoxication following ingestion of *Peganum harmala* seed extract. *Forensic Sci Int* 179: e37–43.

Gao, H., Y.N. Huang, B. Gao, P. Li, C. Inagaki, and J. Kawabata. 2008. Inhibitory effect on α-glucosidase by *Adhatoda vasica* Nees. *Food Chem* 108(3): 965–72.

Gao, P., N. Tao, Q. Ma, W.X. Fan, C. Ni, H. Wang, and Z.H. Qin. 2014. DH332, a synthetic β-carboline alkaloid, inhibits B cell lymphoma growth by activation of the caspase family. *Asian Pac J Cancer Prev* 15(9): 3901–6.

Gautam, C.S., and P.L. Sharma. 1982. Potentiation of oxytocin evoked responses by (+) sotalol deoxysotalol and Vasicine HCl on isolated rat and rabbit uterus. *Indian J Med Res* 76 Suppl: 107–14.

Ghani, A., Z. Ali, T. Islam, S. Sanaullah, and S. Saeed. 2014. Report: Nutrient evaluation and elemental analysis of four selected medicinal plants of Soon Valley Khushab, Punjab, Pakistan. *Pak J Pharm Sci* 27(3): 597–600.

Gibbs, B.F. 2009. Differential modulation of IgE-dependent activation of human basophils by ambroxol and related secretolytic analogues. *Int J Immunopathol Pharmacol* 22(4): 919–27.

Glick, S.D., M.E. Kuehne, J. Raucci, T.E. Wilson, D. Larson, R.W. Keller Jr, and J.N. Carlson. 1994. Effects of iboga alkaloids on morphine and cocaine self-administration in rats: Relationship to tremorigenic effects and to effects on dopamine release in nucleus accumbens and striatum. *Brain Res* 657(1–2): 14–22.

Golovchenko, V.V., D.S. Khramova, A.S. Shashkov, D. Otgonbayar, A. Chimidsogzol, and Y.S. Ovodov. 2012. Structural characterisation of the polysaccharides from endemic Mongolian desert plants and their effect on the intestinal absorption of ovalbumin. *Carbohydr Res* 356: 265–72.

Gong, Y., J. Wu, and S.T. Li. 2015. Immuno-enhancement effects of *Lycium ruthenicum* Murr. polysaccharide on cyclophosphamide-induced immunosuppression in mice. *Int J Clin Exp Med* 8(11): 20631–7.

Gonzalez, M.M., M. Vignoni, M. Pellon-Maison, M.A. Ales-Gandolfo, M.R. Gonzalez-Baro, R. Erra-Balsells, B. Epe et al. 2012. Photosensitization of DNA by β-carbolines: Kinetic analysis and photoproduct characterization. *Org Biomol Chem* 10(9): 1807–19.

Grange, J.M., and N.J. Snell. 1996. Activity of bromhexine and ambroxol, semi-synthetic derivatives of vasicine from the Indian shrub *Adhatoda vasica*, against *Mycobacterium tuberculosis in vitro*. *J Ethnopharmacol* 50(1): 49–53.

Groeger, D., and K. Mothes. 1960. On the biogenesis of peganine. *Arch Pharm* 65: 1049–52.

Gu, L.H., L.P. Liao, H.J. Hu, S.W.A. Bligh, C.H. Wang, G.X. Chou, and Z.T. Wang. 2015. A thin-layer chromatography-bioautographic method for detecting dipeptidyl peptidase IV inhibitors in plants. *J Chromatogr A* 1411: 116–22.

Gupta, O.P., K.K. Anand, B.J. Ghatak, and C.K. Atal. 1978. Vasicine, alkaloid of *Adhatoda vasica*, a promising uterotonic abortifacient. *Indian J Exp Biol* 16(10): 1075–7.

Gupta, O.P., M.L. Sharma, B.J. Ghatak, and C.K. Atal. 1977. Potent uterine activity of alkaloid vasicine. *Indian J Med Res* 66(5): 865–71.

Gutiérrez-Gonzálvez, M.G., M.J. Hazen, and R.H. Espelosín. 1988. Harmine as a substitute for 33258 Hoechst in the FPG technique. *Histochemistry* 89(2): 199–200.

Hamedi, A., S. Farjadian, M.R. Karami. 2015. Immunomodulatory properties of *Trehala manna* decoction and its isolated carbohydrate macromolecules. *J Ethnopharmacol* 162: 121–6.

Hamsa, T.P., and G. Kuttan. 2010. Harmine inhibits tumour specific neo-vessel formation by regulating VEGF, MMP, TIMP and pro-inflammatory mediators both in vivo and *in vitro*. *Eur J Pharmacol* 649(1–3): 64–73.

Han, Y.S., J.M. Kim, J.S. Cho, C.S. Lee, and D.E. Kim. 2005. Comparison of the protective effect of indole beta-carbolines and R-(-)-deprenyl against nitrogen species-induced cell death in experimental culture model of Parkinson's disease. *J Clin Neurol* 1(1): 81–91.

Handforth, A. 2012. Harmaline tremor: Underlying mechanisms in a potential animal model of essential tremor. *Tremor Other Hyperkinet Mov (N Y)* 2: 02-92-769-1.

Handforth, A. 2016. Linking essential tremor to the cerebellum-animal model evidence. *Cerebellum* 15(3): 285–98.

Hara, E.S., M. Ono, S. Kubota, W. Sonoyama, Y. Oida, T. Hattori, T. Nishida, T. Furumatsu, T. Ozaki, M. Takigawa, and T. Kuboki. 2013. Novel chondrogenic and chondroprotective effects of the natural compound harmine. *Biochimie* 95(2): 374–81.

Harrington, N. 2012. Harmala alkaloids as bee-signaling chemicals. *Journal of Student Research* 1: 23–32.

Hart, S., O.M. Fischer, and A. Ullrich. 2004. Cannabinoids induce cancer cell proliferation via tumor necrosis factor alpha-converting enzyme (TACE/ADAM17)-mediated trans-activation of the epidermal growth factor receptor. *Cancer Res* 64(6): 1943–50.

Hazen, M.J., A.I. Pérez-Gorroño, N. Ortiz-Movilla, and M.L. Molero. 2002. An *in vitro* study on the kinetics, subcellular distribution and phototoxicity of harmine in human tumor cells. *Boll Chim Farm* 141(2): 154–7.

He, D., H. Wu, Y. Wei, W. Liu, F. Huang, H. Shi, B. Zhang, X. Wu, and C. Wang. 2015. Effects of harmine, an acetylcholinesterase inhibitor, on spatial learning and memory of APP/PS1 transgenic mice and scopolamine-induced memory impairment mice. *Eur J Pharmacol* 768: 96–107.

He, H., S. Han, T. Zhang, J. Zhang, S. Wang, and J. Hou. 2012. Screening active compounds acting on the epidermal growth factor receptor from *Radix scutellariae* via cell membrane chromatography online coupled with HPLC/MS. *J Pharm Biomed Anal* 62: 196–202.

He, J., J. Yao, H. Sheng, and J. Zhu. 2015. Involvement of the dual-specificity tyrosine phosphorylation-regulated kinase 1A-alternative splicing factor-calcium/calmodulin-dependent protein kinase IIδ signaling pathway in myocardial infarction-induced heart failure of rats. *J Card Fail* 21(9): 751–60.

Helsley, S., R.A. Rabin, and J.C. Winter. 1998. The effects of beta-carbolines in rats trained with ibogaine as a discriminative stimulus. *Eur J Pharmacol* 345(2): 139–43.

Hemmateenejad, B., M. Shamsipur, F. Samari, T. Khayamian, M. Ebrahimi, and Z. Rezaei. 2012. Combined fluorescence spectroscopy and molecular modeling studies on the interaction between harmalol and human serum albumin. *J Pharm Biomed Anal* 67–68: 201–8.

Herraiz, T., D. González, C. Ancín-Azpilicueta, V.J. Arán, and H. Guillén. 2010. beta-Carboline alkaloids in *Peganum harmala* and inhibition of human monoamine oxidase (MAO). *Food Chem Toxicol* 48: 839–45.

Herrendorff, R., M.T. Faleschini, A. Stiefvater, B. Erne, T. Wiktorowicz, F. Kern, M. Hamburger, O. Potterat, J. Kinter, and M. Sinnreich. 2016. Identification of plant-derived alkaloids with therapeutic potential for myotonic dystrophy type I. *J Biol Chem* Jun 13 [Epub ahead of print].

Hostiuc, S., O. Buda, and D.A. Ion. 2014. Harmine for catatonic schizophrenia: A forgotten experiment. *Schizophr Res* 159(1): 249–50.

Hou, J., S. Wan, G. Wang, T. Zhang, Z. Li, Y. Tian, Y. Yu, X. Wu, and J. Zhang. 2016. Design, synthesis, anti-tumor activity, and molecular modeling of quinazoline and pyrido[2,3-d] pyrimidine derivatives targeting epidermal growth factor receptor. *Eur J Med Chem* 118: 276–89.

House, R.V., P.T. Thomas, and H.N. Bhargava. 1995. Comparison of the hallucinogenic indole alkaloids ibogaine and harmaline for potential immunomodulatory activity. *Pharmacology* 51(1): 56–65.

Howe, G.A., B. Xiao, H. Zhao, K.N. Al-Zahrani, M.S. Hasim, J. Villeneuve, H.S. Sekhon, G.D. Goss, L.A. Sabourin, J. Dimitroulakos, and C.L. Addison. 2016. Focal adhesion kinase inhibitors in combination with erlotinib demonstrate enhanced anti-tumor activity in non-small cell lung cancer. *PLoS One* 11(3): e0150567.

Hu, Y., and H. Xie. 2016. [Progress in study on effect of harmine on bone and cartilage metabolism]. *Zhong Nan Da Xue Xue Bao Yi Xue Ban* 41: 328–32.

Hua, H., M. Cheng, X. Li, and Y. Pei. 2002. A new pyrroloquinazoline alkaloid from *Linaria vulgaris*. *Chem Pharm Bull (Tokyo)* 50(10): 1393–4.

Ibraheem, Z.O., R. Abdul Majid, S. Mohd Noor, H. Mohd Sidek, and R. Basir. 2015. The potential of β carbolin alkaloids to hinder growth and reverse chloroquine resistance in *Plasmodium falciparum*. *Iran J Parasitol* 10(4): 577–83.

Idrizi, R., P. Malcolm, C.S. Weickert, K. Zavitsanou, Suresh Sundram. 2016. Striatal but not frontal cortical up-regulation of the epidermal growth factor receptor in rats exposed to immune activation in utero and cannabinoid treatment in adolescence. *Psychiatry Res* 240: 260–4.

Ignacimuthu, S., and N. Shanmugam. 2010. Antimycobacterial activity of two natural alkaloids, vasicine acetate and 2-acetyl benzylamine, isolated from Indian shrub *Adhatoda vasica* Ness. leaves. *J Biosci* 35(4): 565–70.

Illgner, S., and R. Matusch. 2005. Biotransformation of desoxypeganine by microsomal enzymes of the rabbit liver. *Arch Pharm (Weinheim)* 338(1): 49–52.

Im, J.H., Y.R. Jin, J.J. Lee, J.Y. Yu, X.H. Han, S.H. Im, J.T. Hong, H.S. Yoo, M.Y. Pyo, and Y.P. Yun. 2009. Antiplatelet activity of beta-carboline alkaloids from *Perganum harmala*: A possible mechanism through inhibiting PLCgamma2 phosphorylation. *Vascul Pharmacol* 50(5–6): 147–52.

Jafari, E., M.R. Khajouei, F. Hassanzadeh, G.H. Hakimelahi, and G.A. Khodarahmi. 2016. Quinazolinone and quinazoline derivatives: Recent structures with potent antimicrobial and cytotoxic activities. *Res Pharm Sci* 11(1): 1–14.

Jha, D.K., L. Panda, P. Lavanya, S. Ramaiah, and A. Anbarasu. 2012. Detection and confirmation of alkaloids in leaves of *Justicia adhatoda* and bioinformatics approach to elicit its anti-tuberculosis activity. *Appl Biochem Biotechnol* 168(5): 980–90.

Jiménez, J., L. Riverón-Negrete, F. Abdullaev, J. Espinosa-Aguirre, and R. Rodríguez-Arnaiz. 2008. Cytotoxicity of the beta-carboline alkaloids harmine and harmaline in human cell assays *in vitro*. *Exp Toxicol Pathol* 60: 381–9.

Johne, S., D. Gröger, and G. Richter. 1968. [On the biosynthesis of peganine in *Adhatoda vasica* Nees]. *Arch Pharm Ber Dtsch Pharm Ges* 301(10): 721–7.

Johri, R.K., and U. Zutshi. 2002. Mechanism of action of 6, 7, 8, 9, 10, 12-hexahydro-azepino-[2, 1-b] quinazolin-12-one-(RLX)—A novel bronchodilator. *Indian J Physiol Pharmacol* 44(1): 75–81.

Jones, J.O., E.C. Bolton, Y. Huang, C. Feau, R.K. Guy, K.R. Yamamoto, B. Hann, and M.I. Diamond. 2009. Non-competitive androgen receptor inhibition *in vitro* and in vivo. *Proc Natl Acad Sci U S A* 106(17): 7233–8.

Kaczmarek, A., S. Schneider, B. Wirth, and M. Riessland. 2015. Investigational therapies for the treatment of spinal muscular atrophy. *Expert Opin Investig Drugs* 24(7): 867–81.

Kerperien, J., P.V. Jeurink, T. Wehkamp, A. van der Veer, H.J. van de Kant, G.A. Hofman, E.C. van Esch, J. Garssen, L.E. Willemsen, and L.M. Knippels. 2014. Non-digestible oligosaccharides modulate intestinal immune activation and suppress cow's milk allergic symptoms. *Pediatr Allergy Immunol* 25(8): 747–54.

Khadhr, M., D. Bousta, E.H. Hanane, L. El Mansouri, S. Boukhira, M. Lachkar, B. Jamoussi, and S. Boukhchina. 2016. HPLC and GC-MS analysis of Tunisian *Peganum harmala* seeds oil and evaluation of some biological activities. *Am J Ther* Apr 7 [Epub ahead of print].

Khaliq, T., P. Misra, S. Gupta, K.P. Reddy, R. Kant, P.R. Maulik, A. Dube, and T. Narender. 2009. Peganine hydrochloride dihydrate an orally active antileishmanial agent. *Bioorg Med Chem Lett* 19(9): 2585–6.

Khan, F.A., A. Maalik, Z. Iqbal, and I. Malik. 2013. Recent pharmacological developments in β-carboline alkaloid "harmaline." *Eur J Pharmacol* 721(1–3): 391–4.

Khor, B., J.D. Gagnon, G. Goel, M.I. Roche, K.L. Conway, K. Tran, L.N. Aldrich, T.B. Sundberg, A.M. Paterson, S. Mordecai, D. Dombkowski, M. Schirmer, P.H. Tan, A.K. Bhan, R. Roychoudhuri, N.P. Restifo, J.J. O'Shea, B.D. Medoff, A.F. Shamji, S.L. Schreiber, A.H. Sharpe, S.Y. Shaw, and R.J. Xavier. 2015. The kinase DYRK1A reciprocally regulates the differentiation of Th17 and regulunatory T cells. *Elife* May 22: 4.

Kim, D.H., Y.Y. Jang, E.S. Han, and C.S. Lee. 2001. Protective effect of harmaline and harmalol against dopamine- and 6-hydroxydopamine-induced oxidative damage of brain mitochondria and synaptosomes, and viability loss of PC12 cells. *Eur J Neurosci* 13(10): 1861–72.

Kolla, N.J., B. Matthews, A.A. Wilson, S. Houle, R.M. Bagby, P. Links, A.I. Simpson, A. Hussain, and J.H. Meyer. 2015. Lower monoamine oxidase-A total distribution volume in impulsive and violent male offenders with Antisocial Personality Disorder and high psychopathic traits: An [(11)C] harmine positron emission tomography study. *Neuropsychopharmacology* 40(11): 2596–603.

Kolla, N.J., K. Dunlop, J. Downar, P. Links, R. Michael Bagby, A.A. Wilson, S. Houle, F. Rasquinha, A.I. Simpson, and J.H. Meyer. 2016b. Association of ventral striatum monoamine oxidase-A binding and functional connectivity in antisocial personality disorder with high impulsivity: A positron emission tomography and functional magnetic resonance imaging study. *Eur Neuropsychopharmacol* 26(4): 777–86.

Kolla, N.J., L. Chiuccariello, A.A. Wilson, S. Houle, P. Links, R.M. Bagby, S. McMain, C. Kellow, J. Patel, P.V. Rekkas, S. Pasricha, and J.H. Meyer. 2016a. Elevated monoamine oxidase-A distribution volume in borderline personality disorder is associated with severity across mood symptoms, suicidality, and cognition. *Biol Psychiatr* 79(2): 117–26.

Kondoh, D., S. Yamamoto, T. Tomita, K. Miyazaki, N. Itoh, Y. Yasumoto, H. Oike, R. Doi, and K. Oishi. 2014. Harmine lengthens circadian period of the mammalian molecular clock in the suprachiasmatic nucleus. *Biol Pharm Bull* 37(8): 1422–7.

Kovacova, V., J. Zluvova, B. Janousek, M. Talianova, and B. Vyskot. 2014. The evolutionary fate of the horizontally transferred agrobacterial mikimopine synthase gene in the genera *Nicotiana* and *Linaria*. *PLoS One* 9(11): e113872.

Krsková, Z., J. Martin, and J. Dusek. 2011. The inhibition activity of selected beta-carboline alkaloids on enzymes of acetylcholinesterase and butyrylcholinesterase. *Ceska Slov Farm* 60(3): 125–31.

Küçükbay, F.Z., and E. Kuyumcu. 2014. Determination of elements by atomic absorption spectrometry in medicinal plants employed to alleviate common cold symptoms. *Guang Pu Xue Yu Guang Pu Fen Xi* 34(9): 2548–56.

Kuhn, W., T. Müller, H. Grosse, and H. Rommelspacher. 1995b. Plasma levels of the beta-carbolines harman and norharman in Parkinson's disease. *J Neural Transm (Vienna)* 103(12): 1435–40.

Kuhn, W., T. Müller, H. Grosse, and H. Rommelspacher. 1996. Elevated levels of harman and norharman in cerebrospinal fluid of parkinsonian patients. *J Neural Transm (Vienna)* 103(12): 1435–40.

Kuhn, W., T. Müller, H. Grosse, T. Dierks, and H. Rommelspacher. 1995a. Plasma levels of the beta-carbolines harman and norharman in Parkinson's disease. *Acta Neurol Scand* 92(6): 451–4.

Lal, R., and P.L. Sharma. 1981. Potentiation of prostaglandin evoked contractions of isolated rat uterus by vasicine hydrochloride. *Indian J Med Res* 73: 641–8.

Lala, L.G., P.M. D'Mello, and S.R. Naik. 2004. Pharmacokinetic and pharmacodynamic studies on interaction of "Trikatu" with diclofenac sodium. *J Ethnopharmacol* 91(2–3): 277–80.

Lamchouri, F., M. Zemzami, A. Jossang, A. Abdellatif, Z.H. Israili, and B. Lyoussi. 2013. Cytotoxicity of alkaloids isolated from *Peganum harmala* seeds. *Pak J Pharm Sci* 26(4): 699–706.

Lanhers, M.C., I. Bertrand, J. Fleurentin, P.R. Lehr, and J.M. Pelt. 1986. Influence of *Anisotes trisulcus* and *Crepis rueppellii* extracts on sites of bile formation in the rat. *Arzneimittelforschung* 36(5): 826–9.

Lansky, E.P.S. 1975. Consciousness and the Ergotropic and Trophotropic Systems of Arousal. B.A. Thesis, New College, Sarasota, FL.

Lansky, P. 1979. Neurochemistry and the Awakening of Kundalini. In J. White, ed., *Kundalini, Evolution, and Enlightenment*. New York: Anchor Books.

Laviţă, S.I., R. Aro, B. Kiss, M. Manto, and P. Duez. 2016. The role of β-carboline alkaloids in the pathogenesis of essential tremor. *Cerebellum* 15(3): 276–84.

Leclere, L., M. Fransolet, P. Cambier, S. El Bkassiny, A. Tikad, M. Dieu, S.P. Vincent, P. Van Cutsem, and C. Michiels. 2016. Identification of a cytotoxic molecule in heat-modified citrus pectin. *Carbohydr Polym* 137: 39–51.

Lee, C.S., E.S. Han, Y.Y. Jang, J.H. Han, H.W. Ha, and D.E. Kim. 2000. Protective effect of harmalol and harmaline on MPTP neurotoxicity in the mouse and dopamine-induced damage of brain mitochondria and PC12 cells. *J Neurochem* 75(2): 521–31.

Lee, J.Y., Y.M. Lee, G.C. Chang, S.L. Yu, W.Y. Hsieh, J.J. Chen, H.W. Chen, and P.C. Yang. 2011. Curcumin induces EGFR degradation in lung adenocarcinoma and modulates p38 activation in intestine: The versatile adjuvant for gefitinib therapy. *PLoS One* 6(8): e23756.

Leeman-Neill, R.J., Q. Cai, S.C. Joyce, S.M. Thomas, N.E. Bhola, D.B. Neill, J.L. Arbiser, and J.R. Grandis. 2010. Honokiol inhibits epidermal growth factor receptor signaling and enhances the antitumor effects of epidermal growth factor receptor inhibitors. *Clin Cancer Res* 16(9): 2571–9.

Leto, S.M., F. Sassi, I. Catalano, V. Torri, G. Migliardi, E.R. Zanella, M. Throsby, A. Bertotti, and L. Trusolino. 2015. Sustained inhibition of HER3 and EGFR is necessary to induce regression of HER2-amplified gastrointestinal carcinomas. *Clin Cancer Res* 21(24): 5519–31.

Li, S., A. Wang, F. Gu, Z. Wang, C. Tian, Z. Qian, L. Tang, and Y. Gu. 2015. Novel harmine derivatives for tumor targeted therapy. *Oncotarget* 6(11): 8988–9001.

Liu, B. 2011. Study on chemical constituents of *Peganum multisectum*. *Zhong Yao Cai* 34(11): 1719–21.

Liu, H., D. Han, Y. Liu, X. Hou, J. Wu, H. Li, J. Yang, C. Shen, G. Yang, C. Fu, X. Li, H. Che, J. Ai, and S. Zhao. 2013. Harmine hydrochloride inhibits Akt phosphorylation and depletes the pool of cancer stem-like cells of glioblastoma. *J Neurooncol* 112(1): 39–48.

Liu, J., Q. Li, Z. Liu, L. Lin, X. Zhang, M. Cao, and J. Jiang. 2016. Harmine induces cell cycle arrest and mitochondrial pathway-mediated cellular apoptosis in SW620 cells via inhibition of the Akt and ERK signaling pathways. *Oncol Rep* 35(6): 3363–70.

Liu, W., X. Shi, Y. Yang, X. Cheng, Q. Liu, H. Han, B. Yang, C. He, Y. Wang, B. Jiang, Z. Wang, and C. Wang. 2015b. *In vitro* and in vivo metabolism and inhibitory activities of vasicine, a potent acetylcholinesterase and butyrylcholinesterase inhibitor. *PLoS One* 10(4): e0122366.

Liu, W., Y. Wang, D.D. He, S.P. Li, Y.D. Zhu, B. Jiang, X.M. Cheng, Z. Wang, and C.H. Wang. 2015c. Antitussive, expectorant, and bronchodilating effects of quinazoline alkaloids (±)-vasicine, deoxyvasicine, and (±)-vasicinone from aerial parts of *Peganum harmala* L. *Phytomedicine* 22(12): 1088–95.

Liu, Z.L., W.R. Zhu, W.C. Zhou, H.F. Ying, L. Zheng, Y.B. Guo, J.X. Chen, and X.H. Shen. 2014. Traditional Chinese medicinal herbs combined with epidermal growth factor receptor tyrosine kinase inhibitor for advanced non-small cell lung cancer: A systematic review and meta-analysis. *J Integr Med* 12(4): 346–58.

Louis, E.D., K.M. Pellegrino, P. Factor-Litvak, E. Rios, W. Jiang, C. Henchcliffe, and W. Zheng. 2008. Cancer and blood concentrations of the comutagen harmane in essential tremor. *Mov Disord* 23(12): 1747–51.

Ma, Q., W. Chen, and W. Chen. 2016. Anti-tumor angiogenesis effect of a new compound: B-9-3 through interference with VEGFR2 signaling. *Tumour Biol* 37(5): 6107–16.

Ma, X., D. Liu, H. Tang, Y. Wang, T. Wu, Y. Li, J. Yang, J. Yang, S. Sun, and F. Zhang. 2013. Purification and characterization of a novel antifungal protein with antiproliferation and anti-HIV-1 reverse transcriptase activities from *Peganum harmala* seeds. *Acta Biochim Biophys Sin (Shanghai)* 45(2): 87–94.

Ma, Y., and M. Wink. 2010. The beta-carboline alkaloid harmine inhibits BCRP and can reverse resistance to the anticancer drugs mitoxantrone and camptothecin in breast cancer cells. *Phytother Res* 24(1): 146–9.

Mabey, R. 1996. *Flora Britannica*. London: Sinclair-Stevenson Publisher.

Maschauer, S., A. Haller, P.J. Riss, T. Kuwert, O. Prante, and P. Cumming. 2015. Specific binding of [(18)F]fluoroethyl-harmol to monoamine oxidase A in rat brain cryostat sections, and compartmental analysis of binding in living brain. *J Neurochem* 135(5): 908–17.

McIsaac, W.M., P.A. Khairallah, and I.H Page. 1961. 10-Methoxyharmalan, a potent serotonin antagonist which affects conditioned behavior. *Science* 134(3480): 674–5.

Meinguet, C., B. Masereel, and J. Wouters. 2015a. Preparation and characterization of a new harmine-based antiproliferative compound in complex with cyclodextrin: Increasing solubility while maintaining biological activity. *Eur J Pharm Sci* 77: 135–40.

Meinguet, C., C. Bruyère, R. Frédérick, V. Mathieu, C. Vancraeynest, L. Pochet, J. Laloy, J. Mortier, G. Wolber, R. Kiss, B. Masereel, and J. Wouters. 2015b. 3D-QSAR, design, synthesis and characterization of trisubstituted harmine derivatives with *in vitro* antiproliferative properties. *Eur J Med Chem* 94: 45–55.

Mennenga, S.E., J.E. Gerson, T. Dunckley, and H.A. Bimonte-Nelson. 2015. Harmine treatment enhances short-term memory in old rats: Dissociation of cognition and the ability to perform the procedural requirements of maze testing. *Physiol Behav* 138: 260–5.

Miralles, A., S. Esteban, A. Sastre-Coll, D. Moranta, V.J. Asensio, and J.A. García-Sevilla. 2005. High-affinity binding of beta-carbolines to imidazoline I2B receptors and MAO-A in rat tissues: Norharman blocks the effect of morphine withdrawal on DOPA/noradrenaline synthesis in the brain. *Eur J Pharmacol* 518(2–3): 234–42.

Misra, P., T. Khaliq, A. Dixit, S. SenGupta, M. Samant, S. Kumari, A. Kumar, P.K. Kushawaha, H.K. Majumder, A.K. Saxena, T. Narender, and A. Dube. 2008. Antileishmanial activity mediated by apoptosis and structure-based target study of peganine hydrochloride dihydrate: An approach for rational drug design. *J Antimicrob Chemother* 62(5): 998–1002.

Miwa, H. 2007. Rodent models of tremor. *Cerebellum* 6(1): 66–72.

Molero, M.L., M.J. Hazen, A.I. Pérez Gorroño, and J.C. Stockert. 1995. Simple beta-carboline alkaloids as nucleic acids fluorochromes. *Acta Histochem* 97(2): 165–73.

Moura, D.J., C. Rorig, D.L. Vieira, J.A. Henriques, R. Roesler, J. Saffi, and J.M. Boeira. 2006. Effects of beta-carboline alkaloids on the object recognition task in mice. *Life Sci* 79(22): 2099–104.

Moura, D.J., M.F. Richter, J.M. Boeira, J.A. Pêgas Henriques, and J. Saffi. 2007. Antioxidant properties of beta-carboline alkaloids are related to their antimutagenic and antigenotoxic activities. *Mutagenesis* 22(4): 293–302.

Moussa, T.A., and O.A. Almaghrabi. 2016. Fatty acid constituents of *Peganum harmala* plant using Gas Chromatography-Mass Spectroscopy. *Saudi J Biol Sci* 23(3): 397–403.

Mucke, H.A. 2005. The European Association of Addiction Therapy: Inaugural Conference. *IDrugs* 8(10): 816–7.

Murthy, R., K. Erlandsson, D. Kumar, R. Van Heertum, J. Mann, and R. Parsey. 2007. Biodistribution and radiation dosimetry of 11C-harmine in baboons. *Nucl Med Commun* 28(9): 748–54.

Nafisi, S., M. Bonsaii, P. Maali, M.A. Khalilzadeh, and F. Manouchehri. 2010b. Beta-carboline alkaloids bind DNA. *J Photochem Photobiol B* 100: 84–91.

Nafisi, S., Z.M. Malekabady, and M.A. Khalilzadeh. 2010a. Interaction of β-carboline alkaloids with RNA. *DNA Cell Biol* 29(12): 753–61.

Nasehi, M., E. Mashaghi, F. Khakpai, and M.R. Zarrindast. 2013a. Suggesting a possible role of CA1 histaminergic system in harmane-induced amnesia. *Neurosci Lett* 556: 5–9.

Nasehi, M., M. Piri, M. Abdollahian, and M.R. Zarrindast. 2013b. Involvement of nitrergic system of CA1 in harmane induced learning and memory deficits. *Physiol Behav* 109: 23–32.

Nasehi, M., M. Piri, M. Nouri, D. Farzin, T. Nayer-Nouri, and M.R. Zarrindast. 2010. Involvement of dopamine D1/D2 receptors on harmane-induced amnesia in the step-down passive avoidance test. *Eur J Pharmacol* 634(1–3): 77–83.

Nasehi, M., S. Sharifi, and M.R. Zarrindast. 2012. Involvement of the cholinergic system of CA1 on harmane-induced amnesia in the step-down passive avoidance test. *Psychopharmacol* 26(8): 1151–61.

Nilani, P., N. Kasthuribai, B. Duraisamy, P. Dhamodaran, S. Ravichandran, K. Ilango, and B. Suresh. 2009. *In vitro* antioxidant activity of selected antiasthmatic herbal constituents. *Anc Sci Life* 28(4): 3–6.

Noguchi, T., K. Lo, T. Diemer, and D.K. Welsh. 2016. Lithium effects on circadian rhythms in fibroblasts and suprachiasmatic nucleus slices from Cry knockout mice. *Neurosci Lett* 619: 49–53.

Noll, C., A. Tlili, C. Ripoll, L. Mallet, J.L. Paul, J.M. Delabar, and N. Janel. 2012. Dyrk1a activates antioxidant NQO1 expression through an ERK1/2-Nrf2 dependent mechanism. *Mol Genet Metab* 105(3): 484–8.

Owaisat, S., R.B. Raffa, M. Rawl, and S.M. Trips. 2012. In vivo comparison of harmine efficacy against psychostimulants: Preferential inhibition of the cocaine response through a glutamatergic mechanism. *Neurosci Lett* 525(1): 12–6.

Pahwa, G.S., U. Zutshi, and C.K. Atal. 1987. Chronic toxicity studies with vasicine from *Adhatoda vasica* Nees. in rats and monkeys. *Indian J Exp Biol* 25(7): 467–70.

Park, S.Y., Y.H. Kim, Y.H. Kim, G. Park, and S.J. Lee. 2010. Beta-carboline alkaloids harmaline and harmalol induce melanogenesis through p38 mitogen-activated protein kinase in B16F10 mouse melanoma cells. *BMB Rep* 43(12): 824–9.

Patanè, S. 2015. Insights into cardio-oncology: Polypharmacology of quinazoline-based α1-adrenoceptor antagonists. *World J Cardiol* 7(5): 238–42.

Patel, K., M. Gadewar, R. Tripathi, S.K. Prasad, and D.K. Patel. 2012. A review on medicinal importance, pharmacological activity and bioanalytical aspects of beta-carboline alkaloid "Harmine." *Asian Pac J Trop Biomed* 2(8): 660–4.

Paul, B.K., and N. Guchhait. 2011a. Exploring the strength, mode, dynamics, and kinetics of binding interaction of a cationic biological photosensitizer with DNA: Implication on dissociation of the drug-DNA complex via detergent sequestration. *J Phys Chem B* 115(41): 11938–49.

Paul, B.K., and N. Guchhait. 2011b. Modulation of prototropic activity and rotational relaxation dynamics of a cationic biological photosensitizer within the motionally constrained bio-environment of a protein. *J Phys Chem B* 115(34): 10322–34.

Paul, B.K., and N. Guchhait. 2012. Differential interactions of a biological photosensitizer with liposome membranes having varying surface charges. *Photochem Photobiol Sci* 11(4): 661–73.

Paul, B.K., D. Ray, and N. Guchhait. 2012. Binding interaction and rotational-relaxation dynamics of a cancer cell photosensitizer with various micellar assemblies. *J Phys Chem B* 116(32): 9704–17.

Pfau, W., and K. Skog. 2004. Exposure to beta-carbolines norharman and harman. *J Chromatogr B Analyt Technol Biomed Life Sci* 802(1): 115–26.

Philippe, C., M. Zeilinger, M. Mitterhauser, M. Dumanic, R. Lanzenberger, M. Hacker, and W. Wadsak. 2015. Parameter evaluation and fully-automated radiosynthesis of [(11)C] harmine for imaging of MAO-A for clinical trials. *Appl Radiat Isot* 97: 182–7.

Pozo, N., C. Zahonero, P. Fernández, J.M. Liñares, A. Ayuso, M. Hagiwara, A. Pérez, J.R. Ricoy, A. Hernández-Laín, J.M. Sepúlveda, and P. Sánchez-Gómez. 2013. Inhibition of DYRK1A destabilizes EGFR and reduces EGFR-dependent glioblastoma growth. *J Clin Invest* 123(6): 2475–87.

Pulpati, H., Y.S. Biradar, and M. Rajani. 2008. High-performance thin-layer chromatography densitometric method for the quantification of harmine, harmaline, vasicine, and vasicinone in *Peganum harmala*. *J AOAC Int* 91(5): 1179–85.

Racle, J.P., M. Girard, J. Delage, and B. Constantin. 1976. Clinical and anatomopathologic effect of Bisolvon in respiratory resuscitation. *Ann Anesthesiol Fr* 17(1): 51–8.

Ravez, S., O. Castillo-Aguilera, P. Depreux, and L. Goossens. 2015. Quinazoline derivatives as anticancer drugs: A patent review (2011–present). *Expert Opin Ther Pat* 25(7): 789–804.

Rayees, S., U. Mabalirajan, W.W. Bhat, S. Rasool, R.A. Rather, L. Panda, N.K. Satti, S.K. Lattoo, B. Ghosh, and G. Singh. 2015. Therapeutic effects of R8, a semi-synthetic analogue of Vasicine, on murine model of allergic airway inflammation via STAT6 inhibition. *Int Immunopharmacol* 26(1): 246–56.

Rekkas, P.V., A.A. Wilson, V.W. Lee, P. Yogalingam, J. Sacher, P. Rusjan, S. Houle, D.E. Stewart, N.J. Kolla, S. Kish, L. Chiuccariello, and J.H. Meyer. 2014. Greater monoamine oxidase a binding in perimenopausal age as measured with carbon 11-labeled harmine positron emission tomography. *JAMA Psychiatry* 71(8): 873–9.

Ren, J.W., K.M. Chan, P.K. Lai, C.B. Lau, H. Yu, P.C. Leung, K.P. Fung, W.F. Yu, and C.H. Cho. 2012. Extracts from *Radix Astragali* and *Radix Rehmanniae* promote keratinocyte proliferation by regulating expression of growth factor receptors. *Phytother Res* 26(10): 1547–54.

Réus, G.Z., R.B. Stringari, C.L. Gonçalves, G. Scaini, M. Carvalho-Silva, G.C. Jeremias, I.C. Jeremias, G.K. Ferreira, E.L. Streck, J.E. Hallak, A.W. Zuardi, J.A. Crippa, and J. Quevedo. 2012. Administration of harmine and imipramine alters creatine kinase and mitochondrial respiratory chain activities in the rat brain. *Depress Res Treat* 987397.

Réus, G.Z., R.B. Stringari, C.L. Gonçalves, G. Scaini, M. Carvalho-Silva, G.C. Jeremias, I.C. Jeremias, G.K. Ferreira, E.L. Streck, J.E. Hallak, A.W. Zuardi, J.A. Crippa, and J. Quevedo. 2012. Administration of harmine and imipramine alters creatine kinase and mitochondrial respiratory chain activities in the rat brain. *Depress Res Treat* 987397.

Richers, M.T., I. Deb, A.Y. Platonova, C. Zhang, and D. Seidel. 2013. Facile access to ring-fused aminals via direct α-amination of secondary amines with *ortho*-aminobenzaldehydes: Synthesis of vasicine, deoxyvasicine, deoxyvasicinone, mackinazolinone and ruteacarpine. *Synthesis (Stuttg)* 45(13): 1430–1748.

Rüben, K., A. Wurzlbauer, A. Walte, W. Sippl, F. Bracher, and W. Becker. 2015. Selectivity profiling and biological activity of novel β-carbolines as potent and selective DYRK1 kinase inhibitors. *PLoS One* 10(7): e0132453.

Sacher, J., E.A. Rabiner, M. Clark, P. Rusjan, A. Soliman, R. Boskovic, S.J. Kish, A.A. Wilson, S. Houle, and J.H. Meyer. 2012. Dynamic, adaptive changes in MAO-A binding after alterations in substrate availability: An in vivo [(11)C]-harmine positron emission tomography study. *J Cereb Blood Flow Metab* 32(3): 443–6.

Sacher, J., P.V. Rekkas, A.A. Wilson, S. Houle, L. Romano, J. Hamidi, P. Rusjan, I. Fan, D.E. Stewart, and J.H. Meyer. 2015. Relationship of monoamine oxidase-A distribution volume to postpartum depression and postpartum crying. *Neuropsychopharmacology* 40(2): 429–35.

Saeki, M., and H. Egusa. 2014. Novel strategies for the development of anabolic agents for treatment of osteoporosis. *Nihon Yakurigaku Zasshi* 144(6): 277–80.

Sarkar, S., and K. Bhadra. 2014. Binding of alkaloid harmalol to DNA: Photophysical and calorimetric approach. *J Photochem Photobiol B* 130: 272–80.

Sarkar, S., P. Pandya, and K. Bhadra. 2014. Sequence specific binding of beta carboline alkaloid harmalol with deoxyribonucleotides: Binding heterogeneity, conformational, thermodynamic and cytotoxic aspects. *PLoS One* 9(9): e108022.

Scheipl, S., M. Barnard, L. Cottone, M. Jorgensen, D.H. Drewry, W.J. Zuercher, F. Turlais, H. Ye, A.P. Leite, J.A. Smith, A. Leithner, P. Möller, S. Brüderlein, N. Guppy, F. Amary, R. Tirabosco, S.J. Strauss, N. Pillay, and A.M. Flanagan. 2016. EGFR inhibitors identified as a potential treatment for chordoma in a focused compound screen. *J Pathol* 239(3): 320–34.

Schenberg, E.E., J.F. Alexandre, R. Filev, A.M. Cravo, J.R. Sato, S.D. Muthukumaraswamy, M. Yonamine, M. Waguespack, I. Lomnicka, S.A. Barker, and D.X. da Silveira. 2015. Acute biphasic effects of Ayahuasca. *PLoS One* 10(9): e0137202.

Schieferstein, H., M. Piel, F. Beyerlein, H. Lüddens, N. Bausbacher, H.G. Buchholz, T.L. Ross, and F. Rösch. 2015. Selective binding to monoamine oxidase A: *In vitro* and in vivo evaluation of (18)F-labeled β-carboline derivatives. *Bioorg Med Chem* 23(3): 612–23.

Schreiber, K., O. Aurich, and K. Pufahl. 1962. Isolation of peganine from goat's-rue, *Galega officinalis* L. *Arch Pharm* 67: 271–5.

Seifert, A., L.A. Allan, and P.R. Clarke. 2008. DYRK1A phosphorylates caspase 9 at an inhibitory site and is potently inhibited in human cells by harmine. *FEBS J* 275(24): 6268–80.

Shahinas, D., G. Macmullin, C. Benedict, I. Crandall, and D.R. Pillai. 2012. Harmine is a potent antimalarial targeting Hsp90 and synergizes with chloroquine and artemisinin. *Antimicrob Agents Chemother* 56(8): 4207–13.

Shahwar, D., M.A. Raza, S. Tariq, M. Riasat, and M. Ajaib. 2012. Enzyme inhibition, antioxidant and antibacterial potential of vasicine isolated from *Adhatoda vasica* Nees. *Pak J Pharm Sci* 25(3): 651–6.

Shang, X., X. Guo, B. Li, H. Pan, J. Zhang, Y. Zhang, and X. Miao. 2016. Microwave-assisted extraction of three bioactive alkaloids from *Peganum harmala* L. and their acaricidal activity against *Psoroptes cuniculi in vitro*. *J Ethnopharmacol* 192: 350–361.

Shankaraiah, N., C. Jadala, S. Nekkanti, K.R. Senwar, N. Nagesh, S. Shrivastava, V.G. Naidu, M. Sathish, and A. Kamal. 2016. Design and synthesis of C3-tethered 1,2,3-triazolo-β-carboline derivatives: Anticancer activity, DNA-binding ability, viscosity and molecular modeling studies. *Bioorg Chem* 64: 42–50.

Shanon, B. 2010. *The Antipodes of the Mind: The Phenomenology of the Ayahuasca Experience*. Oxford, UK: Oxford University Press.

Sharma, S., M. Kumar, S. Sharma, O.S. Nayal, N. Kumar, B. Singh, and U. Sharma. 2016. Microwave assisted synthesis of phenanthridinones and dihydrophenanthridines by vasicine/KOtBu promoted intramolecular C-H arylation. *Org Biomol Chem* 14(36): 8536–44.

Sharma, S., M. Kumar, V. Kumar, and N. Kumar. 2014. Metal-free transfer hydrogenation of nitroarenes in water with vasicine: Revelation of organocatalytic facet of an abundant alkaloid. *J Org Chem* 79(19): 9433–9.

Shevyakov, S.V., O.I. Davydova, A.S. Kiselyov, D.V. Kravchenko, A.V. Ivachtchenko, and M. Krasavin. 2006. Natural products as templates for bioactive compound libraries. Part 2. Novel modifications of vasicine (peganine) core via efficient and regioselective generation of 3-lithiodeoxyvasicine and its stereoselective addition to aliphatic ketones section sign. *Nat Prod Res* 20(9): 871–81.

Shi, C.C., J.F. Liao, and C.F. Chen. 2001. Comparative study on the vasorelaxant effects of three harmala alkaloids *in vitro*. *Jpn J Pharmacol* 85(3): 299–305.

Shi, C.C., S.Y. Chen, G.J. Wang, J.F. Liao, and C.F. Chen. 2000. Vasorelaxant effect of harman. *Eur J Pharmacol* 390(3): 319–25.

Shi, Z., Z.J. Wang, H.L. Xu, Y. Tian, X. Li, J.K. Bao, S.R. Sun, and B.S. Yue. 2013. Modeling, docking and dynamics simulations of a non-specific lipid transfer protein from *Peganum harmala* L. *Comput Biol Chem* 47: 56–65.

Shingu, T., L. Holmes, V. Henry, Q. Wang, K. Latha, A.E. Gururaj, L.A. Gibson, T. Doucette, F.F. Lang, G. Rao, L. Yuan, E.P. Sulman, N.P. Farrell, W. Priebe, K.R. Hess, Y.A. Wang, J. Hu, and O. Bögler. 2016. Suppression of RAF/MEK or PI3K synergizes cyto-toxicity of receptor tyrosine kinase inhibitors in glioma tumor-initiating cells. *J Transl Med* 14: 46.

Shinozaki, H., K. Hirate, and M. Ishida. 1987. Modification of drug-induced tremor by systemic administration of kainic acid and quisqualic acid in mice. *Neuropharmacology* 26(1): 9–17.

Singh, B., and R.A. Sharma. 2013. Anti-inflammatory and antimicrobial properties of pyrroloquinazoline alkaloids from *Adhatoda vasica* Nees. *Phytomedicine* 20(5): 441–5.

Singh, V.K., V. Mishra, S. Tiwari, T. Khaliq, M.K. Barthwal, H.P. Pandey, G. Palit, and T. Narender. 2013. Anti-secretory and cyto-protective effects of peganine hydrochloride isolated from the seeds of *Peganum harmala* on gastric ulcers. *Phytomedicine* 20(13): 1180–5.

Smythies, J.R., and F. Antun. 1969. Binding of tryptamine and allied compounds to nucleic acids. *Nature* 223(5210): 1061–3.

Sobhani, A.M., S.A. Ebrahimi, and M. Mahmoudian. 2002. An *in vitro* evaluation of human DNA topoisomerase I inhibition by *Peganum harmala* L. seeds extract and its beta-carboline alkaloids. *J Pharm Pharm Sci* 5(1): 19–23.

Soliman, A.M., and S.R. Fahmy. 2011. Protective and curative effects of the 15 KD isolated protein from the *Peganum harmala* L. seeds against carbon tetrachloride induced oxidative stress in brain, tests and erythrocytes of rats. *Eur Rev Med Pharmacol Sci* 15(8): 888–99.

Soliman, A.M., H.S. Abu-El-Zahab, and G.A. Alswiai. 2013. Efficacy evaluation of the protein isolated from *Peganum harmala* seeds as an antioxidant in liver of rats. *Asian Pac J Trop Med* 6(4): 285–95.

Sompalle, R., S.M. Roopan, N.A. Al-Dhabi, K. Suthindhiran, G. Sarkar, and M.V. Arasu. 2016. 1,2,4-Triazolo-quinazoline-thiones: Non-conventional synthetic approach, study of solvatochromism and antioxidant assessment. *J Photochem Photobiol B* 162: 232–9.

Song, Z.Y., J.R. Liu, X.L. Lu, and L.J. Wang. 2006. Harmine induces apoptosis in human SGC-7901 cells. *Zhong Yao Cai* 29(6): 571–3.

Spindler, A., K. Stefan, and M. Wiese. 2016. Synthesis and investigation of Tetrahydro-β-carboline derivatives as inhibitors of the Breast Cancer Resistance Protein (ABCG2). *J Med Chem* Jun 17 [Epub ahead of print].

Suárez-Arroyo, I.J., T.J. Rios-Fuller, Y.R. Feliz-Mosquea, M. Lacourt-Ventura, D.J. Leal-Alviarez, G. Maldonado-Martinez, L.A. Cubano, and M.M. Martínez-Montemayor. 2016. *Ganoderma lucidum* combined with the EGFR tyrosine kinase inhibitor, Erlotinib synergize to reduce inflammatory breast cancer progression. *J Cancer* 7(5): 500–11.

Sun, K., X.H. Tang, and Y.K. Xie. 2015. Paclitaxel combined with harmine inhibits the migration and invasion of gastric cancer cells through downregulation of cyclooxygenase-2 expression. *Oncol Lett* 10(3): 1649–54.

Sun, P., S. Zhang, Y. Li, and L. Wang. 2014. Harmine mediated neuroprotection via evaluation of glutamate transporter 1 in a rat model of global cerebral ischemia. *Neurosci Lett* 583: 32–6.

Tarozzi, A., C. Marchetti, B. Nicolini, M. D'Amico, N. Ticchi, L. Pruccoli, V. Tumiatti, E. Simoni, A. Lodola, M. Mor, A. Milelli, and A. Minarini. 2016. Combined inhibition of the EGFR/AKT pathways by a novel conjugate of quinazoline with isothiocyanate. *Eur J Med Chem* 117: 283–91.

Tascón, M., F. Benavente, N.M. Vizioli, and L.G. Gagliardi. 2016. A rapid and simple method for the determination of psychoactive alkaloids by CE-UV: Application to *Peganum harmala* seed infusions. *Drug Test Anal* Jul 5 [Epub ahead of print].

Tetrud, J.W., and J.W. Langston. 1987. R-(-)-deprenyl as a possible protective agent in Parkinson's disease. *J Neural Transm Suppl* 25: 69–79.

Tian, R., Y. Li, and M. Gao. 2015. Shikonin causes cell-cycle arrest and induces apoptosis by regulating the EGFR-NF-κB signalling pathway in human epidermoid carcinoma A431 cells. *Biosci Rep* 35(2): e00189.

Truman, P., P. Grounds, and K.A. Brennan. 2017. Monoamine oxidase inhibitory activity in tobacco particulate matter: Are harman and norharman the only physiologically relevant inhibitors? *Neurotoxicology* 59: 22–6.

Tuliaganov, N., F.S. Sadritdinov, and G.A. Suleǐmanova. 1986. Pharmacological characteristics of desoxypeganine hydrochloride. *Farmakol Toksikol* 49(3): 37–40.

Vachnadze, V., T. Suladze, N. Vachnadze, L. Kintsurashvili, and J. Novikova. 2015. Alkaloids of *Peganum harmala* and their biological activity. *Georgian Med News* 243: 79–81.

VandenBrink, B.M., R.S. Foti, D.A. Rock, L.C. Wienkers, and J.L. Wahlstrom. 2012. Prediction of CYP2D6 drug interactions from *in vitro* data: Evidence for substrate-dependent inhibition. *Drug Metab Dispos* 40(1): 47–53.

Vaziri, Z., H. Abbassian, V. Sheibani, M. Haghani, M. Nazeri, I. Aghaei, and M. Shabani. 2015. The therapeutic potential of Berberine chloride hydrate against harmaline-induced motor impairments in a rat model of tremor. *Neurosci Lett* 590: 84–90.

Vignoni, M., F.A. Rasse-Suriani, K. Butzbach, R. Erra-Balsells, B. Epe, and F.M. Cabrerizo. 2013. Mechanisms of DNA damage by photoexcited 9-methyl-β-carbolines. *Org Biomol Chem* 11(32): 5300–9.

Vignoni, M., R. Erra-Balsells, B. Epe, and F.M. Cabrerizo. 2014. Intra- and extra-cellular DNA damage by harmine and 9-methyl-harmine. *J Photochem Photobiol B* 132: 66–71.

Vyas, T., R.P. Dash, S. Anandjiwala, and M. Nivsarkar. 2011. Formulation and pharmacokinetic evaluation of hard gelatin capsule encapsulating lyophilized Vasa Swaras for improved stability and oral bioavailability of vasicine. *Fitoterapia* 82(3): 446–53.

Wang, C., Z. Zhang, Y. Wang, and X. He. 2015. Cytotoxic indole alkaloids against human leukemia cell lines from the toxic plant *Peganum harmala*. *Toxins (Basel)* 7(11): 4507–18.

Wang, C., Z. Zhang, Y. Wang, and X. He. 2016. Cytotoxic constituents and mechanism from *Peganum harmala*. *Chem Biodivers* 13(7): 961–8.

Wang, C.H., H. Zeng, Y.H. Wang, C. Li, J. Cheng, Z.J. Ye, and X.J. He. 2015. Antitumor quinazoline alkaloids from the seeds of *Peganum harmala. J Asian Nat Prod Res* 17(5): 595–600.

Wang, K.B., Y.T. Di, Y. Bao, C.M. Yuan, G. Chen, D.H. Li, J. Bai, H.P. He, X.J. Hao, Y.H. Pei, Y.K. Jing, Z.L. Li, and H.M. Hua. 2014. Peganumine A, a β-carboline dimer with a new octacyclic scaffold from *Peganum harmala. Org Lett* 16(15): 4028–31.

Wang, P., J.C. Alvarez-Perez, D.P. Felsenfeld, H. Liu, S. Sivendran, A. Bender, A. Kumar, R. Sanchez, D.K. Scott, A. Garcia-Ocaña, and A.F. Stewart. 2015. A high-throughput chemical screen reveals that harmine-mediated inhibition of DYRK1A increases human pancreatic beta cell replication. *Nat Med* 21(4): 383–8.

Wang, Z., C.L. Wang, J.L. Li, N. Zhang, Y.N. Sun, Z.L. Liu, Z.S. Tang, and J.L. Liu. 2015. Synthesis of new 4-anilinoquinazoline analogues and evaluation of their EGFR inhibitor activity. *Yao Xue Xue Bao* 50(12): 1613–21.

Wang, Z., X. Wu, L. Wang, J. Zhang, J. Liu, Z. Song, and Z. Tang. 2016. Facile and efficient synthesis and biological evaluation of 4-anilinoquinazoline derivatives as EGFR inhibitors. *Bioorg Med Chem Lett* 26(11): 2589–93.

Weng, Q., J. Huang, Y. Zeng, Y. Deng, and M. Hu. 2012. Synthesis and herbicidal activity evaluation of novel β-carboline derivatives. *Molecules* 17(4): 3969–80.

Wu, Z.Y., P.H. Raven, and D.Y. Hong, eds. 2008. *Flora of China. Vol. 11, Oxalidaceae through Aceraceae*. Beijing: Science Press, and St. Louis, MO: Missouri Botanical Garden Press.

Wu, J., F. Zuo, J. Du, P.F. Wong, H. Qin, and J. Xu. 2013. Icariside II induces apoptosis via inhibition of the EGFR pathways in A431 human epidermoid carcinoma cells. *Mol Med Rep* 8(2): 597–602.

Xu, S.W., B.Y. Law, S.W. Mok, E.L. Leung, X.X. Fan, P.S. Coghi, W. Zeng, C.H. Leung, D.L. Ma, L. Liu, and V.K. Wong. 2016. Autophagic degradation of epidermal growth factor receptor in gefitinib-resistant lung cancer by celastrol. *Int J Oncol* 49(4): 1576–88.

Yamazaki, Y., and Y. Kawano. 2010. Inhibitory effect of hydroxyindoles and their analogues on human melanoma tyrosinase. *Z Naturforsch C* 65(1–2): 49–54.

Yang, X., W. Wang, J.J. Qin, M.H. Wang, H. Sharma, J.K. Buolamwini, H. Wang, R. Zhang. 2012. JKA97, a novel benzylidene analog of harmine, exerts anti-cancer effects by inducing G1 arrest, apoptosis, and p53-independent up-regulation of p21. *PLoS One* 7(4): e34303.

Yang, Y., X. Cheng, W. Liu, G. Chou, Z. Wang, and C. Wang. 2015. Potent AChE and BChE inhibitors isolated from seeds of *Peganum harmala* Linn by a bioassay-guided fractionation. *J Ethnopharmacol* 168: 279–86.

Yang, Z., and K.Y. Tam. 2016. Anti-cancer synergy of dichloroacetate and EGFR tyrosine kinase inhibitors in NSCLC cell lines. *Eur J Pharmacol* 789: 458–67.

Ye, F., Y. Che, E. McMillen, J. Gorski, D. Brodman, D. Saw, B. Jiang, and D.Y. Zhang. 2009. The effect of *Scutellaria baicalensis* on the signaling network in hepatocellular carcinoma cells. *Nutr Cancer* 61(4): 530–7.

Yu, A.M., J.R. Idle, K.W. Krausz, A. Küpfer, and F.J. Gonzalez. 2003. Contribution of individual cytochrome P450 isozymes to the O-demethylation of the psychotropic beta-carboline alkaloids harmaline and harmine. *J Pharmacol Exp Ther* 305(1): 315–22.

Yue, P., C. Wang, J. Dan, W. Liu, Z. Wu, and M. Yang. 2015. The importance of solidification stress on the redispersibility of solid nanocrystals loaded with harmine. *Int J Pharm* 480(1–2): 107–15.

Zabeer, A., A. Bhagat, O.P. Gupta, G.D. Singh, M.S. Youssouf, K.L. Dhar, O.P. Suri, K.A. Suri, N.K. Satti, B.D. Gupta, and G.N. Qazi. 2006. Synthesis and bronchodilator activity of new quinazolin derivative. *Eur J Med Chem* 41(3): 429–34.

Zayed, R., and M. Wink. 2005. Beta-carboline and quinoline alkaloids in root cultures and intact plants of *Peganum harmala*. *Z Naturforsch C* 60 (5–6): 451–8.

Zeng, Y., Y. Zhang, Q. Weng, M. Hu, and G. Zhong. 2010. Cytotoxic and insecticidal activities of derivatives of harmine, a natural insecticidal component isolated from *Peganum harmala*. *Molecules* 15(11): 7775–91.

Zhang, H., K. Sun, J. Ding, H. Xu, L. Zhu, K. Zhang, X. Li, and W. Sun. 2014. Harmine induces apoptosis and inhibits tumor cell proliferation, migration and invasion through down-regulation of cyclooxygenase-2 expression in gastric cancer. *Phytomedicine* 21(3): 348–55.

Zhang, L., F. Zhang, W. Zhang, L. Chen, N. Gao, Y. Men, X. Xu, and Y. Jiang. 2015. Harmine suppresses homologous recombination repair and inhibits proliferation of hepatoma cells. *Cancer Biol Ther* 16(11): 1585–92.

Zhang, P., C.R. Huang, W. Wang, X.K. Zhang, J.J. Chen, J.J. Wang, C. Lin, and J.W. Jiang. 2016. Harmine hydrochloride triggers G2 phase arrest and apoptosis in MGC-803 cells and SMMC-7721 cells by upregulating p21, activating Caspase-8/Bid, and downregulating ERK/Bad pathway. *Phytother Res* 30(1): 31–40.

Zhao, L., and M. Wink. 2013. The β-carboline alkaloid harmine inhibits telomerase activity of MCF-7 cells by down-regulating hTERT mRNA expression accompanied by an accelerated senescent phenotype. *PeerJ* 1: e174.

Zheng, X.Y., Z.J. Zhang, G.X. Chou, T. Wu, X.M. Cheng, C.H. Wang, and Z.T. Wang. 2009. Acetylcholinesterase inhibitive activity-guided isolation of two new alkaloids from seeds of *Peganum nigellastrum* Bunge by an *in vitro* TLC-bioautographic assay. *Arch Pharm Res* 32(9): 1245–51.

Zhong, H.J., K.H. Leung, S. Lin, D.S. Chan, Q.B. Han, S.L. Chan, D.L. Ma, and C.H. Leung. 2015. Discovery of deoxyvasicinone derivatives as inhibitors of NEDD8-activating enzyme. *Methods* 71: 71–6.

Zhong, H.J., V.P. Ma, Z. Cheng, D.S. Chan, H.Z. He, K.H. Leung, D.L. Ma, and C.H. Leung. 2012. Discovery of a natural product inhibitor targeting protein neddylation by structure-based virtual screening. *Biochimie* 94(11): 2457–60.

Zhong, Z., Y. Tao, and H. Yang. 2015. Treatment with harmine ameliorates functional impairment and neuronal death following traumatic brain injury. *Mol Med Rep* 12(6): 7985–91.

Zutshi, U., P.G. Rao, A. Soni, O.P. Gupta, and C.K. Atal. 1980. Absorption and distribution of vasicine a novel uterotonic. *Planta Med* 40(4): 373–7.

7 Modern Pharmacognostic Investigation of Harmal

Pharmacognosy is the branch of pharmacy that a century or so ago was the most important, comprising the area of academic concentration of over half of advanced degree pharmacists, though today it is only a small fraction of that. Pharmacognosy is usually defined as the "pharmacology of crude drugs," and as its name *-gnosy* implies, there is also an element in pharmacognosy of the arcane, the esoteric, the gnostic. Divination may come to mind. The emphasis is on the inner knowledge of plants obtained not so much through scientific investigation, but through pondering, apprehending, and communing... the inner knowledge of plants is obtained paradoxically through an inner knowledge of oneself, and later through the impact of that inner awareness to the consciousness of the plant. Sounds way out? Well, it is. But such true, mystically obtained pharmacognosy, tempered with thousands of years of trial and error and practical application, is the ground from which modern pharmacognosy is inspired, influenced, and impelled.

Practically speaking, modern pharmacognosy focuses on two major arenas: (1) phytochemistry (structure) and (2) function as applied to medical treatment or nutritional benefit to mammals, and occasionally also for the benefit of invertebrates or for other plants. Technically, pharmacognosy may encompass not only derivatives from plants, but also from any and all natural materials, including mineral, animal, and even fungal, as in medicinal mushrooms. However, the largest concern in pharmacognosy is with the active chemical components of plants, and their use or potential use in human hygiene, medicine, and daily life. All this work derives directly or indirectly from the herbalists of yesteryear and even for them, from the folk knowledge of the people.

In order to make some order of the vast gamut of pharmacognostical research on *Peganum harmala* as well as on any other relevant *Peganum* species known, several general categories from the list below have been considered. Antioxidant properties are not traversed in a separate section, though they do underlie most of the other categories, particularly as they apply to inflammation in general, and to inflammation in specific contexts.

In short, increasing oxidative tension leads to inflammation and insulin resistance, and reduction in oxidative tension leads to reduction of inflammation and lowering of insulin resistance. These are very general effects applicable to the role of oxidative tension and inflammation in many diseases, from neurodegenerative (e.g., Parkinson's, Alzheimer's), to metabolic (e.g., metabolic syndrome, diabetes, obesity), to neoplastic (cancers of all types). The extent to which infective etiologies also figure in the genesis of chronic illnesses is ever under investigation, and usually controversial. Combating drug resistance is also an important consideration and a generator of research.

The list below subsumes the modern pharmacognositical studies, usually in human cells, both normal and cancer cells, in glass, i.e., *in vitro*, and in live animals—and technically also in other live plants (*in vivo* or *in situ*). In some instances, the modern pharmacognostical research also includes studies in human beings, i.e., case studies and clinical trials. NOTE WELL: All of the studies in this chapter pertain to investigations of *P. harmala* or other *Peganum* species (see also Chapter 2), and their extracts, and even complex fractions of those extracts, but not to their known principal components, i.e., not to pure compounds. Pharmacognostical studies of the pure chemicals known to *Peganum* were considered in Chapter 6 of this book, "Phytochemistry of Harmal."

The functional arenas of modern pharmacognostical findings for *Peganum* sp. are

1. Antidiabetic, metabolic
2. Anti-Parkinson's, cerebroprotective
3. Anti-infective
4. Anti-infestational
5. Uterotonic
6. Anodyne
7. Anticancer
   a. Topoisomerase type 1 inhibition (prevents DNA unwinding)
   b. Antiangiogenic
8. Diuretic
9. Hepatoprotective
10. Antitussive
11. Anti-inflammatory, anti-aging
12. Immunopotentive
13. Cardiovascular
14. Miscellaneous

Below is a survey of the studies extant on the pharmacognosy of *Peganum* and *P. harmala* according to these categories. Each of them is followed by a short overview of the modern ethnomedical uses of *Peganum harmala* all over the world, based on the wide scientific and descriptive literature on the subject and in particular on the extremely useful databases NAPRALERT and Dr. Duke's Phytochemical and Ethnobotanical Databases. This was felt to be necessary, as the results of modern pharmacognostical research in many cases can show the rationale for the traditional uses of plants, and in other cases, traditional and folk medicine can suggest new leads for research. The first overview, Table 7.1, presents the material that does not appear in the more specific tables later in this chapter. The reader is referred to them, especially to those discussing inflammation.

WARNING Although *Peganum harmala* can be a singularly efficacious intervention in numerous medical scenarios, it may cause, and has caused, major damage to

TABLE 7.1

Contemporary Ethnomedical Uses of *Peganum harmala*

Indication	Geography	Part Used	References
(A) Modern Ethnomedical Uses of *Peganum harmala*, General			
Alterative	Iraq	Unspecified	Al-Rawi and Chakravarty 1964
Alterative	Unspecified	Unspecified	Steinmetz 1957; Uphof 1968
Alterative	Pakistan	Seed	Ghafoor 1985
Analgesic	Algeria/North Africa	Unspecified	Hammiche and Merad 1997
Analgesic, oral, hot H_2O extract	Iraq	Seed, dried	Rashan et al. 1989
Analgesic, externally, as a cataplasm	North Africa	Seed, powdered	Boulos 1983
Analgesic, oral, hot H_2O extract	India	Seed	Schipper and Volk 1960
Analgesic, oral, hot H_2O extract	Saudi Arabia	Seed, dried	Al-Yahya 1986
Anodyne	Unspecified	Unspecified	Duke 1992–2016, 2002
Antalgic, oral	Morocco	Seed	Bellakhdar et al. 1991
Alleviation of pain	Unspecified	Seed	Granot 1994
Pain, against	Morocco	Unspecified	MEDUSA n.d.
Pain-killers	Unspecified	Whole plant	Burkill 1985
Depurative	Algeria/North Africa	Unspecified	Hammiche and Merad 1997
Depurative	Libya	Dried entire plant	Hussain and Tobji 1997
Depurative, oral, decoction in oil taken first thing in the morning	North Africa	Plant	Boulos 1983
Diaphoretic, oral, hot H_2O extract	Greece	Seed	Dragendorff 1898
Medicinal use	Middle East	Unspecified	CDFA
Medicine, traditional	Algeria/North Africa	Unspecified	Hammiche and Merad 1997
Medicine: Generally healing	Unspecified	Unspecified	Burkill 1985
Medicine, all-cure	(Muslim population)	Seed	Hammiche and Merad 1997
Medicines	Unspecified	Plant	Burkill 1985
Remedy	Unspecified	Seed	Hanelt et al. 2001
Medicinal: source of a magical medicinal oil (Zet el harmal)	Unspecified	Seed	Hanelt et al. 2001

(Continued)

TABLE 7.1 (CONTINUED)
Contemporary Ethnomedical Uses of *Peganum harmala*

Indication	Geography	Part Used	References
Purifying agent	Egypt	Seed	Uphof 1959; Usher 1974; Kunkel 1984; Facciola 1990 via MEDUSA n.d.
Revulsive	North Africa	Fresh branches	Boulos 1983
Stimulant	Iraq	Unspecified	Al-Rawi and Chakravarty 1964
Stimulant	Unspecified	Unspecified	Steinmetz 1957
Stimulant	Egypt	Seed	MEDUSA n.d.
Stimulant	Pakistan	Seeds, decoction	Ghafoor 1985
Stimulant, as an alcoholic drink	Unspecified	Unspecified	Burkill 1985
Sudorific	Algeria/North Africa	Unspecified	Hammiche and Merad 1997
Sudorific	Iraq	Unspecified	Al-Rawi and Chakravarty 1964
Sudorific	Unspecified	Unspecified	Duke 1992–2016, 2002; Steinmetz 1957
Sudorific	North Africa	Seed	Boulos 1983
Weakness, general, for	Israel	Unspecified	Wild Flowers of Israel. n.d.

(B) Modern Ethnomedical Uses of *Peganum harmala*, Skin and Hair Problems

Abscesses, externally	Pakistan	Dried flower + leaf	Leporatti and Lattanzi 1994
Antiseptic, externally on skin	Algeria/North Africa	Unspecified	Hammiche and Merad 1997
Burns	Algeria/North Africa	Unspecified	Hammiche and Merad 1997
Dermatitis (eczema)	Algeria/North Africa	Unspecified	Hammiche and Merad 1997
Folliculitis	Unspecified	Unspecified	Granot 1994
Purifying agent	Egypt	Seed	Uphof 1959; Usher 1974; Kunkel 1984; Facciola 1990 via MEDUSA n.d.
Skin disease	Libya	Dried entire plant	Hussain and Tobji 1997
Skin diseases	Unspecified	Unspecified	Granot 1994
Skin problems, mostly externally	Israel	Unspecified	Wild Flowers of Israel. n.d.
Skin diseases	North Africa	Seeds, oil extracted from	Boulos 1983
Alopecia	North Africa	Seed, powdered; boiled in olive oil; external: massage	Boulos 1983
Alopecia	Algeria/North Africa	Unspecified	Hammiche and Merad 1997

(Continued)

TABLE 7.1 (CONTINUED)
Contemporary Ethnomedical Uses of *Peganum harmala*

Indication	Geography	Part Used	References
Baldness	Egypt	Seed; externally	Emboden 1979; Bown 1995; Phillips and Rix 1991 via MEDUSA n.d.
Hair-care	Morocco	Seed; externally	Bellakhdar et al. 1991
Makes hair thicker and stronger	North Africa	Seed, powdered; boiled in olive oil	Boulos 1983
Strengthens hair roots	Israel	Unspecified; mostly externally	Granot 1994; Wild Flowers of Israel. n.d.

(C) Modern Ethnomedical Uses of *Peganum harmala*, Eye Problems

Blepharitis	Algeria/North Africa	Unspecified	Hammiche and Merad 1997
Conjunctivitis, purulent	North Africa	Dried powdered plant	Boulos 1983
Conjunctivitis, suppurative	Algeria/North Africa	Unspecified	Hammiche and Merad 1997
Eye complaints	Egypt	Seed	Ayensu 1978; Al-Awdat and Laham 1994; Boulos 1983 via MEDUSA n.d.
Eye conditions	Tunisia	Seed, dried	Boukef et al. 1982
Eye diseases	Unspecified	Unspecified	Granot 1994
Eye diseases	Unspecified	Seed	Hanelt et al. 2001
Eye diseases, infectious	North Africa	Seeds, oil extracted from	Boulos 1983
Eye inflammation	Israel	Unspecified, mostly externally	Wild Flowers of Israel. n.d.
Eyes, for	Unspecified	Unspecified	Uphof 1968
Eyesight, for	Unspecified	Unspecified	Duke 1992–2016, 2002
Sensory system disorders	Morocco	Unspecified	Pelt 1971, Bellakhdar 1997 via MEDUSA n.d.

(D) Modern Ethnomedical Uses of *Peganum harmala*, Gastrointestinal Problems

Anthelmintic	India/North Africa	Seed	Abrol and Chopra 1962; Boulos 1983
Anthelminthic	Unspecified	Seed	Granot 1994; Hanelt et al. 2001
Anthelmintic	Morocco	Seed; oral	Bellakhdar et al. 1991
Anthelmintic	Greece	Seed; oral; hot H_2O extract	Dragendorff 1898
Anthelmintic	Iraq/Saudi Arabia	Seed, dried; oral; hot H_2O extract	Rashan et al. 1989; Al-Yahya 1986
Anthelmintic (Ascaris, Taenia)	Algeria/North Africa	Unspecified	Hammiche and Merad 1997

(Continued)

TABLE 7.1 (CONTINUED)
Contemporary Ethnomedical Uses of *Peganum harmala*

Indication	Geography	Part Used	References
Anthelmintic against tapeworms	Pakistan	Seed powder	Ghafoor 1985
Expels tapeworms	Pakistan	Seed, dried; oral	Said 1984
Tapeworm	Egypt	Seed, dried; oral; hot H_2O extract	Ross et al. 1980
Vermifuge	Egypt	Seed	Ayensu 1978; Al-Awdat and Laham 1994; Boulos 1983 via MEDUSA n.d.
Vermifuge	India/Pakistan	Unspecified	Hassan 1967
Vermifuge	Iraq	Unspecified	Al-Rawi and Chakravarty 1964
Vermifuge	Unspecified	Unspecified	Duke 1992–2016, 2002; Steinmetz 1957; Uphof 1968
Bad digestion	Rabat	Powdered roasted seeds; oral after meals	Boulos 1983
Colic	Algeria/North Africa	Unspecified	Hammiche and Merad 1997
Colic	Unspecified	Unspecified	Duke 1992–2016, 2002
Colic	Egypt	Seed	Ayensu 1978; Al-Awdat and Laham 1994; Boulos 1983 via MEDUSA n.d.
Colic	Pakistan	Seed powder	Ghafoor 1985
Digestive	Egypt	Fruit and seed	Emboden 1979; Bown 1995; Phillips and Rix 1991 via MEDUSA n.d.
Digestive disorders	Algeria/North Africa	Unspecified	Hammiche and Merad 1997
Digestive system disorders	Morocco	Unspecified	Pelt 1971; Bellakhdar 1997 via MEDUSA n.d.
Emetic	Unspecified	Unspecified	Duke 1992–2016, 2002
Emetic	North Africa	Plant	Boulos 1983
Emetic	Egypt	Seed	Ayensu 1978; Al-Awdat and Laham 1994; Boulos 1983 via MEDUSA n.d.
Emetic	Pakistan	Seed	Ghafoor 1985
Emetic	India	Seed; oral; hot H_2O extract	Schipper and Volk 1960
Emetic	Saudi Arabia	Seed, dried; oral; hot H_2O extract	Al-Yahya 1986
Indigestion	Northern Balochistan	Seeds	Burkill 1909, p. 17

(Continued)

TABLE 7.1 (CONTINUED)
Contemporary Ethnomedical Uses of *Peganum harmala*

Indication	Geography	Part Used	References
Relieves stomach gas and pain	Balochistan	Seed; whole seeds swallowed with water	Goodman and Ghafoor 1992
Relieves stomachache and gas	Balochistan	Seeds ground to powder; oral; taken with water	Goodman and Ghafoor 1992
Stomach complaints	Egypt	Fruit and seed	Emboden 1979; Bown 1995; Phillips and Rix 1991 via MEDUSA n.d.
Stomachache	Turkey	Seed; oral	Sezik et al. 1997
(E) Modern Ethnomedical Uses of *Peganum harmala*, Fever			
Antipyretic	Algeria/North Africa	Unspecified	Hammiche and Merad 1997
Fever	North Africa	Seed	Ayensu 1978; Al-Awdat and Laham 1994; Boulos 1983 via MEDUSA n.d.
Fever	India/Pakistan	Unspecified	Hassan 1967
Relieves fever	Balochistan	Seed, burned; inhalation	Goodman and Ghafoor 1992
Diaphoretic	Greece	Seed; oral; hot H_2O extract	Dragendorff 1898
Sudorific	North Africa	Seed	Boulos 1983
Sudorific	Algeria/North Africa; Iraq	Unspecified	Al-Rawi and Chakravarty 1964; Hammiche and Merad 1997
Sudorific	Unspecified	Unspecified	Duke 1992–2016, 2002; Steinmetz 1957
Malaria	Egypt	Seed	Ayensu 1978; Al-Awdat and Laham 1994; Boulos 1983 via MEDUSA n.d.
Malaria	Algeria/North Africa; Iraq	Unspecified	Al-Rawi and Chakravarty 1964; Hammiche and Merad 1997
Malaria	Unspecified	Unspecified	Duke 1992–2016, 2002, Steinmetz 1957
Malaria, chronic, reducing temperature	Pakistan	Seed powder	Ghafoor 1985
Antiperiodic	Pakistan	Seed	Ghafoor 1985
(F) Modern Ethnomedical Uses of *Peganum harmala*, Poison Related			
Poison, as	Unspecified	Unspecified	Duke 1992–2016, 2002
Poisonous/repellant	Unspecified	Unspecified	Burkill 1985

(Continued)

TABLE 7.1 (CONTINUED)
Contemporary Ethnomedical Uses of *Peganum harmala*

Indication	Geography	Part Used	References
Poisoning, against	Algeria/North Africa; Morocco	Unspecified	Hammiche and Merad 1997; Pelt 1971; Bellakhdar 1997 via MEDUSA n.d.
Snake venoms, against	Algeria/North Africa	Unspecified	Hammiche and Merad 1997
Toxic	Morocco	Seed; oral	Bellakhdar et al. 1991
Toxic (claimed to be)	Egypt	Petals; oral	Kamel et al. 1970
Toxic (claimed to be)	Morocco	Seed, dried; oral	Merzouki et al. 2000
(G) Modern Ethnomedical Uses of *Peganum harmala*, Veterinary			
Diuretic for camels	Unspecified	Seed	Granot 1994
Donkey fodder	Israel	Dry stems	Granot 1994
Fodder	Unspecified	Unspecified	Burkill 1985

internal organs and DEATH from improper use. Please treat this plant with extreme respect and be forewarned of its awesome power. Caution and prudence should always be exercised in its employment.

ANTIDIABETIC AND METABOLIC EFFECTS

Much attention has been given to the antidiabetic effects of harmal, apparently owing to its traditional folk use for this purpose (Table 7.2). Three recent studies conducted by Iranian research groups illustrate the heuristic potential of harmal as a medication for diabetes.

Abedi Gaballu et al. (2015) created a "triplex mixture" of equal parts of aqueous extracts of *P. harmala*, *Rhus coriaria*, and *Urtica dioica* on metabolic and histological parameters in alloxan-induced diabetic rats. *R. coriaria* is the well-known, prized, and tart-tasting Sicilian sumac commonly sold in its powdered form in Middle Eastern spice markets. *U. dioica* is the common, or "stinging," nettle, a tasty wild

TABLE 7.2
Modern Ethnomedical Uses of *Peganum harmala*: Antidiabetic

Indication	Geography	Part Used	References
Diabetes	Rabat	Powdered roasted seeds; oral; after meals	Boulos 1983
Diabetes	Algeria/North Africa	Unspecified	Hammiche and Merad 1997

green when cooked as a vegetable, and medicinally esteemed as a genito-urinary remedy, or as a treatment for rheumatisms. The stinging phenomenon obtains from the ability of hair-like projections on the stems and leaves to inject compounds such as histamine and serotonin into the unwitting mammalian encroacher resulting in the eponymous urticaria, the medical term for the characteristic burning skin irritation, which is derived from the name of the genus. Interestingly, the sumac can also elicit a high Ig-E allergic-like dermatitis in sensitive individuals, complete with symptoms of urticaria. The researchers found activity in each of the three aqueous extracts (ED50 = 11.5 $\pm$ 2.57 mg/ml) that led to significant decreases in alkaline phosphatase (1.39–2.23-fold, $p < 0.05$), low-density lipoprotein cholesterol (1.79–3.26-fold, $p < 0.05$), and blood glucose (1.27–4.16-fold, $p < 0.05$) in the diabetic rodents, with the best result devolving from the triplex combination. Also, only the nettles and the triplex were effective in lowering serum triglycerides, but not *Rhus* alone. This trial yielded a classic example showing therapeutic synergies between the extracts of different plants in an intentionally created, traditional style herbal formula, in this case lowering blood sugar and improving lipid profile, and simultaneously, through likely antioxidant and anti-inflammatory mechanisms, also attenuating diabetes-related hepatic and renal compromise.

In a similar study, Komeili et al. (2016) focused exclusively on a hydroalcoholic extract of *P. harmala* seeds at three different doses: 30, 60, and 120 mg/kg in streptozotocin-induced diabetic rats, daily for 4 weeks. Relative to the pharmaceutically induced diabetic rat controls, the animals with diabetes that also received the harmal showed significant, dose-dependent reduction in the abnormally elevated glucose, lipid abnormalities, malondialdehyde—a marker of oxidative stress, alanine transaminase—whose elevation is associated with hepatic stress, aspartate transaminase (also known as SGOT), gamma-glutamyltranspeptidase—an enzyme associated with hyper-oxidation states, bilirubin—the product of red blood cell lysis, and glycated hemoglobin, while the total antioxidant capacity in the harmal-treated groups increased. The results showed the antidiabetic effects of harmal to encompass serum glucose reduction, enhanced antioxidant free radical scavenging, prevention of glycation products, and hepatoprotection, suggesting multifaceted heuristic applications for diabetology.

The pancreatic Islets of Langerhans β cells are responsible for secreting insulin in response to serum glucose. Their malfunction or sluggishness in executing this function is closely tied to both Type 1 (juvenile onset) and Type 2 (adult onset) diabetes mellitus. Additionally, the Islets are poor in supply of antioxidant enzymes, rendering said β cells more sensitive to oxidative damage. Rahimifard et al. (2014) devised an ingenious method for examining the antioxidant, protective effects of a panel of Iranian medicinal herbs, namely *P. harmala*, *Glycyrrhiza glabra* (licorice), *Satureja hortensis* (summer savory), *Rosmarinus officinalis* (rosemary), *Teucrium fruticans*-a member of the genus of plants commonly known as germanders and recently the subject of investigations pertaining to their neuroprotective potential (Simonyan and Chavushyan 2016), *Aloe vera* (Harlev et al. 2012), *Zingiber officinale* (ginger), *Silybum marianum* (the hepatoprotective milk thistle, the so-called wild artichoke, also known by its synonym, *Carduus marianus*), and St. John's wort, *Hypericum perforatum*.

Postlaparotomy, Rahimifard et al. (2014) removed the pancreases from the rats and isolated and incubated the Islets *in vivo* for 24 h. Logarithmic doses of plant materials were added and incubated an additional 24 h, following which cell viability, reactive oxygen species (ROS) production, and insulin secretion in response to glucose challenge were determined. All extracts extended Islet cell survival, their mitochondrial activity, and insulin levels, while reducing ROS production, according to the order (1) *P. harmala*, (2) *G. glabra*, (3) *S. hortensis*, (4) *R. officinalis*, (5) *T. scordium*, (6) *Aloe vera*, (7) *Z. officinale*, (8) *S. marianum*, and (9) *H. perforatum* at logarithmically variable doses of 10, 10(3), 10(4), 10, 10(2), 10(2), 10(–1), 10, and 10(3) μg/mL, respectively.

ANTI-PARKINSON AND CEREBROPROTECTIVE EFFECTS

Harmal can putatively help Parkinson's disease in two interconnected ways. First, harmal modulates neurotransmitters by its inhibitory action on the enzymes that deactivate them, for example by inhibiting acetylcholinesterase, which is normally in part responsible for deactivating neurotransmitter acetylcholine, and second, by selectively inhibiting monoamine oxidase type A (MAO-A), but not inhibiting MAO-B, thus causing a relief of some of the system for deactivating serotonin. Inhibition of catechol-O-methyltransferase (COMT) will have analogous effects on noradrenaline and on dopamine. Overall, the balance and tuning of the neurotransmitter milieu is extremely complex and eludes our understanding at this time (Table 7.3). Discussion of neurotransmitter modulation throughout the body is again considered in Chapter 9.

TABLE 7.3

Modern Ethnomedical Uses of *Peganum harmala*: Neurological and Mental Problems

Indication	Geography	Part Used	References
Anticonvulsant	Jordan	Shade dried aerial parts (seeds); smoked	Alkofahi et al. 1996
Anticonvulsant	Bulgaria	Dried aerial parts; oral	Ivanovska et al. 1997
Anticonvulsive	Jordan	Seed: smoke from burning seeds; inhalation	Al-Khalil 1995
Anti-paralytic	Saudi Arabia	Seed, dried; oral; hot H_2O extract	Al-Yahya 1986
Antispasmodic	Pakistan	Seed	Ghafoor 1985
Antispasmodic	Saudi Arabia	Seed, dried; oral; hot H_2O extract	Al-Yahya 1986
CNS depressant	India/Pakistan	Unspecified	Hassan 1967
CNS stimulant	India/Pakistan	Unspecified	Hassan 1967
CNS stimulant	Saudi Arabia	Seed, dried; oral; hot H_2O extract	Al-Yahya 1986
Epilepsy	Egypt	fruit and seed	Emboden 1979; Bown 1995; Phillips and Rix 1991 via MEDUSA n.d.

(Continued)

TABLE 7.3 (CONTINUED)
Modern Ethnomedical Uses of *Peganum harmala*: Neurological and Mental Problems

Indication	Geography	Part Used	References
Hallucinatory, social	Turkey	Seed	Ertug 2000 via MEDUSA n.d.
Headache	Pakistan	Entire plant; smoked	Leporatti and Lattanzi 1994
Headache	North Africa	Plant, burnt; inhalation	Boulos 1983
Hiccup	Unspecified	Unspecified	Duke 1992–2016, 2002
Hypnotic	India	Seed; oral; hot H_2O extract	Schipper and Volk 1960
Hypotonic	Pakistan	Seed	Ghafoor 1985
Hysteria	Unspecified	Unspecified	Duke 1992–2016, 2002
Hysteria	Egypt	Seed	Ayensu 1978; Al-Awdat and Laham 1994; Boulos 1983 via MEDUSA n.d.
Intoxicant	Unspecified	Unspecified	Font Quer 1979
Mental disorders	Morocco	Unspecified	Pelt 1971; Bellakhdar 1997 via MEDUSA n.d.
Mental illnesses	Egypt	Fruit and seed	Emboden 1979; Bown 1995; Phillips and Rix 1991 via MEDUSA n.d.
Narcotic	Unspecified	Unspecified	Duke 1992–2016, 2002
Narcotic	India/Pakistan	Unspecified	Hassan 1967
Narcotic	Morocco; Egypt; Pakistan	Seed	Pelt 1971; Bellakhdar 1997 via MEDUSA n.d.; Ayensu 1978; Al-Awdat and Laham 1994; Boulos 1983; Ghafoor 1985 via MEDUSA n.d.
Narcotic	India	Seed; oral; hot H_2O extract	Schipper and Volk 1960
Narcotic	India	Seed, dried; oral powder; may be smoked with tobacco; compound drug	Navchoo and Buth 1990
Narcotic	Saudi Arabia	Seed, dried; oral; hot H_2O extract	Al-Yahya 1986
Narcotic, ritual	Unspecified	Seed	Hanelt et al. 2001
Narcotic, social	Egypt	Plant and leaf	Ayensu 1978; Al-Awdat and Laham 1994; Boulos 1983 via MEDUSA n.d.
Nerves, for	Unspecified	Unspecified	Uphof 1968
Nervous breakdown	Israel	Unspecified; oral; hot H_2O extract	Friedman et al. 1986
Nervous disease	Morocco	Seed; oral	Bellakhdar et al. 1991

(*Continued*)

TABLE 7.3 (CONTINUED)
Modern Ethnomedical Uses of *Peganum harmala*: Neurological and Mental Problems

Indication	Geography	Part Used	References
Nervous disorders	Unspecified	Seed	Granot 1994
Nervous illnesses	Egypt	Fruit and seed	Emboden 1979; Bown 1995; Phillips and Rix 1991 via MEDUSA n.d.
Nervous system disorders	Morocco	Unspecified	Pelt 1971; Bellakhdar 1997 via MEDUSA n.d.
Neuralgia	Unspecified	Unspecified	Duke 1992–2016, 2002
Neuralgia	Egypt	Seed	Ayensu 1978/Al-Awdat & Laham 1994, Boulos 1983 via MEDUSA n.d.
Neurotic pains	North Africa	Plant, burnt; inhalation	Boulos 1983
Parkinsonism	Unspecified	Unspecified	Steinmetz 1957
Parkinsonism	Iraq	Unspecified	Al-Rawi and Chakravarty 1964
Parkinsonism	Egypt	Seed	Ayensu 1978; Al-Awdat & Laham 1994; Boulos 1983 via MEDUSA n.d.
Psychoactive	Morocco	Seed	Pelt 1971; Bellakhdar 1997 via MEDUSA n.d.
Relaxant	Jordan	Seed; inhalation	Al-Khalil 1995
Sedative	Unspecified	Seed	Hanelt et al. 2001
Sedative	Egypt	Seed	Ayensu 1978; Al-Awdat and Laham 1994, Boulos 1983 via MEDUSA n.d.
Sedative	Jordan	Shade dried aerial parts (seeds); smoked	Alkofahi et al. 1996
Sedative	Jordan	Seed; inhalation	Al-Khalil 1995
Sedative	Unspecified	Whole plant	Burkill 1985
Sensory system disorders	Morocco	Unspecified	Pelt 1971; Bellakhdar 1997 via MEDUSA n.d.
Soporific	Algeria/North Africa	Unspecified	Hammiche and Merad 1997
Soporific	Unspecified	Unspecified	Uphof 1968
Soporific	Egypt	Seed	Ayensu 1978, Al-Awdat and Laham 1994; Boulos 1983 via MEDUSA n.d.
Spasmolytic	India	Seed; oral; hot H_2O extract	Schipper and Volk 1960
Stupefacient	Iraq	Unspecified	Al-Rawi and Chakravarty 1964
Stupefacient	Unspecified	Unspecified	Steinmetz 1957, Uphof 1968
Tetanus, neonatal	Algeria/North Africa	Unspecified	Hammiche and Merad 1997

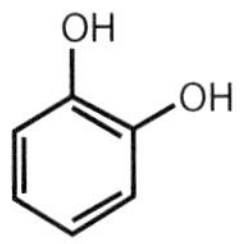

FIGURE 7.1   Catechol.

The enzyme catechol-O-methyltransferase catalyzes the degradation of dopamine as well as that of noradrenalin, notably among the neurotransmitters. Both dopamine and noradrenalin are catecholamines, while serotonin is an indoleamine. Catechol as a pure compound is simply a benzene ring with two hydroxyl groups attached to the ring at adjacent, i.e., *ortho-*, positions (Figure 7.1).

A quick look at the catecholamines dopamine and noradrenalin clearly reveals the presence of the catechol component. Although technically not a catecholamine, the ghost of the catechol can be seen as well in the structure of the entheogen mescaline. Also, note the homology with tyrosine, the dietary amino acid the body uses to produce catecholamines. This is the basis for restriction of tyrosine-rich foods such as chocolate, beans, fermented cheeses, pickled herring, red wine, etc., in patients taking chemical inhibitors of MAO (MAOIs), as the flooding of the bloodstream with catecholamines may result in potentially fatal consequences such as malignant hypertension. The remote possibility of such interactions occurring with harmine ingestion following eating tyrosine rich foods will be discussed also in Chapter 8, related to clinical management of the harmal-treated patient.

Deactivation of dopamine and noradrenalin *in situ* is accomplished in part by COMT, which catalyzes the mobilization of the hydrogen (H) from one of the hydroxyl groups (OH) on the catechol nucleus, and transfers it to the vacancy left by the vacating H to a methyl group (CH3). See, for example, the catalysis of noradrenalin to normetanephrine (Figure 7.2).

However, in addition to the catecholamines, neurotransmitters, there are other catechol combinations within the human physiology. In the case of catecholestrogens, inactivation by COMT may have a beneficial antiestrogen and anticancer effect (Zhu et al. 2009) in addition to the benefit of diminishing dopamine levels and benefitting Parkinson's disease.

The second way harmal is believed to exert putative benefit in Parkinson's disease is through its antioxidant action. Exactly how this works is also undefined, though it may, curiously, involve some of the same compounds responsible for neurotransmitter modulation. However, again the interest in this chapter is limited to studies of complexes derived from harmal, and not to any pure compounds discussed above. It is a facile matter to assume that the action of a complex plant extract is due to this or that principal component, but one should not be overly quick to do so, as minor components, such as γ tocopherol (Shen et al. 2011) in harmal, may contribute or synergize in some as yet unappreciated or incompletely appreciated manner, with alkaloids such as harmine, harmaline, or peganine.

In 2009, Yalcin and Bayraktar segregated their extracts of three plants, *P. harmala*, *Cistus parviflorus*, and *Vitex agnus-castus*, into alkaloid-rich and flavonoid-rich fractions, and tested them as possible inhibitors of COMT employing a fluorometric

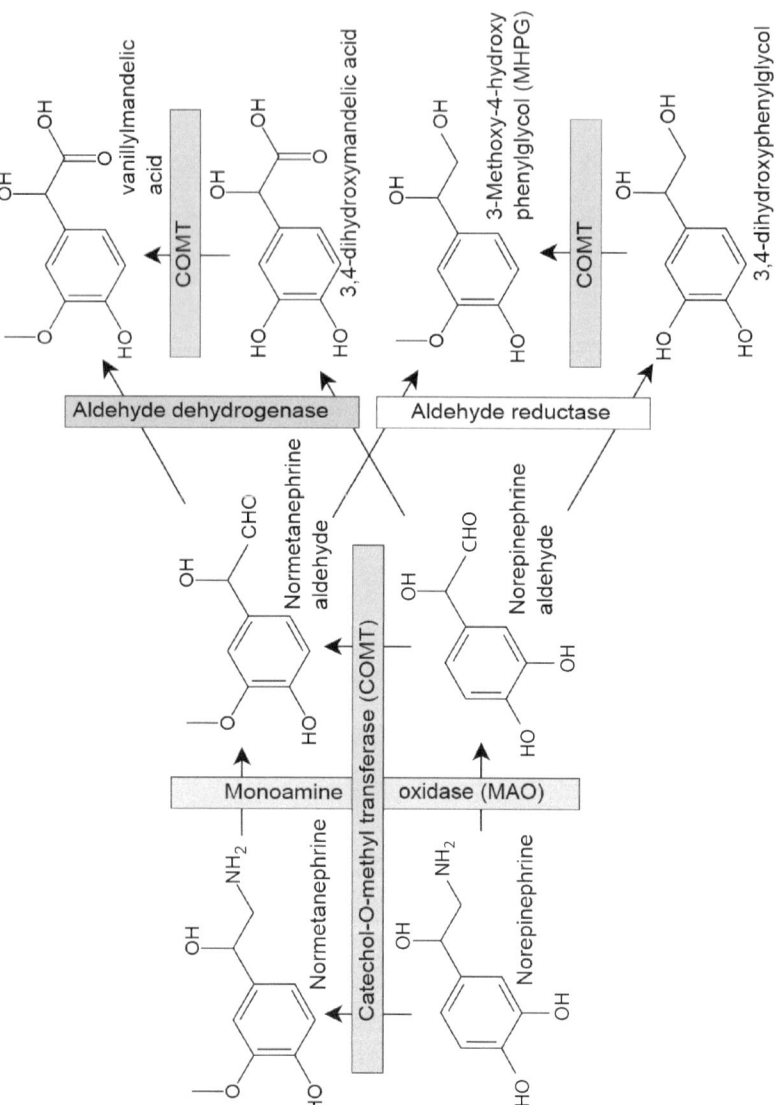

FIGURE 7.2 Catalysis of noradrenalin to normetanephrine. (By Mikael Häggström, used with permission. After Inkscape Reference: Rod Flower. From Rang, H.P. et al. *Rang & Dale's Pharmacology*, 6th ed., Churchill Livingstone, Edinburgh, 2007.)

enzyme assay, utilizing S-adenosylmethionine (SAM) and aesculetin (ES) as methyl donor and acceptor substrates, respectively. Of the other two herbs besides *P. harmala* that were investigated, *Cistus parviflorus*, a species of the *Cistus* genus, members of which are widely known as "rockrose," is used in traditional medicines as a remedy for colds and flu, diarrhea, as a reliable nervine, and, like harmal, as an emmenagogue. *Ladanum*, a resin secreted by the glandular trichomes, contains polyphenolics, terpenoids, and alkaloids possessing antioxidant, antibacterial, antifungal, and anticancer properties, while total leaf aqueous extracts kill or weaken influenza virus (Papaefthimiou et al. 2014). The other plant, *V. agnus-castus*, is the large "chaste berry" tree, which may also be known as "monk's pepper." Indeed, the fruits, as well as the leaves and shoots, have a reputation in herbal medicine as an anaphrodisiac for both sexes, and a history of employment for postmenstrual syndrome, mastalgia, headache, and as a galactogogue (Rani and Sharma 2013). The hormonally active compounds and further pharmacological uses are actively being pursued (Ahangarpour et al. 2016; Kikalishvili et al. 2016; Yao et al. 2016).

Of these three plants studied by Yalcin and Bayraktar (2009), inhibition of COMT was greatest in the alkaloid fraction extracted from *P. harmala*.

The cerebroprotective influence of a harmala alkaloid-rich fraction from *P. harmala* seeds was quantified in mice subjected to sodium nitrite-induced hypoxia and ethanol-induced neurodegeneration by Biradar et al. (2013). Their study was a good example of the antioxidant properties, specifically neuro-antioxidant properties, of harmal and how and why they may provide a theoretical underpinning that supports application of *P. harmala* extracts for prevention and treatment of Parkinson's and Alzheimer's diseases.

Specifically, Birader's group in India found their harmal extract at the doses employed to be without toxicity. At 5, 2.5, and 1.25 mg/kg p.o., the extract significantly ($p < 0.001$) and dose-dependently inhibited sodium nitrite induced memory impairment by decreasing time needed to "find the bottle" in a specially designed test. The findings on memory preservation co-occurred with inhibition of acetylcholinesterase ($p < 0.01$) and MAO-A ($p < 0.001$), increasing glutathione, a powerful *in vitro* natural antioxidant ($p < 0.001$), and decreasing thiobarbituric acid reactive substances (TBARS) ($p < 0.001$), an index of lipid peroxidation, in whole brain. Finally, 5 mg/kg of the harmal extract protected against neuronal DNA fragmentation in brain frontotemporal cortex following sodium nitrite assault.

In the newest study by Rezaei et al. (2016) from Iran, Parkinson's disease was induced in rats by direct injection *in situ* of 6-hydroxydopamine into the substantia nigra (Figure 7.3). Intraperitoneal injections of an aqueous harmal extract were given at time points −144 h, −120 h, −96 h, −72 h, −48 h, −24 h, −2 h, 4 h, and 24 h before and after the 6-hydroxydopamine. The authors have reported that compared to their controls, the rats that received the *harmal* injections had significantly less muscle stiffness and one-direction rotational behavior with significantly less global brain acetylcholinesterase activity, and histologically, diminished degeneration of dopaminergic neurons. The authors also discuss *P. harmala*'s known inhibition of angiotensin converting enzyme in terms of its beneficial potential for treatment of Parkinson's disease.

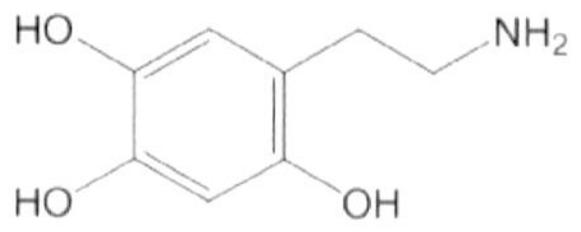

FIGURE 7.3   6-Hydroxydopamine. Used to induce Parkinson's disease in rats. (Rezaei, M. et al. 2016.)

FIGURE 7.4   Ripe seed capsules of *Peganum harmala*. The seeds have been used through history for the treatment of different neurological and mental problems. (7/20/2015, Judaean Desert, by side of the Inn of the Good Samaritan, Israel: by Helena Paavilainen.)

Herraiz et al. (2010) from Madrid, Spain showed *P. harmala* seed extract to potently and reversibly cause a competitive inhibition of human monoamine oxidase (MAO-A) with an IC (50) of 27 µg/l, while the root extract of the same plant inhibited MAO-A with an IC(50) of 159 µg/l. Stems and leaf extract showed considerably less activity. None of the extracts showed significant inhibition of MAO-B. It was assumed that these actions were related to the presence of β-carbolines within the extracts, and that these actions were implicated in *harmal*'s reported antidepressant activity (Figure 7.4).

ANTI-INFECTIVE

"Anti-infective" means attack upon microbial or viral invaders. For clarity, we will now discuss the following considered subcategories: antiprotozoal, antibacterial, antifungal, and antiviral (Table 7.4).

TABLE 7.4
Modern Ethnomedical Uses of *Peganum harmala*: Infectious Diseases

Indication	Geography	Part Used	References
Abscesses	Pakistan	Dried flower + leaf (external)	Leporatti and Lattanzi 1994
Anthelmintic	India; North Africa	Seed	Abrol and Chopra 1962; Boulos 1983
Anthelmintic	Unspecified	Seed	Granot 1994; Hanelt et al. 2001
Anthelmintic	Morocco	Seed; oral	Bellakhdar et al. 1991
Anthelmintic	Greece	Seed; oral; hot H_2O extract	Dragendorff 1898
Anthelmintic	Iraq; Saudi Arabia	Seed, dried; oral; hot H_2O extract	Al-Yahya 1986; Rashan et al. 1989
Anthelmintic against tapeworms	Pakistan	Seed powder	Ghafoor 1985
Expels tapeworms	Pakistan	Seed, dried; oral	Said 1984
Vermifuge	Egypt	Seed	Ayensu 1978; Al-Awdat and Laham 1994; Boulos 1983 via MEDUSA n.d.
Vermifuge	India/Pakistan; Iraq	Unspecified	Al-Rawi and Chakravarty 1964; Hassan 1967
Vermifuge	Unspecified	Unspecified	Duke 1992–2016, 2002; Steinmetz 1957; Uphof 1968
Anti-inflammatory	Pakistan	Entire plant; smoked	Leporatti and Lattanzi 1994
Anti-inflammatory	Saudi Arabia	Seed, dried; oral; hot H_2O extract	Al-Yahya 1986
Antimicrobial	Iraq	Seed, dried; oral; hot H_2O extract	Rashan et al. 1989
Antiperiodic	Pakistan	Seed	Ghafoor 1985
Antiseptic	Unspecified	Seed; incense	Hanelt et al. 2001
Antiseptic	Pakistan	Seeds and leaves; external; smoke	Ghafoor 1985
Antitussive	Pakistan	Entire plant; smoked	Leporatti and Lattanzi 1994
Bactericide	India	Seed	Abrol and Chopra 1962
Blepharitis	Algeria/North Africa	Unspecified	Hammiche and Merad 1997
Conjunctivitis, purulent	North Africa	Dried powdered plant	Boulos 1983
Conjunctivitis, suppurative	Algeria/North Africa	Unspecified	Hammiche and Merad 1997

(Continued)

TABLE 7.4 (CONTINUED)
Modern Ethnomedical Uses of *Peganum harmala*: Infectious Diseases

Indication	Geography	Part Used	References
Dermatitis (eczema)	Algeria/North Africa	Unspecified	Hammiche and Merad 1997
Eye inflammation	Israel	Unspecified; mostly externally	Wild Flowers of Israel n.d.
Infectious eye diseases	North Africa	Seed, oil extracted from	Boulos 1983
Fever	North Africa	Seed	Ayensu 1978; Al-Awdat and Laham 1994; Boulos 1983 via MEDUSA n.d.
Fever	Balochistan	Seeds, burned; inhalation	Goodman and Ghafoor 1992
Fever	India/Pakistan	Unspecified	Hassan 1967
Folliculitis	Unspecified	Unspecified	Granot 1994
Gingivitis	Morocco	Roots; macerated with vinegar; gargarism	Bellakhdar 1997 via MEDUSA n.d.
Laryngitis	Pakistan	Seed, decoction; oral	Ghafoor 1985
Laryngitis	Unspecified	Unspecified	Duke 1992–2016, 2002
Malaria	Egypt	Seed	Ayensu 1978; Al-Awdat and Laham 1994; Boulos 1983 via MEDUSA n.d.
Malaria	Algeria/North Africa; Iraq	Unspecified	Al-Rawi and Chakravarty 1964; Hammiche and Merad 1997
Malaria	Unspecified	Unspecified	Duke 1992–2016, 2002; Steinmetz 1957
Malaria, chronic, reducing temperature in	Pakistan	Seed powder	Ghafoor 1985
Mumps	Algeria/North Africa	Unspecified	Hammiche and Merad 1997
Otitis	Tunisia	Seed, dried; external	Boukef et al. 1982
Pediculicide	India	Root; external: hot H_2O extract	Schipper and Volk 1960
Pediculicide	Unspecified	Unspecified	Duke 1992–2016, 2002
Kills head lice	India	Root; external	Nayar 1955
Rhinitis	Unspecified	Seed	Granot 1994
Syphilis	India	Seed	Ayensu 1978; Al-Awdat and Laham 1994; Boulos 1983 via MEDUSA n.d.

(Continued)

TABLE 7.4 (CONTINUED)
Modern Ethnomedical Uses of *Peganum harmala*: Infectious Diseases

Indication	Geography	Part Used	References
Tapeworm	Egypt	Seed, dried; oral; hot H_2O extract	Ross et al. 1980
Tetanus, neonatal	Algeria/North Africa	Unspecified	Hammiche and Merad 1997
Throat inflammation	Pakistan	Dried flower and leaf; external	Leporatti and Lattanzi 1994
Wounds	Pakistan	Seeds and leaves; external; fumigation by burning seeds and leaves	Ghafoor 1985

ANTIPROTOZOAL

Coccidia is a large subclass of obligate parasitic, single-celled protozoa that spend their lives and reproduce within animal cells (Figure 7.5). Infected animals, which may include any mammal, many fish, and birds, spread spores called oocysts in their feces. Coccidiosis is an economically important disease of chickens, and may also affect humans. Symptoms can include bloody diarrhea and weakness or disfiguring skin involvement, and the infection is possibly fatal in the young or weak.

Tanweer et al. (2014), from the Gomal College of Veterinary Sciences in Pakistan, divided 200 1-week-old broiler chickens into five groups: negative control, positive control receiving a standard dose of **coccidiosis** with no *P. harmala*, and three groups of chickens receiving the coccidiosis challenge plus 200, 250, or 300 mg/l of a methanolic extract of *P. harmala* seeds in their drinking water. The untreated positive controls developed the symptoms of coccidiosis with weight loss and depression of the feed conversion ratio. The birds receiving the harmal in days 14–35 showed diminution or reversal of these symptoms at a rate commensurate with the concentration of the *P. harmala* extract. Since the market increasingly eschews chickens treated with antibiotics—the standard treatment for coccidiosis is sulfa-drugs—*P. harmala* represents an attractive alternative.

Leishmaniasis is a common protozoal parasitic disease affecting millions of people each year in over a hundred countries. The parasite is spread by sand flies, and can result in both visceral and cutaneous forms. There is substantial morbidity and a significant number of fatalities worldwide. The extent of the scourge has provoked renewed interest in natural products as treatments in lieu of the antifungal amphotericin B and pentavalent antimonial drugs and their attendant side effects, complicated by growing global concern over drug resistance developing among the parasites (Cheuka et al. 2016). Accordingly, *harmal* has been investigated as a potential alternative, natural treatment.

One Iranian study pitted aqueous and ethanolic *harmal* extracts against Glucantime, a proprietary pentavalent antimonial, anti-leishmaniasis drug, in BALB/C mice.

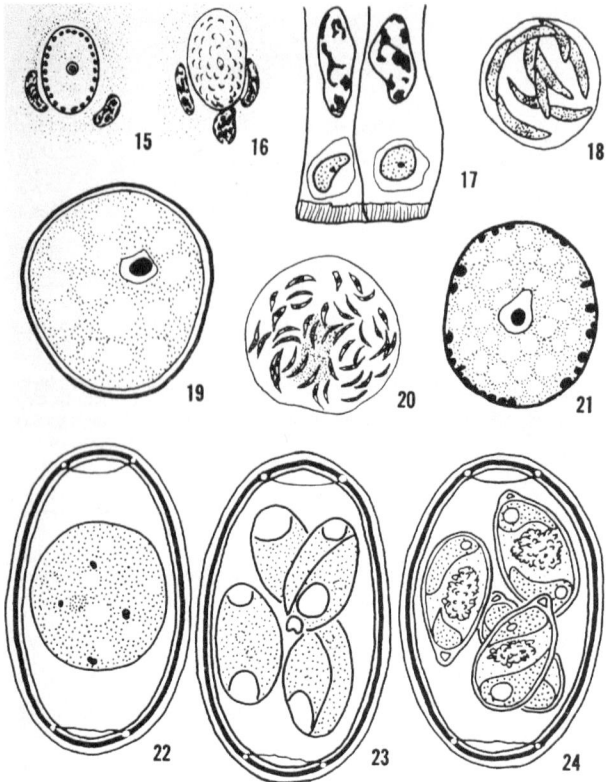

FIGURE 7.5 *Coccidia oocysts.* (From Levine, N.D., and Ivens, V., *The coccidian parasites* (Protozoa, Sporozoa) *of rodents,* University of Illinois Press, Urbana, 1965. Internet Archive Book Images, https://www.flickr.com/photos/internetarchivebookimages/20474596349/.)

Mice were inoculated with *Leishmania major* to develop cutaneous lesions. The ethanolic extract of *P. harmala* exhibited superior activity relative to the aqueous extract in significantly preventing or reducing the impact of the cutaneous lesions measured by size and parasitic load. Its action was not inferior to that of the pentavalent antimonial drug (Khoshzaban et al. 2014).

A companion Iranian look at *harmal* and *Leishmania major* (Mirzaei et al. 2007) similarly went *in vitro* to explore its therapeutic power relative to the pentavalent antimonial. They grew the promastigote form of *L. major,* which is the flagellated stage that lives in sandflies, as opposed to the nonmotile, a-flagellate amastigote stage, which nevertheless expertly establishes itself in mammals (Kima 2007). Mirzaei et al. (2007) and they showed a model to measure inhibition of *L. major* promastigotes. Though their standard *P. harmala* extract prepared according to the method of Manske and Holmes (1952) was effective against the parasite in a manner comparable to that of the antimonial, it was so at approximately 100 times the antimonial dose. Overall, however, the *harmal* was adjudged to possess comparable action to the trivalent Sb compound, albeit at a higher dose. This differential in dose (IC 50) is to be expected whenever a crude drug is compared to a pure compound.

Potassium antimonyl tartrate [Sb (III)], the trivalent antimonial historic standard drug against leishmaniasis (Figure 7.6). Antimony potassium tartrate, or tartar emetic, is known from ancient times. These days, newer pentavalent antimonial compounds are favored (Figure 7.7).

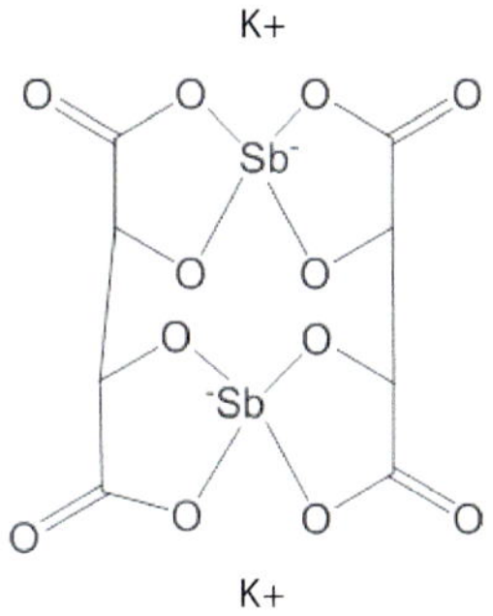

FIGURE 7.6   Antimony potassium tartrate.

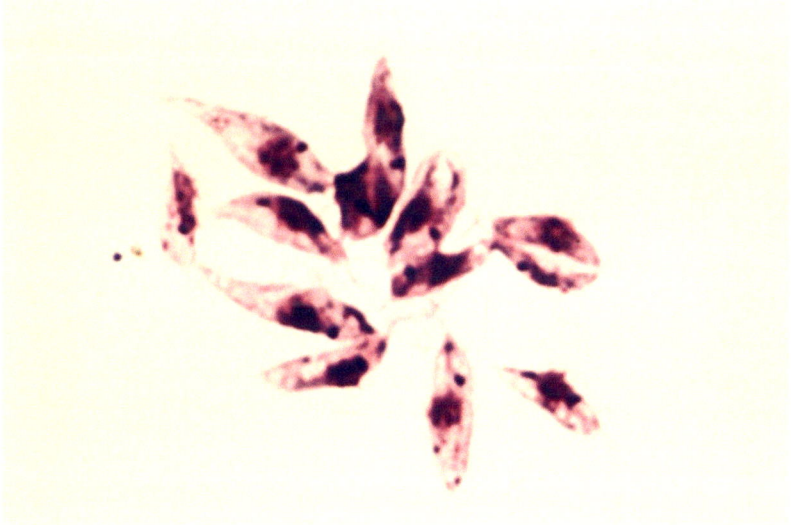

FIGURE 7.7   *Leishmania tropica.* Under the acellular culture condition, the protozoa transforms into the form of promastigote, a flagellated and elongated morphology seen in the mid-gut of the vector. (By Yutaka Tsutsumi, M.D., Professor, Department of Pathology, Fujita Health University School of Medicine, http://info.fujita-hu.ac.jp/~tsutsumi/photo/photo176-7 .htm, retrieved 3/3/2017.)

ANTIBACTERIAL

Writing in the *Pakistan Journal of Biological Sciences*, Irshaid et al. (2014) reported upon their investigations on both antibacterial and antioxidant virtues of their methanolic extracts of the aerial parts of *P. harmala*, *Artemisia sieberi*, green flowered *Rosmarinus officinalis*, and *Sarcopoterium spinosum*, an ethnobotanical member of the rose family, highly regarded in Israel and proven *in vitro* active against diabetes mellitus (Bachrach 2007; Rosenzweig et al. 2007). According to Irshaid's study, *P. harmala* possessed significant antioxidant activity measured by free radical scavenging *in silico*, and was also antibiotic against the several species of bacteria available, but the most potent antibiotic and antioxidant, as well as the plant with the highest holder of polyphenols, was *S. spinosum*. Though in this series there appeared to be some correlation between polyphenolic levels and antibiotic activity, an etiologic relationship between antibacterial action and polyphenol content could be established, suggesting possible roles of nonpolyphenolic compounds in the antibiotic effect.

Excellent activity of *P. harmala* ethanolic extract against *Streptococcus mutans*, the most important causative pathogen in dental caries, was found by researchers at the Dental School of Shiraz University (Motamedifar et al. 2016), home of the legendary eponymous red wine, the wine a mighty source of resveratrol, creating its own antibacterial synergy with its ethanol (Cirano et al. 2016). Motamedifar's agar diffusion and micro-broth dilution systems were used to quantify the effect of their *harmal* tincture relative to a 0.2% chlorhexidine solution. In this case, the tincture was four times as potent against *S. mutans* growth as the 0.2% chlorhexidine; however, toxicity of the *harmal* solution against green monkey kidney epithelial cells, i.e., Vera cells, led them to conclude that the *harmal* tincture would be inappropriate for daily human use as a topical dental applique.

In Tunisia and France, researchers studied the antibiotic effects of ethyl acetate, chloroform, butanol, and methanol extracts of *P. harmala* leaves in solid media and found the chloroform and methanolic extracts preferentially toxic to gram positive as opposed to gram negative bacteria. Antioxidant and antiviral effects of the extract were also noted (see the following Antiviral section; Edziri et al. 2010) (Figure 7.8).

ANTIFUNGAL

In Iran again, Aboualigalehdari et al. (2016) collected vaginal samples of *Candida albicans* from 27 different women suffering from vaginitis and studied the effects of their *harmal* extract against the 27 different *C. albicans* samples in generating a biofilm, i.e., a slimy adherent composed of an assemblage of microorganisms in an organic matrix. They confirmed that 12 µg/ml easily inhibited biofilm in strong biofilm formation, while 10 and 6 µg/ml inhibited biofilm formation in moderate and weak biofilm formation of *C. albicans* strains, respectively. The authors propose *in vivo* evaluation to further investigate the possibility and suitability of *P. harmala* extracts as treatment for common *C. albicans* vaginitis.

FIGURE 7.8 Leaves of *Peganum harmala* are antioxidant, antibacterial, and antiviral. (3/23/2015, Mount Scopus Botanical Garden, Jerusalem, Israel: by Helena Paavilainen.)

ANTIVIRAL

The work cited above by Edziri et al. (2010) also showed their methanolic extract of *P. harmala* leaves to have good activity against human cytomegalovirus (HCMV) strain AD-169 (ATCC Ref. VR 538) utilizing their "shell-vial" culture methodology. HCMV is frequently associated with the salivary glands, and may be life threatening in immunocompromised patients. The methanolic leaf extract also showed significant antioxidant activity *in silico*.

Kiani et al. (2008) communicated to the *Iranian Journal of Virology* the ability of *P. harmala* seed extract to prevent replication of the human herpes simplex virus type 1 (HSV-1) in Vero cells. Different concentrations (100, 500, 571, 667, 800, and 1000 µg/ml) of the extract were employed to Vero cells up to 667 µg/ml. These non-toxic doses nonetheless inhibited replication of HSV-1, indicating the potential of *harmal* extract as an anti-HSV-1 medicament. There is a great need for such a relatively nontoxic substance, since HSV-1 infection may be life-threatening in immuno-compromised individuals such as organ transplant recipients, patients with acquired immunodeficiency syndrome, and premature infants.

Aldhebiani et al. (2015) describe in the *Saudi Journal of Biological Sciences* something of the background regarding viral diseases of fig trees, most especially *Ficus carica*, the fig of commerce (Lansky et al. 2008, 2010). A sign of one of these diseases, Fig leaf mottle-associated virus-1 (FLMaV-1), is *chlorosis* (as in bleaching): the appearance of white spots, streaks, blots, rings, or feathery wisps throughout the leaf. These chlorotic areas can be measured and the severity of disease in leaves appearing 3 weeks after a graft graded. Extracts of four plants, *Thuja* leaf, ginger roots, *harmal* seeds, and turmeric rhizome (*Curcuma longa*), were investigated as possible inhibitors of FLMaV-1. Pretreatment for days prior to grafting of FLMaV-1 infected rootlets with either the *Zingiber* or the *Curcuma* provided protection against the development of viral symptoms in the leaves, and against infection-associated higher total photosynthesis pigments, total soluble intracellular proteins, and total phenolics. As protectors, *harmal* and *Thuja* leaf had no effect. But as potential treatments following the outbreak of leaf signs, the ginger and turmeric extracts were without effect, but the *harmal* and *Thuja* leaf extracts were effective as antiviral agents, resulting in significant reduction in the infection-associated parameters.

ANTI-INFESTATIONAL

Whereas the previous section dealt with microbes, this section concerns infestation with more complex organisms, either as parasites within or on the body, or as pests outside the body or in the environment (Table 7.5). A Moroccan study (Jbilou et al. 2008) examined the insecticidal activity of extracts of *P. harmala*, and of six other plants, namely *Centaurium erythraea*, *Ajuga iva*, *Aristolochia baetica*, *Pteridium aquilinum*, and *Raphanus raphanistrum*, on larval development, α-amylase activity, and offspring production of *Tribolium castaneum*, the red flour beetle, a worldwide pest that consumes stored cereal grains and their products (Figure 7.9).

Of the seven plants, *C. erythraea*, or European centaury, a plant renowned as a medicine for gastric and hepatic maladies (Tuluce et al. 2011) and appreciated for its antioxidant, free radical scavenging capabilities (Siler et al. 2014; Valentão et al. 2003), had the strongest insecticidal effect against growth of the larvae of *T. castaneum* with a 63% mortality 10 days after treatment, compared to 58% for the second place *P. harmala*. Larvae fed on treated diets had lower α-amylase (the enzyme that breaks down complex starches to glucose and mannose) than larvae fed on untreated diet, suggesting a decrease in the larva's ability or involvement in digesting grain following treatment with *harmal* and/or other herbs.

TABLE 7.5
Modern Ethnomedical Uses of *Peganum harmala*: Antiparasitic and Repellent

Indication	Geography	Part Used	References
Anthelmintic	India; North Africa	Seed	Abrol and Chopra 1962; Boulos 1983
Anthelmintic	Unspecified	Seed	Granot 1994; Hanelt et al. 2001
Anthelmintic	Morocco	Seed; oral	Bellakhdar et al. 1991
Anthelmintic	Greece	Seed; oral; hot H_2O extract	Dragendorff 1898
Anthelmintic	Iraq/Saudi Arabia	Seed, dried; oral; hot H_2O extract	Al-Yahya 1986; Rashan et al. 1989
Anthelmintic (*Ascaris, Taenia*)	Algeria/North Africa	Unspecified	Hammiche and Merad 1997
Anthelmintic against tapeworms	Pakistan	Seed powder	Ghafoor 1985
Expels tapeworms	Pakistan	Seed, dried; oral	Said 1984
Antiperiodic	Pakistan	Seed	Ghafoor 1985
Insect repellant	Iraq	Unspecified	Al-Rawi and Chakravarty 1964
Insect repellant	Unspecified	Unspecified	Duke 1992–2016, 2002
Insecticide	India	Seed	Abrol and Chopra 1962; Schipper and Volk 1960
Malaria	Egypt	Seed	Ayensu 1978; Al-Awdat and Laham 1994; Boulos 1983 via MEDUSA n.d.
Malaria	Algeria/North Africa; Iraq	Unspecified	Al-Rawi and Chakravarty 1964; Hammiche and Merad 1997
Malaria	Unspecified	Unspecified	Duke 1992–2016, 2002; Steinmetz 1957
Malaria, chronic, reducing temperature	Pakistan	Seed powder	Ghafoor 1985
Mosquito repellent	India	Entire plant; external	Nayar 1955
Pediculicide	India	Root; external: hot H_2O extract	Schipper and Volk 1960
Pediculicide	Unspecified	Unspecified	Duke 1992–2016, 2002
Kills head lice	India	Root; external	Nayar 1955
Poisonous/repellant	Unspecified	Unspecified	Burkill 1985
Protects woolen garments from moths	India	Seed; external	Nayar 1955
Tapeworm	Egypt	Seed, dried; oral; hot H_2O extract	Ross et al. 1980

(Continued)

TABLE 7.5 (CONTINUED)
Modern Ethnomedical Uses of *Peganum harmala*: Antiparasitic and Repellent

Indication	Geography	Part Used	References
Vermifuge	Egypt	Seed	Ayensu 1978; Al-Awdat and Laham 1994; Boulos 1983 via MEDUSA n.d.
Vermifuge	India/Pakistan; Iraq	Unspecified	Al-Rawi and Chakravarty 1964; Hassan 1967
Vermifuge	Unspecified	Unspecified	Duke 1992–2016, 2002; Steinmetz 1957; Uphof 1968

FIGURE 7.9   *Tribolium castaneum*, the red flour beetle. (By Eric Day, Virginia Tech, Blacksburg, Va. http://www.genome.gov/pressDisplay.cfm?photoID=87, courtesy of National Human Genome Research Institute, https://www.genome.gov, via Wikimedia Commons.)

Dastagir and Hussain (2014) of the Pharmacognosy Lab, Department of Botany, University of Peshawar, Pakistan found their n-hexane extract of *P. harmala* to have a higher toxicity (i.e., lower LD 50) against brine shrimp than their methanolic extract. Overall, the LD 50 was lower than that for other herbs tested, including *Fagonia cretica*, *Chrozophora tinctoria*, and *Ricinus communis*. While not a pest, but rather a valuable food source used in aquaculture, brine shrimps nevertheless remain a favorite for testing drug toxicity, so the effect of *harmal* on these organisms is included here (Figure 7.10).

Shang et al. (2016) on the other hand, chose a true pest, *Psoroptes cuniculi*, commonly known as the rabbit ear mite, causing havoc in many other domestic animals

FIGURE 7.10   Brine shrimp (*Artemia monica*) from Mono Lake. (9/9/2008, California: by djpmapleferryman via http://www.flickr.com.)

and of course generating economic losses. They employed microwave extraction of *P. harmala* seeds to produce an extract, fractionated it by chromatography, and found the extract and the principal components (peganine, harmine, harmaline) acaricidal against *P. cuniculi* (Figure 7.11).

Thus, *harmal* could be considered of great heuristic interest for the treatment of ascarides in domestic animals.

FIGURE 7.11   *Psoroptes cuniculi*, the rabbit ear mite. (From Diseases of Research Animals—DORA, University of Missouri—Comparative Medicine Program and IDEXX-BioResearch, 2013. http://dora.missouri.edu/rabbits/psoroptes-cuniculi/ retrieved 3/2/2017.)

UTEROTONIC

One of the most common folk usages for *P. harmala* seeds is as an abortifacient, and consequently the effect of the drug on the uterus has been investigated (Table 7.6). Fathiazada et al. (2006) found their hydroalcoholic extract of *P. harmala* seeds to promote contractions in *ex vivo* rat uterus and endometrium-free myometrium (stripped myometrium) preparations significantly more than the solvent control,

TABLE 7.6
Modern Ethnomedical Uses of *Peganum harmala*: Gynecological/Urogenital

Indication	Geography	Part Used	References
Abortifacient	Israel	Dried aerial parts; oral; decoction	Shapira et al. 1989
Abortifacient	India	Entire plant; oral	Saha et al. 1961
Abortifacient	India	Entire plant; oral; infusion	Kakrani and Saluja 1993
Abortifacient	India	Fresh entire plant; oral	Singh et al. 1996
Abortifacient	Egypt	Seed	Ayensu 1978; Al-Awdat and Laham 1994; Boulos 1983 via MEDUSA n.d.
Abortifacient	Unspecified	Seed	Granot 1994
Abortifacient	India	Seed; oral	Abrol and Chopra 1962; Chopra 1933; Saha et al. 1961
Abortifacient	India (by Indian midwives)	Seed; oral	Chopra et al. 1949
Abortifacient	India	Seed; oral; hot H_2O extract	Malhi and Trivedi 1972
Abortifacient	Arabic countries (Unani medicine)/India	Seed, dried; oral; hot H_2O extract	Kamboj 1988; Nayar 1954; Razzack 1980
Abortifacient	Arabic countries (Unani medicine)	Seed, dried/plant?; vaginal (pessary)	Razzack 1980
Abortifacient	Pakistan	Seeds, decoction	Ghafoor 1985
Abortifacient	Algeria/North Africa; India/Pakistan	Unspecified	Hammiche and Merad 1997; Hassan 1967
Abortifacient at 2–3 months	Morocco	Seed, dried; oral; handful of seeds	Merzouki et al. 2000
Antifertility agents	Morocco	Seed	Pelt 1971; Bellakhdar 1997 via MEDUSA n.d.
Aphrodisiac	Egypt	Seed	Ayensu 1978; Al-Awdat and Laham 1994; Boulos 1983 via MEDUSA n.d.
Aphrodisiac	Unspecified	Seed	Hanelt et al. 2001

(Continued)

TABLE 7.6 (CONTINUED)
Modern Ethnomedical Uses of *Peganum harmala*: Gynecological/Urogenital

Indication	Geography	Part Used	References
Aphrodisiac	India/Iraq/Middle East	Unspecified	Al-Rawi and Chakravarty 1964; Granot 1994; Hassan 1967
Aphrodisiac	Unspecified	Unspecified	Uphof 1968
Aphrodisiac, male	India/Iraq/Kuwait	Seed; oral	Alami et al. 1976; Al-Rawi and Chakravarty 1964; Chopra et al. 1949
Aphrodisiac, male	Egypt	Seed; oral; hot H_2O extract	Lewis and Elvin-Lewis 1977
Cancer, uterine	Unspecified	Unspecified	Hartwell 1967–1971
Dysmenorrhea	India	Seed; oral; hot H_2O extract	Schipper and Volk 1960
Dysmenorrhea	Unspecified	Unspecified	Duke 1992–2016, 2002
Emmenagogue	Israel	Dried aerial parts; oral; decoction	Shapira et al. 1989
Emmenagogue	India	Entire plant; oral	Saha et al. 1961
Emmenagogue	India	Fresh entire plant; oral	Singh et al. 1996
Emmanagogue	Egypt/North Africa	Seed	Boulos 1983; Ayensu 1978; Al-Awdat and Laham 1994; Boulos 1983 via MEDUSA n.d.
Emmenagogue	India; Pakistan	Seed, oral	Ahmad 1957; Chopra 1933; Chopra et al. 1949; Saha et al. 1961
Emmenagogue	Greece/India/Iraq	Seed; oral; hot H_2O extract	Al-Rawi and Chakravarty 1964; Dragendorff 1898; Malhi and Trivedi 1972; Schipper and Volk 1960
Emmenagogue	Arabic countries (Unani medicine)/ India	Seed, dried; oral; hot H_2O extract	Kamboj 1988; Razzack 1980
Emmenagogue	Algeria/North Africa; India/ Pakistan; Iraq	Unspecified	Al-Rawi and Chakravarty 1964; Hammiche and Merad 1997; Hassan 1967
Emmenagogue	Unspecified	Unspecified	Steinmetz 1957
Emmenagogue, mild	India	Seed, dried; oral; hot H_2O extract	Nayar 1954
Galactagogue	Pakistan	Seed	Ghafoor 1985
Galactagogue	India	Seed; oral	Chopra et al. 1949
Galactagogue	India/Pakistan; Iraq	Unspecified	Al-Rawi 1964; Hassan 1967
Galactagogue	Unspecified	Unspecified	Steinmetz 1957; Uphof 1968

(Continued)

TABLE 7.6 (CONTINUED)
Modern Ethnomedical Uses of *Peganum harmala*: Gynecological/Urogenital

Indication	Geography	Part Used	References
Increases the flow of milk	Pakistan	Seeds, decoction of	Ghafoor 1985
Genito-urinary system disorders	Morocco	Unspecified	Pelt 1971; Bellakhdar 1997 via MEDUSA n.d.
Infertility, female	Balochistan	Seed; fumes from burning seeds introduced into the vagina by a special pipe	Goodman and Ghafoor 1992
Infertility, female	Algeria/North Africa	Unspecified	Hammiche and Merad 1997
Menstrual problems	Egypt	Seed oil	Emboden 1979; Bown 1995; Phillips & Rix 1991 via MEDUSA n.d.
Pregnancy/birth	Egypt	Fruit & seed	Emboden 1979; Bown 1995; Phillips and Rix 1991 via MEDUSA n.d.
Easing delivery	Kuwait	Leaf and stem; oral; type of extract not specified	Alami et al. 1976
Sexual disorders	Egypt	Seed oil	Emboden 1979; Bown 1995; Phillips and Rix 1991 via MEDUSA n.d.
Syphilis	India	Seed	Ayensu 1978; Al-Awdat and Laham 1994; Boulos 1983 via MEDUSA n.d.
Urogenital agent	India/Pakistan	Unspecified	Hassan 1967
Uterine pain in pregnant women	Balochistan	Seed; fumes from burning seeds introduced into the vagina by a special pipe	Goodman and Ghafoor 1992
Uterus, prolapse of	Egypt	Seed	Ayensu 1978; Al-Awdat and Laham 1994; Boulos 1983 via MEDUSA n.d.

and this effect was unaffected by atropine injections. Further, the *harmal* stimulated contractions in a calcium-free solution in the presence of KCl, leading to speculation that *harmal* may increase calcium influx via voltage-dependent calcium channels.

ANODYNE

Part of *harmal*'s historical usage relates to relief of pain, so antipain properties of *P. harmala* have been examined (Table 7.7). In Morocco, Monsef et al. (2004) set

TABLE 7.7

Modern Ethnomedical Uses of *Peganum harmala*: Pain Alleviation

Indication	Geography	Part Used	References
Analgesic	India	Seed; oral; hot H_2O extract	Schipper and Volk 1960
Analgesic	Iraq/ Saudi Arabia	Seed, dried; oral; hot H_2O extract	Al-Yahya 1986; Rashan et al. 1989
Analgesic	North Africa	Seed, powdered; externally as a cataplasm	Boulos 1983
Analgesic	Algeria/North Africa	Unspecified	Hammiche and Merad 1997
Anodyne	Unspecified	Unspecified	Duke 1992–2016, 2002
Antalgic	Morocco	Seed; oral	Bellakhdar et al. 1991
Pain, alleviation of	Unspecified	Seed	Granot 1994
Pain	Morocco	Unspecified	Pelt 1971; Bellakhdar 1997 via MEDUSA n.d.
Pain-killers	Unspecified	Whole plant	Burkill 1985
Articular pains	Libya	Dried entire plant	Hussain and Tobji 1997
Articular pains	North Africa	Seed, powdered; externally: powdered seeds mixed with honey and ginger rubbed on skin	Boulos 1983
Joint pains	Israel	Unspecified; mostly externally	Wild Flowers of Israel. n.d.
Colic	Egypt	Seed	Ayensu 1978; Al-Awdat and Laham 1994; Boulos 1983 via MEDUSA n.d.
Colic	Pakistan	Seed powder	Ghafoor 1985
Colic	Algeria/North Africa	Unspecified	Hammiche and Merad 1997
Colic	Unspecified	Unspecified	Duke 1992–2016, 2002
Dysmenorrhea	India	Seed; oral; hot H_2O extract	Schipper and Volk 1960
Dysmenorrhea	Unspecified	Unspecified	Duke 1992–2016, 2002
Headache	Pakistan	Entire plant; smoked	Leporatti and Lattanzi 1994
Headache	North Africa	Plant, burnt; inhalation	Boulos 1983
Muscle pains	Unspecified	Seed	Granot 1994
Narcotic	Egypt/Morocco/ Pakistan	Seed	Ghafoor 1985; Ayensu 1978; Al-Awdat and Laham 1994; Boulos 1983 via MEDUSA n.d.; Pelt 1971; Bellakhdar 1997 via MEDUSA n.d.
Narcotic	India	Seed; oral; hot H_2O extract	Schipper and Volk 1960

(Continued)

TABLE 7.7 (CONTINUED)
Modern Ethnomedical Uses of *Peganum harmala*: Pain Alleviation

Indication	Geography	Part Used	References
Narcotic	Saudi Arabia	Seed, dried; oral; hot H_2O extract	Al-Yahya 1986
Narcotic	India	Seed, dried; oral powder; may be smoked with tobacco; compound drug	Navchoo and Buth 1990
Narcotic	India/Pakistan	Unspecified	Hassan 1967
Narcotic	Unspecified	Unspecified	Duke 1992–2016, 2002
Narcotic, ritual	Unspecified	Seed	Hanelt et al. 2001
Neuralgia	Egypt	Seed	Ayensu 1978; Al-Awdat and Laham 1994; Boulos 1983 via MEDUSA n.d.
Neuralgia	Unspecified	Unspecified	Duke 1992–2016, 2002
Neurotic pains	North Africa	Plant, burnt; inhalation	Boulos 1983
Rheumatic pains	North Africa	Seed, oil extracted from	Boulos 1983
Stomachache	Turkey	Seed; oral	Sezik et al. 1997
Stomach gas and pain	Balochistan	Seed; whole seeds swallowed with water	Goodman and Ghafoor 1992
Stomachache and gas	Balochistan	Seed; oral; ground to powder; taken with water	Goodman and Ghafoor 1992
Toothache	North Africa	Root and seed; pounded roots and seeds mixed with tobacco; smoked in pipe	Boulos 1983
Toothache	Unspecified	Seed	Granot 1994
Toothache	Israel	Unspecified	Wild Flowers of Israel n.d.
Tooth pain	Unspecified	Unspecified	Granot 1994
Womb pain in pregnant women	Balochistan	Seed; fumes from burning seeds introduced into the vagina by a special pipe	Goodman and Ghafoor 1992

out to gauge the antinociceptive effects of the total alkaloid extract of *P. harmala*, the abundant Iranian plant, using a formalin-induced pain response in mice. The extracts were administered intraperitoneally (i.p.) 30 min before formalin injection to the mouse paw, and nociception recorded 0–5 min post formalin (early phase) and 15–40 min post formalin (late phase). In both phases, the *harmal* injection resulted in decreased nociception in a dose-dependent manner, with the highest doses of *harmal* (30 mg/kg) resulting in 100% pain suppression compared to controls at both early and late phases.

Farouk et al. (2008), also in Morocco, took the work to another level, proving the nociception suppression from their *P. harmala* seed extract employed both alone and when combined with morphine in three different mouse models of nociception, namely writhing, formalin, and hot plate tests. Again the alkaloid-rich extract of *harmal* at 12.5 and 25 mg/kg dose-dependently and significantly decreased nociception of i.p. injection of acetic acid. Because 1 mg/kg naloxone reversed the effect, it was concluded that both peripheral and central action of *harmal* are likely mediated at least in part through modulation of opioid receptors.

A collaborative effort among several Pakistani scientists (Shoaib et al. 2016) largely confirmed the results of earlier studies, showing their *P. harmala* alkaloid extract to reduce writhings induced by acetic acid injection. The extract was considered safe by this group for the experimental mouse at doses up to 1250 mg/kg. As in the previous report, naloxone injections could reverse the experimental analgesia.

ANTICANCER

DNA TOPOISOMERASE TYPE 1

The pioneering study by Sobhani et al. (2002) on the effect of *P. harmala* seed extract *in vitro* remains the classic modern paper in the new and ancient field of **harmalogy**. The essence of their investigation *in vitro* was of the *harmal* seed extract as well as its principal harmala alkaloids, namely harmine and congeners, all of which showed potency, with harmine the strongest, in inhibiting the relaxation of DNA strands to allow them to replicate themselves. The enzyme that catalyzes that relaxation is called (DNA) topoisomerase type I, and Sobhani found the inhibition reproducible and robust. Also of great interest in their paper was the observation that *harmal* was the most commonly used medicine for treating cancer among folk healers in Iran/Persia and that it was often, or usually, used in combination with another more specifically indigenous herb, *Dracocephalum kotschyi*, and furthermore, that the ratio between the two herbs used was roughly 9:1, that is, nine parts *harmal*, one part *D. kotschyi* (Table 7.8).

Sobhani's classic was followed by Jahaniani et al. (2005) who elucidated the structure of a known flavone, **xanthomicrol** (Stout and Stout 1961), that Jahaniani had found in the leaves of *D. kotschyi* as a principal component and largely causing its powerful and specific cancer cytotoxic effect. Xanthomicrol's antiproliferative selectivity toward malignant cells as opposed to nonmalignant cells was equal to that of **doxorubicin** (Figures 7.12 and 7.13). Jahaniani et al. (2005) also taught that the two plants, *harmal* and *D. kotschyi*, or rather their methanolic extracts, were the principal constituents of a proprietary preparation commonly used in modern Iran for the treatment of all kinds of cancers—*Spinal-Z*.

Xanthomicrol was the subject of a recent review outlining its anticancer, antiplatelet, and antispasmodic properties. It is a methoxylated flavone that accumulates on the surfaces of leaves of certain species of plants. Purportedly, the function

TABLE 7.8
Modern Ethnomedical Uses of *Peganum harmala*: Neoplasia-Related

Indication	Geography	Part Used	References
Abscesses	Pakistan	Dried flower + leaf; external	Leporatti and Lattanzi 1994
Alterative	Pakistan	Seed	Ghafoor 1985
Alterative	Iraq	Unspecified	Al-Rawi and Chakravarty 1964
Alterative	Unspecified	Unspecified	Steinmetz 1957; Uphof 1968
Analgesic	India	Seed; oral; hot H_2O extract	Schipper and Volk 1960
Analgesic	Iraq/Saudi Arabia	Seed, dried; oral; hot H_2O extract	Al-Yahya 1986; Rashan et al. 1989
Analgesic	North Africa	Seed, powdered; externally; as a cataplasm	Boulos 1983
Analgesic	Algeria/North Africa	Unspecified	Hammiche and Merad 1997
Anodyne	Unspecified	Unspecified	Duke 1992–2016, 2002
Antalgic	Morocco	Seed; oral	Bellakhdar et al. 1991
Pain	Morocco	Unspecified	Pelt 1971; Bellakhdar 1997 via MEDUSA n.d.
Pain, alleviation of	Unspecified	Seed	Granot 1994
Pain-killers	Unspecified	Whole plant	Burkill 1985
Anti-inflammatory	Pakistan	Entire plant; smoked	Leporatti and Lattanzi 1994
Anti-inflammatory	Saudi Arabia	Seed, dried; oral; hot H_2O extract	Al-Yahya 1986
Antiperiodic	Pakistan	Seed	Ghafoor 1985
Antipyretic	Algeria/North Africa	Unspecified	Hammiche and Merad 1997
Fever	North Africa	Seed	Ayensu 1978; Al-Awdat and Laham 1994; Boulos 1983 via MEDUSA n.d.
Fever	India/Pakistan	Unspecified	Hassan 1967
Reducing temperature [in chronic malaria]	Pakistan	Seed powder	Ghafoor 1985
Relieves fever	Balochistan	Seed, burned; inhalation	Goodman and Ghafoor 1992
Cancer, uterine	Unspecified	Unspecified	Hartwell 1967–1971
Depurative	Libya	Dried entire plant	Hussain and Tobji 1997
Depurative	North Africa	Plant; oral; decoction in oil; taken first thing in the morning	Boulos 1983
Depurative	Algeria/North Africa	Unspecified	Hammiche and Merad 1997
Dermatitis (eczema)	Algeria/North Africa	Unspecified	Hammiche and Merad 1997
Weakness, general	Israel	Unspecified	Wild Flowers of Israel n.d.

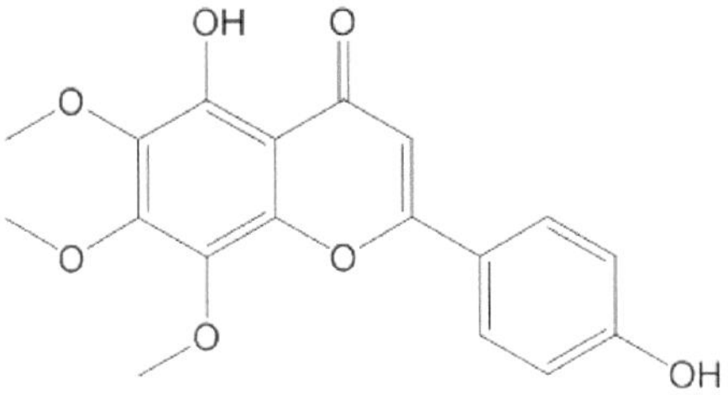

FIGURE 7.12   Xanthomicrol.

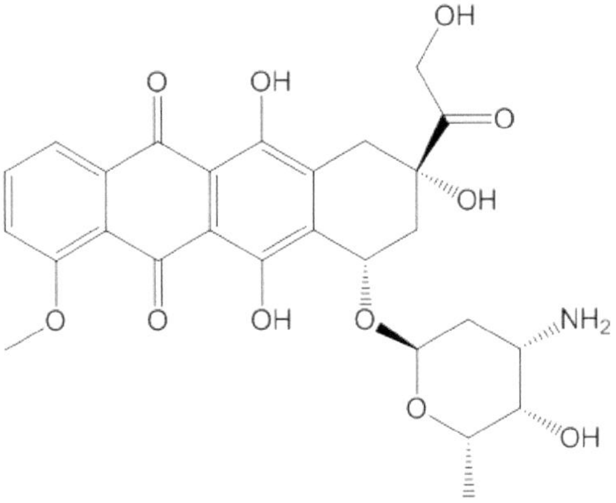

FIGURE 7.13   Doxorubicin (Adriamycin).

for the plant of these methoxylated flavones is protection against ultraviolet light (Fattahi et al. 2014).

Refinements of Sobhani's Topoisomerase I assay were made by Cao et al. (2005) from Guangzhou, China, who marveled at the extraordinary DNA intercalating abilities of the harmala alkaloids. Possibly this intercalating in some way underlay the specific inhibition of topoisomerase type I, as there was no effect of any of these compounds on topoisomerase type II (though this has been challenged—see Chapter 5, "Harmal as Anticancer Agent in Traditional Chinese and Iranian Medicine").

ANGIOGENESIS

Yavari et al. (2015), from the Medical Biology Research Center of the Kermanshah University of Medical Sciences in Iran, studied the effect of a low-pH, stable, hydroalcoholic extract of *P. harmala* seeds on indices related to angiogenesis, such as normal epithelial cell proliferation and expression of vascular epithelial growth factor (VEGF). Their *harmal* extract was found to significantly decrease epithelial cell proliferation—ID50 of ~85 µg/ml, and VEGF secretion at doses greater than 10 μg/ml. Overall the findings show considerable potential of *harmal* as an antiangiogenic agent.

G-Quadruplex: Genome Regulation

G-quadruplexes are secondary three-dimensional structures of DNA that also involve anomalous, base interactions irregular to the Watson and Crick canon. They are now an intensive subject of research as essential genome regulators, and so, highly relevant for cutting edge cancer research (Valton and Prioleau 2016). Because of their general resistance to being victims of drug resistance by the host, and relative profile of higher safety than synthetic chemical agents, natural products are an attractive repository for prospecting after new drugs.

Wang et al. (2016) screened 17 medicinal plants for possible binding to G-quadruplex d(TTGGGTT)4 by (1)H NMR spectroscopy. The most potent binding in the group was noted for the crude extract of the seeds of *P. harmala*. Four β-carboline alkaloids were isolated from their extract, which they dubbed pegaharmines A–D, confirming their structures and absolute configurations with extensive NMR analyses, x-ray crystallography, ECD calculations, and CD exciton chirality approaches (Figure 7.14). Pegaharmine D (4), which showed the strongest G-quadruplex interaction, and the associated significant cytotoxic activity against three cancer cell lines, suggested concurrence of G-quadruplex binding and cytotoxicity to cancer cells, and a novel means for screening of potentially anticancer efficacy of putative therapeutic compounds and preparations. Natural small molecules with power to modulate quadruplex may represent a kind of new frontier of anticancer pharmacognosy (Paulo and Francisco 2016).

Differentiation

Differentiation is a nonlethal approach to anticancer therapeutics. Rather than kill the bad cells, they are reformed. In general, cancers can be seen to represent de-differentiation of normal cells. Nowhere else is this more apparent than in leukemia. Normal white blood cells have clear functions and specific abilities with which to carry out these functions. For example, one function of normal white blood cells is phagocytosis, the ability to eat and digest foreign particles, such as dead cells or bacteria. These cells have clear function, they are differentiated. When cells become de-differentiated, they lose the ability to carry out their clear, ordered function. This is the de-differentiated primitive state of cancer.

Herbal agents that produce differentiation (Kawaii and Lansky 2003) are called differentiating agents. Zaker et al. (2007) created extracts of *P. harmala* and focused on purified harmala alkaloids in their investigations, but overall, their studies suggested strongly that the complex *P. harmala* total alkaloid extracts as well as pure compounds exerted pro-differentiating effects in leukemia cells, providing a possible comparable alternative to conventional pro-differentiating agents such as ATRA (Figure 7.15).

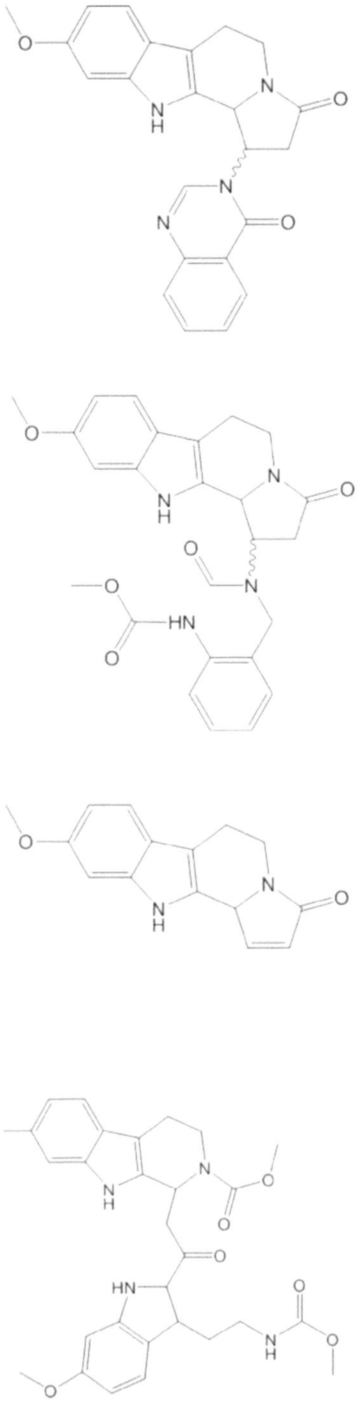

FIGURE 7.14 Pegaharmines A–D (from top to bottom).

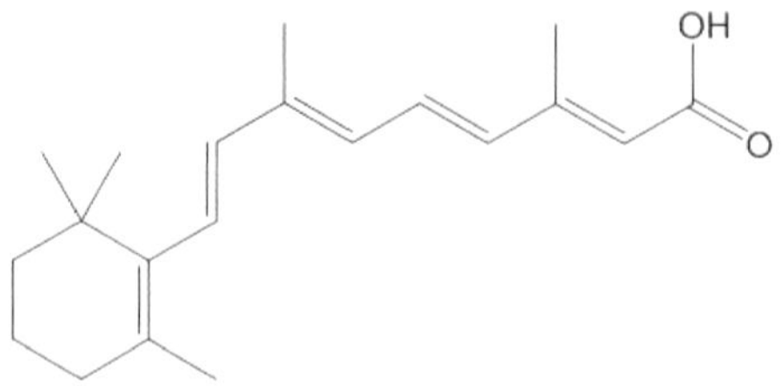

FIGURE 7.15   ATRA (all-trans-retinoic acid).

DIURETIC

Diuresis is often an important pharmacological objective in therapeutics, for example, in reducing swellings or hypertension, or whenever elimination of fluids is desirable. Plants have been chosen for their diuretic effects since ancient days (Table 7.9). As a common plant in the region, *P. harmala* was apparently also employed for this purpose, and Al-Saikhan et al. (2016) set out to prove or disprove the veracity of these traditional usages in controlled, *in vitro* experiments. They compared the magnitude of diuresis of a methanolic extract of *P. harmala* relative to the conventional powerful diuretic furosemide. The *harmal* was used in three different doses: 150, 300, and 450 mg/kg, while the furosemide was used at 10 mg/kg orally. Results showed urine volume at 5 and 24 h and electrolyte excretion (Na+, K+, and Cl-) at 24 h significantly and dose-dependently higher in harmal-treated rats relative to untreated rats ($p < 0.05$). These effects were equivalent to the dose-appropriate effects from furosemide.

TABLE 7.9
Modern Ethnomedical Uses of *Peganum harmala*: Kidney and Bladder

Indication	Geography	Part Used	References
Calculus	Unspecified	Unspecified	Duke 1992–2016, 2002
Diuretic	North Africa	Plant	Boulos 1983
Diuretic	Iraq	Seed, dried; oral; hot H_2O extract	Rashan et al. 1989
Diuretic	India/Pakistan	Unspecified	Hassan 1967
Diuretic for camels	Unspecified	Seed	Granot 1994
Dropsy in the feet	Persia	Unspecified; fomentation	Baillon 1871–1888, p. 451
Dysuria	Turkey	Seed, decocted	Honda et al. 1996; Ertug 2000; Citoglu 1997 via MEDUSA n.d.
Dysuria	Turkey	Seed, dried; oral; decoction	Honda et al. 1996
Urinary disorders	Egypt	Seed oil	Emboden 1979; Bown 1995; Phillips and Rix 1991 via MEDUSA n.d.

HEPATOPROTECTIVE

Hepatoprotection is a key goal in health maintenance, and highly pertinent to the question, "is life worth living?" The answer is—"it depends on the liver." Indeed healthy hepatic function is a fundament of well-being (Table 7.10).

Hamden et al. (2008a) investigated the putative liver-protective qualities of both ethanol and chloroform extracts of *P. harmala* on diseases induced by thiourea in adult male rat (Figure 7.16).

TABLE 7.10

Modern Ethnomedical Uses of *Peganum harmala*: Liver and Spleen

Indication	Geography	Part Used	References
Depurative	North Africa	Plant; oral; decoction in oil taken first thing in the morning	Boulos 1983
Depurative	Libya	Dried entire plant	Hussain and Tobji 1997
Depurative	Algeria/North Africa	Unspecified	Hammiche and Merad 1997
Digestive	Egypt	Fruit and seed	Emboden 1979; Bown 1995; Phillips and Rix 1991 via MEDUSA n.d.
Digestive disorders	Algeria/North Africa	Unspecified	Hammiche and Merad 1997
Digestive system disorders	Morocco	Unspecified	Pelt 1971; Bellakhdar 1997 via MEDUSA n.d.
Bad digestion	Rabat	Powdered roasted seeds; oral; after meals	Boulos 1983
Indigestion	Northern Balochistan	Seeds	Burkill 1909, p. 17
Icterus	India	Seed; oral; hot H_2O extract	Schipper and Volk 1960
Jaundice	Pakistan	Seed powder	Ghafoor 1985
Jaundice	Saudi Arabia	Seed, dried; oral; hot H_2O extract	Al-Yahya 1986
Jaundice	Unspecified	Unspecified	Duke 1992–2016, 2002
Spleen disease	Unspecified	Unspecified	Granot 1994

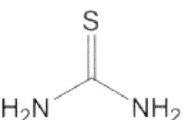

FIGURE 7.16   Thiourea (hepatotoxin).

Thiourea is a carcinogen. Its effects on body weight, thyroid function, and endo-crine cancer parameters have been recorded. Hepatoprotection was related to aspar-tate aminotransferase (AST), alanine aminotransferase (ALT), and serum bilirubin. Since both ethanolic and chloroform *harmal* extracts were able to revert neuron-specific enolase (NSE) and thyroglobulin (TG) levels to normal following their post-thiourea elevation, the hepatoprotective potency of *harmal* was proven.

In an independent investigation conducted in Tunisia, Bourogaa et al. (2015), also evaluated a *harmal* extract as putative hepatoprotector in rat, in this case after expo-sure to ethanol in water 35% (4 g/kg/day) for 6 weeks. Some animals also received the *harmal* extract by i.p. injection, 10 mg/kg, daily, and control rats received i.p. saline. The chronic ethanol exposure increased lipid peroxidation as measured by increased thiobarbituric acid reactive substances (TBARS) in liver, with concurrent disturbances in antioxidant defense, hepatic superoxide dismutase, catalase, and glu-tathione peroxidase. *Harmal* injection blocked the lipid peroxidation and attenuated the antioxidant disturbance.

ANTITUSSIVE

Liu et al. (2015) in Shanghai explored the antitussive, expectorant, and bronchodilat-ing effects of a 50% ethanolic *harmal* extract and alkaloid-rich fraction of the aerial parts of *P. harmala in vitro*, guided by their observation that these are the most important medicinal uses of the plant according to traditional Uighur medicine in their region, especially for asthma and cough (Table 7.11).

Cough models in mice and guinea pigs employed ammonia liquor, capsaicin, and citric acid to induce cough. Phenol red secretion experiments were performed in mice to assess expectoration. The *harmal* extract suppressed cough in these settings as effectively as codeine, suggesting its use as a lead compound for treatment of respiratory disease.

ANTI-INFLAMMATORY, ANTI-AGING

Bremner et al. (2009) scrutinized 64 plant drugs from 61 different plant species orig-inating in southeastern Spain for their anti-inflammatory activity, especially through their effect on the transcription factor NF-κ-B and the cytokine TNF-α. *P. harmala* extract suppressed cytokine TNF-α at a dose of 10 µg/ml (Table 7.12).

TABLE 7.11
Modern Ethnomedical Uses of *Peganum harmala*: Respiratory System

Indication	Geography	Part Used	References
Antitussive	Pakistan	Entire plant; smoked	Leporatti and Lattanzi 1994
Antitussive	Algeria/North Africa	Unspecified	Hammiche and Merad 1997
Bronchodilator	Saudi Arabia	Seed, dried; oral; hot H_2O extract	Al-Yahya 1986

TABLE 7.12
Modern Ethnomedical Uses of *Peganum harmala*: Inflammation-Related

Indication	Geography	Part Used	References
Abscesses	Pakistan	Dried flower and leaf; external	Leporatti and Lattanzi 1994
Anti-inflammatory	Pakistan	Entire plant; smoked	Leporatti and Lattanzi 1994
Anti-inflammatory	Saudi Arabia	Seed, dried; oral; hot H_2O extract	Al-Yahya 1986
Antimicrobial	Iraq	Seed, dried; oral; hot H_2O extract	Rashan et al. 1989
Antiperiodic	Pakistan	Seed	Ghafoor 1985
Antiseptic	Unspecified	Seed; incense	Hanelt et al. 2001
Antiseptic	Pakistan	Seeds & leaves; external; smoke	Ghafoor 1985
Antiseptic on skin	Algeria/North Africa	Unspecified	Hammiche and Merad 1997
Arthritis	Unspecified	Whole plant	Burkill 1985
Articular pains	Libya	Dried entire plant	Hussain and Tobji 1997
Articular pains	North Africa	Powdered seeds; externally; mixed with honey and ginger; rubbed on skin	Boulos 1983
Joint pains	Israel	Unspecified; mostly externally	Wild Flowers of Israel n.d.
Asthma	Egypt	Seed	Ayensu 1978; Al-Awdat and Laham 1994; Boulos 1983 via MEDUSA n.d.
Asthma	India	Seed; oral; hot H_2O extract	Schipper and Volk 1960
Asthma	Algiers	Seed; oral; in small doses)	Boulos 1983
Asthma	Pakistan	Seed powder	Ghafoor 1985
Asthma	India	Seed, dried; oral; powder given with hot water	Singh 1995
Asthma	India/Pakistan	Unspecified	Hassan 1967
Asthma	Unspecified	Unspecified	Duke 1992–2016, 2002
Bactericide	India	Seed	Abrol and Chopra 1962
Blepharitis	Algeria/North Africa	Unspecified	Hammiche and Merad 1997
Burns	Algeria/North Africa	Unspecified	Hammiche and Merad 1997
Cancer, uterine	Unspecified	Unspecified	Hartwell 1967–1971
Colic	Egypt	Seed	Ayensu 1978; Al-Awdat and Laham 1994; Boulos 1983 via MEDUSA n.d.

(Continued)

TABLE 7.12 (CONTINUED)
Modern Ethnomedical Uses of *Peganum harmala*: Inflammation-Related

Indication	Geography	Part Used	References
Colic	Pakistan	Seed powder	Ghafoor 1985
Colic	Algeria/North Africa	Unspecified	Hammiche and Merad 1997
Colic	Unspecified	Unspecified	Duke 1992–2016, 2002
Conjunctivitis, purulent	North Africa	Dried powdered plant	Boulos 1983
Conjunctivitis, suppurative	Algeria/North Africa	Unspecified	Hammiche and Merad 1997
Dermatitis (eczema)	Algeria/North Africa	Unspecified	Hammiche and Merad 1997
Eye inflammation	Israel	Unspecified; mostly external	Wild Flowers of Israel. n.d.
Infectious eye diseases	North Africa	Seed, oil extracted from	Boulos 1983
Folliculitis	Unspecified	Unspecified	Granot 1994
Genito-urinary system disorders	Morocco	Unspecified	Pelt 1971; Bellakhdar 1997 via MEDUSA n.d.
Gingivitis	Morocco	Root; macerated with vinegar; gargarism	Bellakhdar 1997 via MEDUSA n.d.
Headache	Pakistan	Entire plant; smoked	Leporatti and Lattanzi 1994
Headache	North Africa	Plant, burnt; inhalation	Boulos 1983
Hemorrhoids	Tunisia	Plant	Enquête ethnobotanique réalisée en Tunisie [P.N.M. 93] via MEDUSA n.d.
Hemorrhoids	North Africa	Plant; oral; decoction in oil taken first thing in the morning	Boulos 1983
Hemorrhoids	Libya	Dried entire plant; oral	Hussain and Tobji 1997
Hemorrhoids	Turkey	Root; oral: decoction	Tabata et al. 1994
Hemorrhoids	Egypt	Seed; external	Emboden 1979; Bown 1995; Phillips and Rix 1991 via MEDUSA n.d.
Hemorrhoids	Algeria/North Africa	Unspecified	Hammiche and Merad 1997
Hemorrhoids	Israel	Unspecified; mostly external	Wild Flowers of Israel. n.d.
Hemorrhoids	Unspecified	Unspecified	Granot 1994
Laryngitis	Pakistan	Seed, decoction; oral	Ghafoor 1985
Laryngitis	Unspecified	Unspecified	Duke 1992–2016, 2002
Malaria	Egypt	Seed	Ayensu 1978; Al-Awdat and Laham 1994; Boulos 1983 via MEDUSA n.d.

(Continued)

TABLE 7.12 (CONTINUED)
Modern Ethnomedical Uses of *Peganum harmala*: Inflammation-Related

Indication	Geography	Part Used	References
Malaria	Algeria/North Africa; Iraq	Unspecified	Al-Rawi and Chakravarty 1964; Hammiche and Merad 1997
Malaria	Unspecified	Unspecified	Duke 1992–2016, 2002; Steinmetz 1957
Malaria, chronic, reducing temperature in	Pakistan	Seed powder	Ghafoor 1985
Neuralgia	Egypt	Seed	Ayensu 1978; Al-Awdat and Laham 1994; Boulos 1983 via MEDUSA n.d.
Otitis	Tunisia	Seed, dried; external	Boukef et al. 1982
Prostatitis, stopping involuntary urination in	Turkey	Seed, dried; oral	Tabata et al. 1994
Revulsive	North Africa	Fresh branches	Boulos 1983
Rheumatic pains	North Africa	Seed, oil extracted from	Boulos 1983
Rheumatism	India	Leaf; oral; decoction	Singh et al. 1996
Rheumatism	Tunisia	Plant	Enquête ethnobotanique réalisée en Tunisie [P.N.M. 93] via MEDUSA n.d.
Rheumatism	Unspecified	Whole plant	Burkill 1985
Rheumatism	North Africa	Plant, fresh; external; rubbing in sheep's fat	Boulos 1983
Rheumatism	Egypt	Seed	Ayensu 1978; Al-Awdat and Laham 1994; Boulos 1983 via MEDUSA n.d.
Rheumatism	India	Seed; oral	Shah 1982
Rheumatism	India	Seed; oral; hot H_2O extract	Schipper and Volk 1960
Rheumatism	Morocco	Seed; external	Bellakhdar et al. 1991
Rheumatism	Tunisia	Seed, dried; external	Boukef et al. 1982
Rheumatism	North Africa	Powdered seeds; externally; mixed with honey and ginger; rubbed on skin	Boulos 1983
Rheumatism	Unspecified	Unspecified	Duke 1992–2016, 2002; Uphof 1968
Rhinitis	Unspecified	Seed	Granot 1994
Sciatica	North Africa	Seed; infusion	Boulos 1983
Throat inflammation	Pakistan	Dried flower and leaf; external	Leporatti and Lattanzi 1994

(Continued)

TABLE 7.12 (CONTINUED)
Modern Ethnomedical Uses of *Peganum harmala*: Inflammation-Related

Indication	Geography	Part Used	References
Toothache	North Africa	Pounded roots and seeds mixed with tobacco; smoked in pipes	Boulos 1983
Toothache	Unspecified	Seed	Granot 1994
Toothache	Israel	Unspecified	Wild Flowers of Israel. n.d.
Tooth pain	Unspecified	Unspecified	Granot 1994
Urinary disorders	Egypt	Seed oil	Emboden 1979; Bown 1995; Phillips and Rix 1991 via MEDUSA n.d.
Wound healing	Algeria/North Africa	Unspecified	Hammiche and Merad 1997
Wounds & injuries	Morocco	Unspecified	Pelt 1971; Bellakhdar 1997 via MEDUSA n.d.
Wounds	Pakistan	Seeds and leaves; external; fumigation by burning seeds and leaves	Ghafoor 1985

Khadhr et al. (2016) studied the seed oil of Tunisian *P. harmala* and found it to possess γ-tocopherol (573.66 µg/g) and linoleic acid (62.5%) (Figures 7.17 and 7.18). When formulated into a 20% cream and applied to the plantar aspect of rat paw, it effected a reduction in inflammation, 5 h following carrageenan application, superior to that of diclofenac 1%. A small peripheral analgesia from cream applications was also observed.

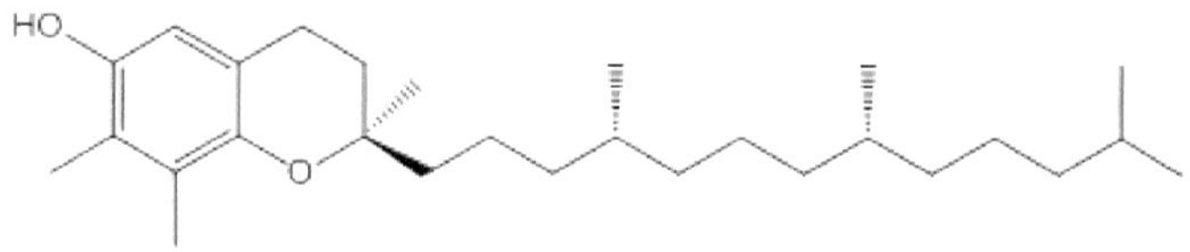

FIGURE 7.17   γ-Tocopherol.

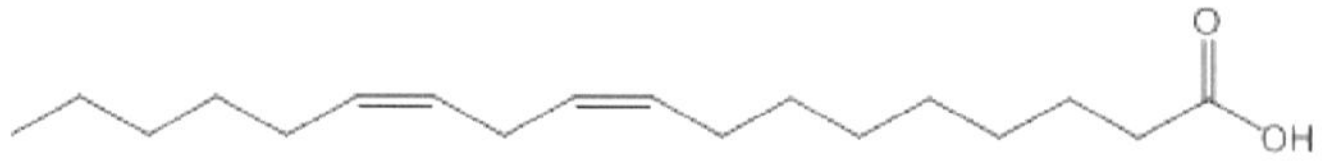

FIGURE 7.18   Linoleic acid.

Bensalem et al. (2014), from Algeria and Belgium, looked at inhibition *in vitro* by harmala alkaloids extract of myeloperoxidase, a lysosomal enzyme released during the inflammatory response from granules of neutrophils. Its inhibition may constitute a mechanism for curbing inflammation. Extracts from *P. harmala* seeds, roots, and aerial parts were tested, with the seed extract being the most potent inhibitor of myeloperoxidase (Figure 7.19).

Mohammadirad et al. (2013) induced aging artificially in male BALB/c mice by daily administration of 500 mg/kg for 6 weeks of D-galactose, a sugar known to cause cognitive and physical symptoms of accelerated aging (Parameshwaran et al. 2010). The animals also received for 4 weeks daily doses of one of the following herbs according to the associated mg/kg: *Zingiber officinale* (250), *Glycyrrhiza glabra* (150), *Rosmarinus officinalis* (300), *Peganum harmala* (50), *Aloe vera* (150), *Satureja hortensis* (200), *Teucrium scordium* (200), *Hypericum perforatum* (135), or *Silybum marianum* (150) or sham. Proinflammatory markers such as tumor necrosis factor-α (TNF-α), interleukin-1β (IL-β), interleukin-6 (IL-6), NF-kappaB (NF-κb), total antioxidant power (TAP), thiobarbituric acid reactive substances (TBARS) as lipid peroxidation (LPO) marker, testosterone, and dehydroepiandrosterone-sulfate (DHEA-S) were quantified in blood at the conclusion of treatment. D-galactose dosing resulted in significant oxidative stress and promoted an inflammatory aging cascade that was largely reversed

FIGURE 7.19 Stems of *Peganum harmala* rising from a woody rhizome. *Peganum* root has been used for toothache and gingivitis. (2/25/2015, Mount Scopus Botanical Garden, Jerusalem, Israel: by Helena Paavilainen.)

by each of the herbal extracts at the doses given, with *Silybum marianum* (wild artichoke) most effective in overturning the aging changes.

Hamden et al. (2008b) investigated the effects of 17β-estradiol (E2), *P. harmala* extract, and 40% caloric restriction on various testis parameters during aging in 12-month-old male rats over a 6-month period. Estrogen (E2) as well as caloric restriction protected the testis from a decrease in testosterone. The *harmal* extract, as well as caloric restriction or E2, increased superoxide dismutase activity and decreased activity of testicular enzymes γ-glutamyl transferase, alkaline phosphatase, and lactate dehydrogenase, while increasing activity of testicular catalase. *Harmal* and caloric restriction also resulted in lower lipid peroxidation and recovery of spermatogenesis in the treated animals. In a subsequent work (Hamden et al. 2009) the group was able to replicate its results and suggest that the anti-aging mechanisms, antioxidant activity, and hepatoprotection of both *harmal* and caloric restriction possibly could be attributed to their ability to increase E2 level, which as an antioxidant acts as a scavenger of ROS.

IMMUNOPOTENTIVE

Immunological reactability is an important index of health. Improving such reactivity is often a key goal of medical intervention. Toghyani et al. (2015) developed a model for assessing immunological strength in broiler chicks, and the effect of *harmal* seed or seed extracts on this index. A total of 350 chicks were allocated to groups of 14, each of which received a dietary treatment such as control, 1 or 2 g/kg *harmal* seed in diet, or 100 or 200 mg/l *harmal* seed extract in water. Broilers received dietary treatments for 1–42 days, after which two birds from each pen were randomly weighed and sacrificed at 42 days of age. *Harmal* treatments did not affect the relative weights of lymphoid organs (bursa of Fabricius and spleen), but antibody titer against Newcastle and influenza viruses, as well as sheep red blood cell antigen, were significantly ($p < 0.05$) enhanced by feeding *harmal* seed or extract. In conclusion, provision of *harmal* seed and extract in diet enhanced immunological competence in broiler chicks.

A related study by Dawood and Qubih (2012) at the University of Mosul, Iraq, studied 140 chicks receiving a vaccine against Marek's disease. The vaccinated animals suffered various insults such as congestion and enlargement of organs and hemorrhage that appeared to be at least partially mollified in the animals receiving also *harmal* seed or extract in the diet.

CARDIOVASCULAR

Although direct pharmacognostical studies of *P. harmala* on cardiovascular parameters are lacking, a survey of folk usage in Morocco, Italy, Tunisia, and Jordan, and modern studies on its principal components, led by inference to the reasonable conclusion that *P. harmala* may help correct hypertension (Table 7.13).

TABLE 7.13

Modern Ethnomedical Uses of *Peganum Harmala*: Heart and Blood Circulation

Indication	Geography	Part Used	References
Cardiac diseases	North Africa	Seed; infusion	Boulos 1983
Dropsy in the feet	Persia	Unspecified; fomentation	Baillon 1871–1888, p. 451
Hypertension, for	Algeria/North Africa	–	Hammiche and Merad 1997
Hypertension, arterial	Tunisia	Plant	Enquête ethnobotanique réalisée en Tunisie [P.N.M. 93] via MEDUSA n.d.

Further, *P. harmala* extracts possess anti-adrenergic, anticholinergic, antihistaminic, and antispasmodic properties, which influenced the cardiovascular system, including also bradycardic effects (Moloudizargari et al. 2013). Ethnobotanical assessments in Morocco also confirmed folk use of *harmal* for treating hypertension and cardiac diseases (Eddouks et al. 2002). Obviously, basic laboratory *in vitro* studies and also clinical research on *harmal* and hypertension are needed.

MISCELLANEOUS

In Jordan, Aqel and Hadidi (1991) showed a direct relaxant effect of a *P. harmala* aqueous seed extract *ex vivo* on isolated muscle contractions of rabbit or guinea pig intestine, trachea, aorta, jejunum, and ileum induced by a 10^{-4} power dose of acetylcholine, histamine, or noradrenaline, thus highlighting the antispasmodic potential of *harmal*.

In Baghdad, Nessrian Ali Hussien (2015) studied concentrations of natural radioactivity in *P. harmala* seeds from throughout Baghdad province using HPGegamma-ray spectroscopy established at low background configuration. Activity concentrations of 40K (1460.81 keV), 75Se (96.73 keV), 85Kr (513.99 keV), 88Y (1836.01 keV), and 134Cs (563.23, 801.93, and 1167.94 keV) were determined. The average activity concentrations of radioactivity of 40K, 75Se, 85Kr, and 134Cs were 425 ± 12.22, 350 ± 10.05, 304 ± 8.73, and 159.17 ± 4.57 Bq/kg, respectively, within international limits. Trace elements in descending order were K > Fe > Ti > Zr > Nb > Sr > Y > Zn > Ga. The findings confirmed the safety of *harmal* as a medicine for the future. The author also pointed out that "[in] the Iraqi tradition different utilizations of *harmal* have been defined, among them, dried grain from this plant are baked in homes to protect against abuse (evil) eye and applied as an incense to sanctify the air as well as the intellect" (Table 7.14).

TABLE 7.14
Modern Ethnomedical Uses of *Peganum harmala*: Miscellaneous

Indication	Geography	Part Used	References
Brings blessing to the inhabitants of the house	Israel	Whole plant; hanged on the entrance of the house	Wild Flowers of Israel. n.d.
Dye	Unspecified	Root and seed and stem	Burkill 1985
Dye, red, production	Middle East	Unspecified	CDFA
Dye ("Turkey red"), produced from the seeds, used for dyeing hats (tarbooshes) and carpets	Unspecified	Seed	FOC; Granot 1994; Hanelt et al. 2001; McLean and Ivimey-Cook 1956 via Hammiche and Merad 1997
Dye, red, for the coloring of carpets	Western Asia/ Egypt	Seed	Ayensu 1978; Al-Awdat and Laham 1994; Boulos 1983 via MEDUSA n.d.; Uphof 1959; Usher 1974; Singh and Kachroo 1976; Emboden 1979 via MEDUSA n.d.
Exorcism agent	Israel	Unspecified	Granot 1994
Food: as spice	Egypt	Seed	Ayensu 1978; Al-Awdat and Laham 1994; Boulos 1983 via MEDUSA n.d.; Uphof 1959; Usher 1974; Kunkel 1984; Facciola 1990 via MEDUSA n.d.
Condiment	Unspecified	Seed	Baillon 1871–1888, p. 451; Hanelt et al. 2001
Food: edible oil	Egypt	Seed oil	Uphof 1959; Usher 1974; Kunkel 1984; Facciola 1990 via MEDUSA n.d.
Oil, edible, similar to cottonseed oil	Unspecified	Seed, dehulled, oil of	CDFA
Frictions and applications (purpose not mentioned)	Morocco	Fresh sap from the branches; externally; Seed, roasted and crushed; alone or in a mixture; internally or externally	Bellakhdar 1997 via MEDUSA n.d.
Fumigant	Unspecified	Unspecified	Duke 1992–2016, 2002

(Continued)

TABLE 7.14 (CONTINUED)
Modern Ethnomedical Uses of *Peganum harmala*: Miscellaneous

Indication	Geography	Part Used	References
Incense	Egypt	Seed	Ayensu 1978; Al-Awdat and Laham 1994; Boulos 1983 via MEDUSA n.d.; Uphof 1959; Usher 1974; Singh and Kachroo 1976; Emboden 1979 via MEDUSA n.d.
Ink	Unspecified	Root and seed and stem	Burkill 1985
Keeps evil eye away	Israel	Whole plant; hanged on the entrance of the house	Wild Flowers of Israel. n.d.
Magic use	Balochistan	Seed; burned over hot coals; inhalation	Goodman and Ghafoor 1992
Magic: source of a magical medicinal oil (Zet el harmal)	Unspecified	Seed	Hanelt et al. 2001
Oil "zit-el-harmel," preparation of	Algiers	Fruit?	Winkler
Magic, prophylaxis against harm caused by	Algeria/North Africa	Unspecified	Hammiche and Merad 1997
Mordant	Unspecified	Root and seed and stem	Burkill 1985
Ritual narcotic	Unspecified	Seed	Hanelt et al. 2001
Seeds, trade item	Unspecified	Seed	Hammiche and Merad 1997; Hanelt et al. 2001
Social: hallucinogens	Turkey	Seed	Ertug 2000 via MEDUSA n.d.
Social: narcotics	Egypt	Plant and leaf	Ayensu 1978; Al-Awdat and Laham 1994; Boulos 1983 via MEDUSA n.d.
Social: religion, superstitions, magic	Unspecified	Plant and seed	Burkill 1985
Social: religious uses	Egypt	Plant and leaf	Ayensu 1978; Al-Awdat and Laham 1994; Boulos 1983 via MEDUSA n.d.
Social: religious uses	Morocco	Seed	Pelt 1971; Bellakhdar 1997 via MEDUSA n.d.
Social: smoking materials	Egypt	Plant and leaf	Ayensu 1978; Al-Awdat and Laham 1994; Boulos 1983 via MEDUSA n.d.

(*Continued*)

TABLE 7.14 (CONTINUED)
Modern Ethnomedical Uses of *Peganum harmala*: Miscellaneous

Indication	Geography	Part Used	References
Social: smoking materials	Morocco/Turkey	Seed	Pelt 1971; Bellakhdar 1997 via MEDUSA n.d.; Ertug 2000 via MEDUSA n.d.
Suppresses excess saliva discharge during sleep	Balochistan	Seed, 2–3 orally before bedtime	Goodman and Ghafoor 1992
Tattoos	Unspecified	Root and seed & stem	Burkill 1985
Zoroastrian drug "haoma," candidate for	Egypt	Seed	Ayensu 1978; Al-Awdat and Laham 1994; Boulos 1983 via MEDUSA n.d.

REFERENCES

Abedi Gaballu, F., Y. Abedi Gaballu, O. Moazenzade Khyavy, A. Mardomi, K. Ghahremanzadeh, B. Shokouhi, and H. Mamandy. 2015. Effects of a triplex mixture of *Peganum harmala*, *Rhus coriaria*, and *Urtica dioica* aqueous extracts on metabolic and histological parameters in diabetic rats. *Pharm Biol* 53(8): 1104–9.

Aboualigalehdari, E., N. Sadeghifard, M. Taherikalani, Z. Zargoush, Z. Tahmasebi, B. Badakhsh, A. Rostamzad, S. Ghafourian, and I. Pakzad. 2016. Anti-biofilm properties of Peganum harmala against Candida albicans. *Osong Public Health Res Perspect* 7(2): 116–8.

Abrol, B.K., and I.C. Chopra. 1962. Some vegetable drug resources of Ladakh (Little Tibet). Part I. *Curr Sci* 31: 324–6.

Ahangarpour, A., S.A. Najimi, and Y. Farbood. 2016. Effects of *Vitex agnus-castus* fruit on sex hormones and antioxidant indices in a d-galactose-induced aging female mouse model. *J Chin Med Assoc* 79(11): 589–96.

Alami, R., A. Macksad, and A.R. El-Gindy. 1976. *Medicinal plants in Kuwait*. Kuwait: Al-Assiriya Printing Press.

Al-Awdat, M., and G. Laham. 1994. *Al-Nabatat al-tibbiyah wa-sti'malatuha*. [Medicinal plants and their uses.] Damascus: Al-Ahally Publ.

Aldhebiani, A.Y., E.K.F. Elbeshehy, A. Areej, A.A. Baeshen, and T. Elbeaino. 2015. Inhibitory activity of different medicinal extracts from Thuja leaves, ginger roots, Harmal seeds and turmeric rhizomes against Fig leaf mottle-associated virus 1 (FLMaV-1) infecting figs in Mecca region. *Saudi J Biol Sci*. http://dx.doi.org/10.1016/j.sjbs.2015.11.005. Available online 10 November 2015.

Al-Khalil, S. 1995. A survey of plants used in Jordanian traditional medicine. *Int J Pharmacog* 33(4): 317–23.

Alkofahi, A., H. Masaadeh, and S. Al-Khalil. 1996. Antimicrobial evaluation of some plant extracts of traditional medicine of Jordan. *Alex J Pharm Sci* 10(2): 123–6.

Al-Rawi, A., and H.L. Chakravarty. 1964. *Medicinal Plants of Iraq*. Min Agr Techn Bull no 15. Baghdad: Govt Press.

Al-Saikhan, F.I., and M.N. Ansari. 2016. Evaluation of the diuretic and urinary electrolyte effects of methanolic extract of *Peganum harmala* L. in Wistar albino rats. *Saudi J Biol Sci* 23(6): 749–53.

Al-Yahya, M.A. 1986. Phytochemical studies of the plants used in traditional medicine of Saudi Arabia. *Fitoterapia* 57(3): 179–82.

Aqel, M., and M. Hadidi. 1991. Direct relaxant effect of *Peganum harmala* seed extract on smooth muscles of rabbit and guinea pig. *Pharm Biol* 29: 176–82.

Ayensu, E.S. 1978. *Medicinal Plants of West Africa*. Algonac: Reference Publications.

Bachrach, Z.Y. 2007. Ethnobotanical studies of *Sarcopoterium spinosum* in Israel. *Israel Journal of Plant Sciences* 55(1): 111–4.

Bellakhdar, J. 1997. *La pharmacopée traditionnelle marocaine: Médecine arabe ancienne et savoirs populaires*. Paris: Ibis Press.

Bellakhdar, J., R. Claisse, J. Fleurentin, and C. Younos. 1991. Repertory of standard herbal drugs in the Moroccan pharmacopoea. *J Ethnopharmacol* 35(2): 123–43.

Bensalem, S., J. Soubhye, I. Aldib, L. Bournine, A.T. Nguyen, M. Vanhaeverbeek, A. Rousseau, K.Z. Boudjeltia, A. Sarakbi, J.M. Kauffmann, J. Nève, M. Prévost, C. Stévigny, F. Maiza-Benabdesselam, F. Bedjou, P. Van Antwerpen, and P. Duez. 2014. Inhibition of myeloperoxidase activity by the alkaloids of *Peganum harmala* L. (*Zygophyllaceae*). *J Ethnopharmacol* 154(2): 361–9.

Biradar, S.M., H. Joshi, and K.C. Tarak. 2013. Cerebroprotective effect of isolated harmine alkaloids extracts of seeds of *Peganum harmala* L. on sodium nitrite-induced hypoxia and ethanol-induced neurodegeneration in young mice. *Pak J Biol Sci* 16(23): 1687–97.

Boukef, K., H.R. Souissi, and G. Balansard. 1982. Contribution to the study on plants used in traditional medicine in Tunisia. *Plant Med Phytother* 16(4): 260–79.

Boulos, L. 1983. *Medicinal Plants of North Africa*. Algonac: Reference Publications.

Bourogaa, E., R.M. Jarraya, M. Damak, and A. Elfeki. 2015. Hepatoprotective activity of *Peganum harmala* against ethanol-induced liver damages in rats. *Arch Physiol Biochem* 121(2): 62–7.

Bown, D. 1995. *Encyclopaedia of Herbs and their Uses*. London: Dorling Kindersley.

Bremner, P., D. Rivera, M.A. Calzado, C. Obón, C. Inocencio, C. Beckwith et al. 2009. Assessing medicinal plants from South-Eastern Spain for potential anti-inflammatory effects targeting nuclear factor-Kappa B and other pro-inflammatory mediators. *J Ethnopharmacol* 124: 295–305.

Burkill, H.M. 1985. *The Useful Plants of West Tropical Africa*. Vol. 5. Kew, UK: Royal Botanic Gardens.

Burkill, I.H. 1909. *A Working List of the Flowering Plants of Baluchistan*. Calcutta: Superintendent Government Printing.

Cao, R., W. Peng, H. Chen, Y. Ma, X. Liu, X. Hou, H. Guan, and A. Xu. 2005. DNA binding properties of 9-substituted harmine derivatives. *Biochem Biophys Res Commun* 338(3): 1557–63.

CDFA. California Department of Food and Agriculture. 2015. [Online database.] https://www.cdfa.ca.gov/plant/ipc/encycloweedia/weedinfo/peganum.htm (accessed May 15 2015).

Cheuka, P.M., G. Mayoka, P. Mutai, and K. Chibale. 2016. The role of natural products in drug discovery and development against neglected tropical diseases. *Molecules* 22(1): E58.

Chopra, R.N. 1933. *Indigenous Drugs of India: Their medical and economic aspects*. Calcutta, India: The Art Press.

Chopra, R.N., R.L. Badhwar, and S. Ghosh. 1949. *Poisonous Plants of India*. Calcutta: Manager of Publications, Government of India Press. Volume 1.

Cirano, F.R., R.C. Casarin, F.V. Ribeiro, M.Z. Casati, S.P. Pimentel, T. Taiete, and M.M. Bernardi. 2016. Effect of Resveratrol on periodontal pathogens during experimental periodontitis in rats. *Braz Oral Res* 30(1): e128.

Dastagir, G., and F. Hussain. 2014. Cytotoxic activity of plants of family *Zygophyllaceae* and *Euphorbiaceae*. *Pak J Pharm Sci* 27(4): 801–5.

Dawood, Z.A., and T.S. Qubih. 2012. Effect of *Peganum harmala* on histological reactions after post Marek's disease vaccination in layer hens. *Iraqi J Vet Sci* 26(4): 339–46.

DORA. Diseases of Research Animals. 2013. [Online database.] Curators of the University of Missouri. University of Missouri. Comparative Medicine Program and IDEXX-BioResearch. http://dora.missouri.edu/ (accessed May 15 2015).

Dragendorff, G. 1898. *Die Heilpflanzen der verschiedenen Volker und Zeiten.* Stuttgart: F. Enke.

Duke, J.A. 1992–2016. Dr. Duke's Phytochemical and Ethnobotanical Databases. [Online database.] U.S. Department of Agriculture, Agricultural Research Service. Home Page, http://phytochem.nal.usda.gov/http://dx.doi.org/10.15482/USDA.ADC/1239279 (accessed March 10 2017).

Duke, J.A. 2002. *Handbook of Phytochemical Constituents of GRAS Herbs and Other Economic Plants.* Boca Raton: CRC Press.

Eddouks, M., M. Maghrani, A. Lemhadri, M.L. Ouahidi, H. Jouad. 2002. Ethnopharmacological survey of medicinal plants used for the treatment of diabetes mellitus, hypertension and cardiac diseases in the south-east region of Morocco (Tafilalet). *J Ethnopharmacol* 82(2–3): 97–103.

Edziri, H., M. Mastouri, M. Matieu, M. Zine, L. Gutman, and M. Aouni. 2010. Biological activities of *Peganum harmala* leaves. *Afr J Biotechnol* 9(48): 8199–205.

Emboden, W. 1979. *Narcotic Plants.* New York: Macmillan.

Ertug, F. 2000. An ethnobotanical study in Central Anatolia (Turkey). *Econ Bot* 54(2): 155–82.

Facciola, S. 1990. *Cornucopia—A Source Book of Edible Plants.* Vista: Kampong Publication.

Farouk, L., A. Laroubi, R. Aboufatima, A. Benharref, and A. Chait. 2008. Evaluation of the analgesic effect of alkaloid extract of *Peganum harmala* L.: Possible mechanisms involved. *J Ethnopharmacol* 115(3): 449–54.

Fathiazada, F., Y. Azarmib, and L. Khodai. 2006. Pharmacological effects of *Peganum harmala* seeds extract on isolated rat uterus. *Iranian Journal of Pharmaceutical Sciences* 2(2): 81–6.

Fattahi, M., R.M. Cusido, A. Khojasteh, M. Bonfill, and J. Palazon. 2014. Xanthomicrol: A comprehensive review of its chemistry, distribution, biosynthesis and pharmacological activity. *Mini Rev Med Chem* 14(9): 725–33.

Font Quer, P. 1979. *Plantas medicinales: El Dioscórides renovado.* Barcelona: Editorial Labor.

Friedman, J., Z. Yaniv, A. Dafni, and D. Palevitch. 1986. A preliminary classification of the healing potential of medicinal plants, based on a rational analysis of an ethnopharmacological field survey among Bedouins in the Negev Desert, Israel. *J Ethnopharmacol* 16(2–3): 275–87.

Ghafoor, A. 1985. Zygophyllaceae. In Flora of Pakistan, eFloras. St. Louis: Missouri Botanical Garden; Cambridge, MA: Harvard University Herbaria. http://www.efloras.org (accessed July 24 2014).

Goodman, S.M., and A. Ghafoor. 1992. *The Ethnobotany of Southern Balochistan, Pakistan: With particular reference to medicinal plants.* Chicago: Field Museum of Natural History.

Granot, Y. 1994. *Medical Plants of the Negev.* Midreshet Sde Boker. Interdisciplinary Education Center. http://www.boker.org.il/learning/?http://www.boker.org.il/meida/negev/meida.htm.

Hamden, K., D. Silandre, C. Delalande, A. Elfeki, and S. Carreau. 2008b. Protective effects of estrogens and caloric restriction during aging on various rat testis parameters. *Asian J Androl* 10(6): 837–45.

Hamden, K., H. Masmoudi, F. Ellouz, A. El Feki, and S. Carreau. 2008a. Protective effects of *Peganum harmala* extracts on thiourea-induced diseases in adult male rat. *J Environ Biol* 29: 73–7.

Hamden, K., S. Carreau, F. Ayadi, H. Masmoudi, and A. El Feki. 2009. Inhibitory effect of estrogens, phytoestrogens, and caloric restriction on oxidative stress and hepato-toxicity in aged rats. *Biomed Environ Sci* 22(5): 381–7.

Hammiche, V., and R. Merad. 1997. *Peganum harmala*. International Programme on Chemical Safety Poisons Information Monograph 402. http://www.inchem.org/docu ments/pims/plant/pim402fr.htm [In French.]

Hanelt, P., R. Büttner, and R. Mansfeld, eds. 2001. *Mansfeld's Encyclopedia of Agricultural and Horticultural Crops (except Ornamentals)*. Berlin: Springer.

Harlev, E., E. Nevo, E.P. Lansky, R. Ofir, and A. Bishayee. 2012. Anticancer potential of aloes: Antioxidant, antiproliferative, and immunostimulatory attributes. *Planta Med* 78(9): 843–52.

Hartwell, J.L. 1967–1971. Plants used against cancer: A survey. *Lloydia* 32(1): 78–107; 32(2): 153–205; 32(3): 247–96; 33(1): 97–194; 33(3): 288–392; 34(1): 103–60; 34(2): 204–55; 34(3): 310–61; 34(4): 386–425.

Hassan, I. 1967. Some folk uses of *Peganum harmala* in India and Pakistan. *Econ Bot* 21(3): 284.

Herraiz, T., D. González, C. Ancín-Azpilicueta, V.J. Arán, and H. Guillén. 2010. beta-Carboline alkaloids in *Peganum harmala* and inhibition of human monoamine oxidase (MAO). *Food Chem Toxicol* 48: 839–45.

Honda, G., E. Yesilada, M. Tabata, E. Sezik, T. Fujita, Y. Takeda, Y. Takaishi, and T. Tanaka. 1996. Traditional medicine in Turkey. VI. Folk medicine in West Anatolia: Afyon, Kutahya, Denizli, Mugla, Aydin Provinces. *J Ethnopharmacol* 53: 75–87.

Hussain, H., and R.S. Tobji. 1997. Antibacterial screening of some Libyan medicinal plants. *Fitoterapia* 68(5): 467–70.

Hussien, N.A. 2015. The concentrations of natural radioactivity in *Pergamum* (sic) *har-mala* seeds. *International Journal of Application or Innovation in Engineering & Management (IJAIEM)* 4(8): 45–50.

Irshaid, F.I., K.A. Tarawneh, J.H. Jacob, and A.M. Alshdefat. 2014. Phenol content, anti-oxidant capacity and antibacterial activity of methanolic extracts derived from four Jordanian medicinal plants. *Pak J Biol Sci* 17(3): 372–9.

Ivanovska, N., S. Philipov, and R. Istatkova. 1997. Evaluation of anti-inflammatory activity of plants used in Bulgarian folk medicines. *Fitoterapia* 68(5): 417–22.

Jahaniani, F., S.A. Ebrahimi, N. Rahbar-Roshandel, and M. Mahmoudian. 2005. Xanthomicrol is the main cytotoxic component of *Dracocephalum kotschyii* and a potential anti-cancer agent. *Phytochemistry* 66(13): 1581–92.

Jbilou, R., H. Amri, N. Bouayad, N. Ghailani, A. Ennabili, and F. Sayah. 2008. Insecticidal effects of extracts of seven plant species on larval development, alpha-amylase activ-ity and offspring production of *Tribolium castaneum* (Herbst) (*Insecta*: *Coleoptera*: *Tenebrionidae*) *Bioresour Technol* 99: 959–64.

Kakrani, H.K., and A.K. Saluja. 1993. Traditional treatment through herbal drugs in Kutch District, Gujarat State, India. Part I. Uterine disorders. *Fitoterapia* 65(5): 463–5.

Kamboj, V.P. 1988. A review of Indian medicinal plants with interceptive activity. *Indian J Med Res* 1988(4): 336–55.

Kamel, S.H., T.M. Ibrahim, A.A. Afifi, and S.M. Hamza. 1970. Chemical studies on the Egyptian plant *Peganum harmala*. *U A R J Vet Sci* 7(1): 61; *Biological Abstracts* 54 9840.

Kawaii, S., and E.P. Lansky. 2004. Differentiation-promoting activity of pomegranate (*Punica granatum*) fruit extracts in HL-60 human promyelocytic leukemia cells. *J Med Food* 7(1): 13–8.

Khadhr, M., D. Bousta, E.H. Hanane, L. El Mansouri, S. Boukhira, M. Lachkar, B. Jamoussi, and S. Boukhchina. 2016. HPLC and GC-MS analysis of Tunisian *Peganum harmala* seeds oil and evaluation of some biological activities. *Am J Ther* Apr 7 [Epub ahead of print].

Khoshzaban, F., F. Ghaffarifar, and H.R. Jamshidi Koohsari. 2014. *Peganum harmala* aqueous and ethanol extracts effects on lesions caused by *Leishmania major* (MRHO/IR/75/ER) in BALB/c mice. *Jundishapur J Microbiol* 7(7): e10992.

Kiani, S.J., M. Shamsi Shahrabadi, A. Ataei, and N. Sajjadi. 2008. *Peganum harmala* seed extract can prevent HSV-1 replication *in vitro*. *Iran J Virol* 1(4): 11–6.

Kikalishvili, B., D. Zurabashvili, Ts. Sulakvelidze, M. Malania, and D. Turabelidze. 2016. Study of lipids seed's oil of *Vitex agnus castus* growing in Georgia. *Georgian Med News* (256–257): 77–81.

Kima, P.E. 2007. The amastigote forms of *Leishmania* are experts at exploiting host cell processes to establish infection and persist. *Int J Parasitol* 37(10): 1087–96.

Komeili, G., M. Hashemi, and M. Bameri-Niafar. 2016. Evaluation of antidiabetic and anti-hyperlipidemic effects of *Peganum harmala* seeds in diabetic rats. *Cholesterol* 2016: 7389864.

Kunkel, G. 1984. *Plants for Human Consumption: An annotated checklist of the edible phanerogams and ferns.* Koenigstein: Koeltz Scientific Books.

Lansky, E.P., H.M. Paavilainen, A.D. Pawlus, and R.A. Newman. 2008. *Ficus* spp. (fig): Ethnobotany and potential as anticancer and anti-inflammatory agents. *J Ethnopharmacol* 119(2): 195–213.

Lansky, E.P., H.M. Paavilainen, and S. Lansky. 2011. *Figs: The Genus* Ficus. Boca Raton, FL: CRC Press.

Leporatti, M.L., and E. Lattanzi. 1994. Traditional phytotherapy on coastal areas of Makran (Southern Pakistan). *Fitoterapia* 65(2): 158–61.

Levine, N.D., and V. Ivens. 1965. *The coccidian parasites (*Protozoa, Sporozoa*) of rodents.* Urbana: University of Ilinois Press, 275. Via Internet Archive Book Images, https://www.flickr.com/photos/internetarchivebookimages/20474596349/.

Lewis, W.H., and M.P.F. Elvin-Lewis. 1977. *Medical botany.* New York: Wiley-Interscience.

Liu, W., X. Cheng, Y. Wang, S. Li, T. Zheng, Y. Gao, G. Wang, S. Qi, J. Wang, J. Ni, Z. Wang, and C. Wang. 2015. In vivo evaluation of the antitussive, expectorant and broncho-dilating effects of extract and fractions from aerial parts of *Peganum harmala* linn. *J Ethnopharmacol* 162: 79–86.

Malhi, B.S., and V.P. Trivedi. 1972. Vegetable antifertility drugs of India. *Q J Crude Drug Res* 12: 1922.

Manske, R.H.F., and H.L. Holmes. 1952. *The Alkaloids: Chemistry and Pharmacology.* New York: Academic Press, p. 393.

MEDUSA. The Medusa Database (http://medusa.maich.gr) and references contained therein. [n.d.]. Home page (accessed November 12 2012).

Merzouki, A., F. Ed Derfoufi, and J.M. Mesa. 2000. Hemp (*Cannabis sativa* L.) and abortion. *J Ethnopharmacol* 73: 501–3.

Mirzaei, M., S.J. Nosratabadi, A. Derakhshanfar, and I. Sharifi. 2007. Antileishmanial activity of *Peganum harmala* extract on the *in vitro* growth of *Leishmania major* promastigotes in comparison to a trivalent antimony drug. *Veterinarski Arhiv* 77: 365–75.

Mohammadirad, A., F. Aghamohammadali-Sarraf, S. Badiei, Z. Faraji, R. Hajiaghaee, M. Baeeri, M. Gholami, and M. Abdollahi. 2013. Anti-aging effects of some selected Iranian folk medicinal herbs-biochemical evidences. *Iran J Basic Med Sci* 16(11): 1170–80.

Moloudizargari, M., P. Mikaili, S. Aghajanshakeri, M.H. Asghari, and J. Shayegh. 2013. Pharmacological and therapeutic effects of *Peganum harmala* and its main alkaloids. *Pharmacogn Rev* 7(14): 199–212.

Monsef, H.R., A. Ghobadi, M. Iranshahi, and M. Abdollahi. 2004. Antinociceptive effects of *Peganum harmala* L. alkaloid extract on mouse formalin test. *J Pharm Pharm Sci* 7: 65–9.

Motamedifar, M., H. Khosropanah, and S. Dabiri. 2016. Antimicrobial activity of *Peganum harmala* L. on *Streptococcus mutans* compared to 0.2% chlorhexidine. *J Dent (Shiraz)* 17(3): 213–8.

Navchoo, I.A., and G.M. Buth. 1990. Ethnobotany of Ladakh, India: Beverages, narcotics, foods. *Econ Bot* 44(3): 318–21.

Nayar, S.L. 1954. Poisonous seeds of India. Part II. *J Bombay Nat Hist Soc* 52(2–3): 1–18.

Nayar, S.L. 1955. Vegetable insecticides. *Bull Natl Inst Sci India* 1955(4): 137–45.

Papaefthimiou, D., A. Papanikolaou, V. Falara, S. Givanoudi, S. Kostas, and A.K. Kanellis. 2014. Genus *Cistus*: A model for exploring labdane-type diterpenes' biosynthesis and a natural source of high value products with biological, aromatic, and pharmacological properties. *Front Chem* 2: 35.

Parameshwaran, K., M.H. Irwin, K. Steliou, and C.A. Pinkert. 2010. D-galactose effectiveness in modeling aging and therapeutic antioxidant treatment in mice. *Rejuvenation Res* 13(6): 729–35.

Paulo, A., and A.P. Francisco. 2016. Oncogene expression modulation in cancer cell lines by DNAG-quadruplex-interactive small molecules. *Curr Med Chem* Aug 29 [Epub ahead of print].

Pelt, J.M. 1971. *Drogues et plantes magiques.* Paris: Horizons de France.

Phillips, R., and M. Rix. 1991. *Perennials.* Vols.1–2. London: Pan Books.

Rahimifard, M., M. Navaei-Nigjeh, N. Mahroui, S. Mirzaei, Z. Siahpoosh, A. Nili-Ahmadabadi, A. Mohammadirad, M. Baeeri, R. Hajiaghaie, and M. Abdollahi. 2014. Improvement in the function of isolated rat pancreatic islets through reduction of oxidative stress using traditional Iranian medicine. *Cell J* 16(2): 147–163.

Rang, H.P., M.M. Dale, J.M. Ritter, and R.J. Flower. 2007. *Rang & Dale's Pharmacology.* 6th edition. Edinburgh: Churchill Livingstone.

Rani, A., and A. Sharma. 2013. The genus *Vitex*: A review. *Pharmacogn Rev* 7(14): 188–98.

Rashan, L.J., M.H. Adaay, and A.L.T. Al-Khazraji. 1989. *In vitro* antiviral activity of the aqueous extract from the seeds of *Peganum harmala*. *Fitoterapia* 60(4): 365–7.

Razzack, H.M.A. 1980. The concept of birth control in Unani medical literature. Unpublished manuscript of the author.

Rezaei, M., S. Nasri, M. Roughani, Z. Niknami, and S.A. Ziai. 2016. *Peganum harmala* L. extract reduces oxidative stress and improves symptoms in 6-hydroxydopamine-induced Parkinson's disease in rats. *Iran J Pharm Res* 15(1): 275–81.

Rosenzweig, T., G. Abitbol, and D. Taler. 2007. Evaluating the anti-diabetic effects of Sarcopoterium spinosum extracts *in vitro*. Israel Journal of Plant Sciences 55(1): 103–9.

Ross, S.A., S.E. Megalla, D.W. Bishay, and A.H. Awad. 1980. Studies for determining antibiotic substances in some Egyptian plants, Part II. Antimicrobial alkaloids from the seeds of *Peganum harmala* L. *Fitoterapia* 51(6): 309–12.

Saha, J.C., E.C. Savini, and S. Kasinathan. 1961. Ecbolic properties of Indian medicinal plants. Part 1. *Indian J Med Res* 49: 130–51.

Said, M. 1984. Potential of herbal medicines in modern medical therapy. *Ancient Sci Life* 4(1): 36–47.

Schipper, A., and O.H. Volk. 1960. The alkaloids of *Peganum harmala*. *Dtsch Apoth Ztg* 100: 255.

Sezik, E., E. Yeşilada, M. Tabata, G. Honda, Y. Takaishi, T. Fujita, T. Tanaka, and Y. Takeda. 1997. Traditional medicine in Turkey. VIII. Folk medicine in East Anatolia: Erzurum, Erzincan, Agri, Kars, Igdir Provinces. *Econ Bot* 51(3): 195–211.

Shang, X., X. Guo, B. Li, H. Pan, J. Zhang, Y. Zhang, and X. Miao. 2016. Microwave-assisted extraction of three bioactive alkaloids from *Peganum harmala* L. and their acaricidal activity against *Psoroptes cuniculi in vitro. J Ethnopharmacol* 192: 350–361.

Shapira, Z., J. Terkel, Y. Egozi, A. Nyska, and J. Friedman. 1989. Abortifacient potential for the epigeal parts of *Peganum harmala. J Ethnopharmacol* 27(3): 319–25.

Shen, Y., K. Lebold, E.P. Lansky, M.G. Traber, and E. Nevo. 2011. "Tocol-omic" diversity in wild barley, Short communication. *Chem Biodivers* 8: 2322–30.

Shoaib, M., S.W. Shah, N. Ali, I. Shah, S. Ullah, M. Ghias, M.N. Tahir, F. Gul, S. Akhtar, A. Ullah, W. Akbar, and A. Ullah. 2016. Scientific investigation of crude alkaloids from medicinal plants for the management of pain. *BMC Complement Altern Med* 16: 178.

Siler, B., S. Zivković, T. Banjanac, J. Cvetković, J. Nestorović Živković, A. Cirić, M. Soković, and D. Mišić. 2014. Centauries as underestimated food additives: Antioxidant and anti-microbial potential. *Food Chem* 147: 367–76.

Simonyan, K.V., and V.A. Chavushyan. 2016. Protective effects of hydroponic *Teucrium polium* on hippocampal neurodegeneration in ovariectomized rats. *BMC Complement Altern Med* 16(1): 415.

Singh, V., B.K. Kapahi, and T.N. Srivastava. 1996. Medicinal herbs of Ladakh especially used in home remedies. *Fitoterapia* 67(1): 38–48.

Sobhani, A.M., S.A. Ebrahimi, and M. Mahmoudian. 2002. An *in vitro* evaluation of human DNA topoisomerase I inhibition by *Peganum harmala* L. seeds extract and its beta-carboline alkaloids. *J Pharm Pharm Sci* 5(1): 19–23.

Steinmetz, E.F. 1957. *Codex vegetabilis.* Amsterdam: [s.n.].

Stout, G.H., and V.F. Stout. 1961. The structure and synthesis of xanthomicrol. *Tetrahedron* 14(3–4): 296–303.

Tanweer, A.J., N. Chand, U. Saddique, C.A. Bailey, and R.U. Khan. 2014. Antiparasitic effect of wild rue (*Peganum harmala* L.) against experimentally induced coccidiosis in broiler chicks. *Parasitol Res* 113(8): 2951–60.

Toghyani, M., A. Ghasemi, and S.A. Tabeidian. 2015. The effect of different levels of seed and extract of harmal (*Peganum harmala* L.) on immune responses of broiler chicks. *International Journal of Biological, Biomolecular, Agricultural, Food and Biotechnological Engineering* 9(1): 31–4.

Tuluce, Y., H. Ozkol, I. Koyuncu, and H. Ine. 2011. Gastroprotective effect of small centaury (*Centaurium erythraea* L) on aspirin-induced gastric damage in rats. *Toxicol Ind Health* 27(8): 760–8.

Uphof, J.C.Th. 1959. *Dictionary of Economic Plants.* Weinheim: Kramer.

Uphof, J.C.Th. 1968. *Dictionary of Economic Plants.* 2nd ed. Lehre: Kramer.

USDA, ARS, National Genetic Resources Program. Germplasm Resources Information Network (GRIN) [Online database]. National Germplasm Resources Laboratory, Beltsville, MD. http://www.ars-grin.gov (accessed March 4 2014).

Usher, G. 1974. *A Dictionary of Plants Used by Man.* London: Constable.

Valentão, P., E. Fernandes, F. Carvalho, P.B. Andrade, R.M. Seabra, and M.L. Bastos. 2003. Hydroxyl radical and hypochlorous acid scavenging activity of small centaury (*Centaurium erythraea*) infusion. A comparative study with green tea (*Camellia sinensis*). *Phytomedicine* 10(6–7): 517–22.

Valton, A.L., and M.N. Prioleau. 2016. G-Quadruplexes in DNA replication: A problem or a necessity? *Trends Genet* 32(11): 697–706.

Wang, K.B., Y.T. Di, Y. Bao, C.M. Yuan, G. Chen, D.H. Li, J. Bai, H.P. He, X.J. Hao, Y.H. Pei, Y.K. Jing, Z.L. Li, and H.M. Hua. 2014. Peganumine A, a β-carboline dimer with a new octacyclic scaffold from *Peganum harmala. Org Lett* 16(15): 4028–31.

Wild Flowers of Israel. [n.d.] Home page. http://www.wildflowers.co.il/hebrew/plant.asp?ID=387 (accessed January 3 2015).

Yalcin, D., and O. Bayraktar. 2009. Inhibition of catechol-O-methyltransferase (COMT) by some plant-derived alkaloids and phenolics. *J Mol Catal* 64: 162–6.

Yao, J.L., S.M. Fang, R. Liu, M.B. Oppong, E.W. Liu, G.W. Fan, and H. Zhang. 2016. A review on the terpenes from genus *Vitex. Molecules* 21(9): E1179.

Yavari, N., F. Emamian, R. Yarani, H. Reza Mohammadi-Motlagh, K. Mansouri, and A. Mostafaie. 2015. *In vitro* inhibition of angiogenesis by heat and low pH stable hydroalcoholic extract of *Peganum harmala* seeds via inhibition of cell proliferation and suppression of VEGF secretion. *Pharm Biol* 53(6): 855–61.

Zaker, F., A. Oody, and A. Arjmand. 2007. A study on the antitumoral and differentiation effects of *Peganum harmala* derivatives in combination with ATRA on leukaemic cells. *Arch Pharm Res* 30: 844–9.

Zhu, B.T., P. Wang, M. Nagai, Y. Wen, and H.W. Bai. 2009. Inhibition of human catechol-O-methyltransferase (COMT)-mediated O-methylation of catechol estrogens by major polyphenolic components present in coffee. *J Steroid Biochem Mol Biol* 113(1–2): 65–74.

8 Clinical Anticancer Use of Harmal: Two Cases

CASE 1. OLIGOASTROCYTOMA

A 29-year-old white Jewish male, of mixed Ashkenazi and Sephardi parentage, approached ESL in August 2012 seeking herbal therapy for his malignant brain tumor of two years. The tumor first presented itself in 2010, when while sitting at his computer, he suddenly noted a malignant glowing green line across the screen, screamed out to his wife nearby, and awoke hours later in the hospital. He had a total tonic-clonic seizure, the first of his life, and after an MRI was told he had a "golf-ball sized" right occipito-parietal lobe brain tumor and was scheduled for consultation with the country's leading neurosurgeon. The doctor told the patient that he had two alternatives: (1) do nothing specific, just take antiseizure medication for the rest of his life, or (2) have it removed surgically. The patient then asked that physician what *he* would do had *he* to choose. The physician reported that he "would probably have it taken out."

Surgery was scheduled and the operation was declared a success. From the patient's own report, he was singing a *piyut* (a devotional melody) when he felt the anaesthesia taking effect and he was "going under," and when he awoke from the surgery he immediately picked up in the melody from where he had left off. The others in the room confirmed this to be so. Pathology revealed a stage 3, anaplastic (poorly differentiated and rapidly growing), primary brain tumor of the type "oligoastrocytoma," which is actually a rare hybrid of oligodendroma and astrocytoma, with an occurrence of about 2.5% of all brain tumors, themselves about 1% of all cancers.

A few short months post-op, a follow-up MRI revealed again the presence of tumor. Surgery was again scheduled, performed, declared a success, and again, a follow-up MRI mere months later revealed again that the tumor was present. With some trepidation for the frequency of craniotomies, a third surgery was scheduled, performed, declared a success, and short months later, the follow-up MRI declared again a return of tumor.

At that point, the patient was sent for radiotherapy. The sessions, mounting to a total of 60 Gy radiation, were given 5 days a week for 6 weeks. An MRI just 6 weeks following the last session revealed that the tumor had gotten larger.

The patient was then offered the standard medical chemotherapy used for brain tumors, but when he read the circular about this drug's effect on sperm quality, and that he would need to freeze some sperm prior to beginning therapy, decided not to take the temozolomide, and began researching the Internet for alternative treatments. He discovered the compelling anecdotes of individuals who had apparently cured themselves of various cancers by ingesting an extract of flowers of *Cannabis indica* going by the eponymous name of the developer and promoter: Rick Simpson oil (RSO) (Singh and Bali 2013).

The patient was begun on a regimen of the RSO prepared under license and from a generous gift of Tikun Olam, Ltd., a government-approved grower of medical *Cannabis* in Israel. He was started on a ketogenic diet based on organic fruits and vegetables, raw salmon, and medicinal foods including concentrated red wine, fenugreek, garlic, cabbage juice, capers, kimchi, beef, sheep, shark cartilage, fish oil, medicinal mushrooms, Pomegranate Emulsion (Rimonest Ltd., Haifa, Israel), and walnuts. He continued to take Vitamin D 5,000 units twice a day, and antiseizure medication. He was also given extracts prepared from up to 20 different medicinal herbs for which preclinical studies had shown efficacy *in vitro* or in animal models against glioma and/or epileptiform seizures. In addition, he received treatment with other complementary modalities, including acupuncture, shiatsu massage, and intercessory prayer. He was given instruction in and practiced medical *qigong*, especially "marrow washing *qigong*," a kind of imaginal, meditative form of gentle movement.

After 3 months of the above regimen, it was decided to add treatment with *P. harmala*. All ingestions of *harmal* were taken under the direct supervision of the physician, for a period of not less than 4 h postingestion. Because of the known MAO inhibition properties of *harmal*, the patient was advised to avoid tyrosine-rich foods according to the MAO-I diets that are commonly prescribed (e.g., avoiding red wine, sardines, anchovies, pickled herring, hard cheese, chocolate, peanuts, miso, tempeh), while still adhering to the ongoing ketogenic and organic principles, for a period of 12 h pretreatment and 24 h post-treatment.

At the time of the first ingestion, the patient was given 50 mg of seeds to consciously and thoroughly chew. They were very bitter. The patient and doctor waited for any noticeable effect. The patient was encouraged, then and at other times, to perceive his expected healing as an exciting "adventure." The effect of this first ingestion was unremarkable. There was some mild exhilaration and feeling of well-being after about 20 min. The patient even appeared a bit calmer than baseline, and reported having a kind of "spiritual" feeling. The four hours or so were spent sitting and gazing outside from a balcony, and also doing some walking in the neighborhood of the house where the session was taking place. There were no untoward effects whatsoever during or in the days following the initial experience, and no hallucinations or bizarre perceptions, or inappropriate or unusual behaviors of any kind.

It was decided to continue with this approximate procedure on a once weekly basis for some coming weeks, all the while gradually monitoring the effects of the intervention clinically, and gradually increasing the amount of seeds to chew. So the second session was 80 mg seeds, the next 120 mg, and so on for quite a few weeks. No week was missed for at least 5 months. The patient continued with the dietary regimen described above during this period, including daily ingestion of up to 20 different herbal and mushroom extracts from the group of *Acorus calamus, Angelica sinensis, Bacopa monnieri, Boswellia serrata, Cannabis indica, C. sativa, Centella asiatica, Curcuma longa, Glycyrrhiza glabra, Ganoderma lucidum, Grifola frondosa, Hydrastis canadensis, Ligusticum wallachi, Ligustrum lucidum, Lycium* sp., *Olea europea, Passiflora incarnata, Scutellaria baicalensis, S. barbata, S. lateriflora, Withania somnifera*, and *Zingiber officinale*, all recognized in the medical literature for their specific anti-glioma and/or anti-seizure effects, except for *Scutellaria barbata*, which is recognized neither against glioma nor seizure, but is recognized as an important anticancer herb in general. However, one herb was withheld on the day of treatment with *harmal* and on the 2 days immediately preceding and immediately following the treatment day: *Hypericum perfoliatum*, which like *harmal* also contains potential photoactivated toxins, and has MAO-related effects, and therefore it was felt there may be relative contraindications to using *H. perfoliatum* and *P. harmala* together or too proximal in time. So *H. perfoliatum* was taken only 4 days a week, i.e., the days most distant from the 3 days off for *harmal*. On both *Hypericum* days, and on the *harmal* day, the patient was encouraged to sit facing the sun at a high sun hour for at least 15 min, 1–2 h post ingestion of the flourochrome (*Hypericum* or *Peganum*), with more or less sun exposure depending on season, with more exposure indicated in winter, and less in summer.

Because authorities like Kapoor (2000), writing on the Ayurvedic use of *harmal*, claimed that the therapeutic dose was 3–6 g per day, the dose was conscientiously increased gradually to an eventual maximum dose of 1.8 g, once weekly for a total of 6 months. Although there was considerable trepidation about using it, *harmal*, wildcrafted in the desert, it was gradually ameliorated by consulting with experts on ethnomedical *harmal* use and harmala alkaloid pharmacology, respectively. Caution always seemed justified. *P. harmala* is classified everywhere as a poisonous plant, and the literature is not devoid of fatalities associated with its ingestion. All manner of physiologic mischief may be inferred from what is known of the pharmacology of this plant. Doses high enough to induce vomiting, traditionally often indicated as a *sine qua non* for *harmal*'s putative efficacy as an anticancer drug, were never reached. Consistently, the *harmal* had no negative effects whatsoever, and many positive ones. Salubrious sequellae, associated with its use, as perceived by the patient, included

- Improved appetite
- Better sleep
- Better psychological disposition
- Feeling of being more spiritual and connected to nature

Eventually, however, untoward effects possibly related to the *harmal* did begin to emerge that did prove dysphoric and led to eventual discontinuance of the *harmal* as a treatment for this patient. In short, the patient had begun to feel, according to his own comfort level, *too* spiritual. The old moniker for harmine, i.e., telepathine, first known as an unidentified compound in *Banisteriopsis caapi*, the "vine of the soul," that is the principal component of Ayahuasca, became appropriate and comprehensible.

It is not unusual for individuals in some sort of therapy to "internalize" their therapist, so that they can almost hear his or her voice guiding or advising them in many life occurrences that occur outside of the formal hours of therapy. This phenomenon is natural, and may be considered part of the "positive transference" of a healing encounter. There is also such a thing recognized among psychoanalysts in particular that is called a "counter transference," whereby the patient somehow becomes internalized in the therapist. Such phenomena are well-known to shamans, such as the ones who work with Ayahuasca, and also in the field of hypnosis, where they may be known, according to Dr. Udi Bonshtein, an Israeli psychologist licensed to hypnotize, as "tranceference" and "countertranceference" (Bonshtein 2012). Although formal hypnosis was never used in this case, as it was in the case to follow, the concepts are nevertheless germane.

Anyway, the patient began to complain that the doctor (ESL) was right there in his thoughts and consciousness, even at such times when the patient perhaps preferred privacy. This became disconcerting and dysphoric. On one occasion, the physician at home in bed had a dream possibly very indirectly involving the patient but highly vivid, lucid, and evocative. The doctor rose from bed to record the dream at the computer. His habit is to record dreams and send them to himself by email, so he had to first enter the email. The time was 7:03 in the morning, and within the previous 3 min, an email from this patient had arrived. This was extraordinarily unusual, because this patient, in spite of his own efforts and those of others, is an absolute night owl. He would be fast asleep every night by 5:00 am, unless something very, very out of the ordinary had occurred. In fact, it had never occurred in many months, even years, of treatment. The doctor opened this untimely email, and it was the patient reporting a dream that had just awoken him a minute earlier. Need more be said here of this? It was an uncanny and quasi-numinous experience. The doctor's notice of the synchronous dream was never shared with the patient, for fear (justified or not!) that to do so might further unnecessarily exacerbate his ongoing dysphoria with the putatively *harmal*-provoked telepathy.

Eventually, the harmal was discontinued, the telepathy receded, and in lieu of an MRI, concerning which the doctor had reservations, a specialized ultrasound of the adult human brain was sought and received by neurosurgeon/electrical engineer, David Michaeli, MD, PhD, who claimed that compared to the MRI that the patient had had 2 weeks prior to commencing the herbal and other complementary treatments, the tumor had shrunk 75%. The same, decidedly experimental, echoencephalography of an adult human brain *in situ* was repeated on four more occasions during an additional 11 months, following which it was finally agreed by all parties for the patient to receive an MRI after a full 20 months without one. The results revealed what is shown in the figures: after 20 months of complementary intervention in the absence of any conventional anticancer treatments aside from the ketogenic diet, and employing weekly therapeutic *harmal*, from 50 mg to 1.8 g of dried seeds each time over a 4–6 month period, and daily use of other medicinal herbs including pomegranate (*Punica granatum*), turmeric (*Curcuma longa*) and *Cannabis sativa* or *C. indica* (high CBD, low THC varieties, namely Raphael [*sativa*] or Avidekel [*indica*]; kind donations of Tikun Olam, Ltd., Tel Aviv, Israel)—after this 20-month period without MRI, with greatly curtailed exposure to electronic devices, especially smart phones and tablets, and with elimination of all toxic fluids from the home, such as cleaning solutions, switching to only ecological preparations, the MRI results more than corroborated the echoencephalographic impressions. Professor John Moshe Gomori, head of MRI and neuroradiology in Hadassah Hospital, Jerusalem, noted that in that MRI after 20 months one no longer saw any sign of tumor, only scar tissue remaining from previous surgeries and radiation. Fifteen months later, another MRI was obtained which continued to not show tumor. Four and a half years since beginning complementary therapy, and over 2 years since the last *harmal* treatment, the patient continued to be free of obvious pathology or of seizures relating to his previous illness/underlying dyscrasia (Figures 8.1 through 8.4).

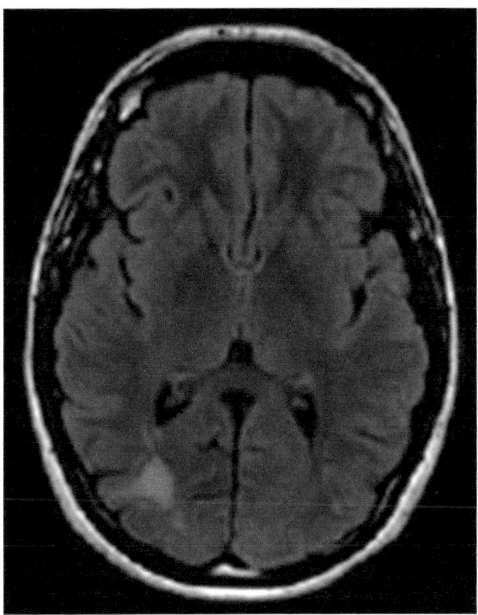

FIGURE 8.1   MRI image. 2/28/2012. The tumor was the white area in the left lower corner of the image.

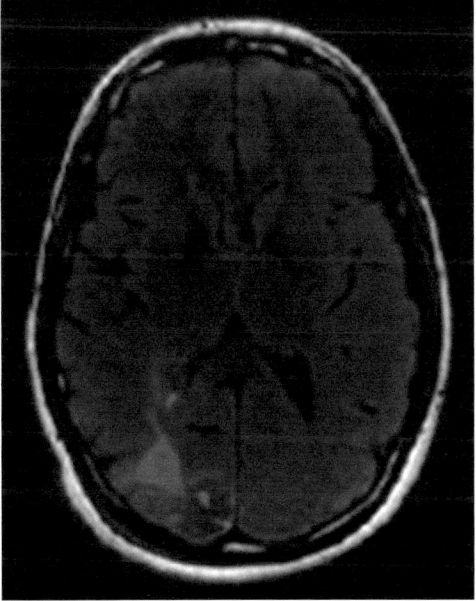

FIGURE 8.2   MRI image. 5/3/2012. Following Surgical Excision. The tumor appeared again in the same place. It was the light grayish area.

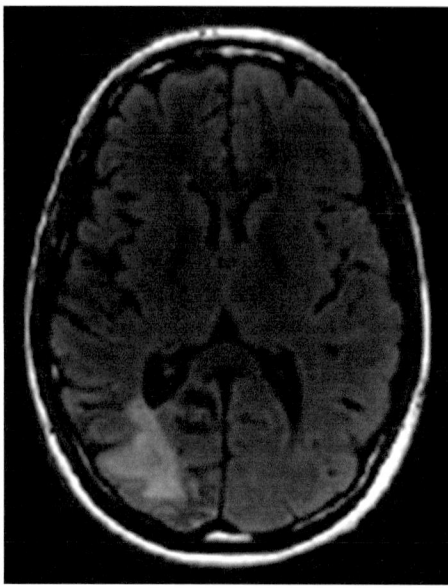

FIGURE 8.3   MRI image. 8/14/2012. Following 6 weeks of four times a week, total 60 Gy radiotherapy, the tumor appeared *bigger*?

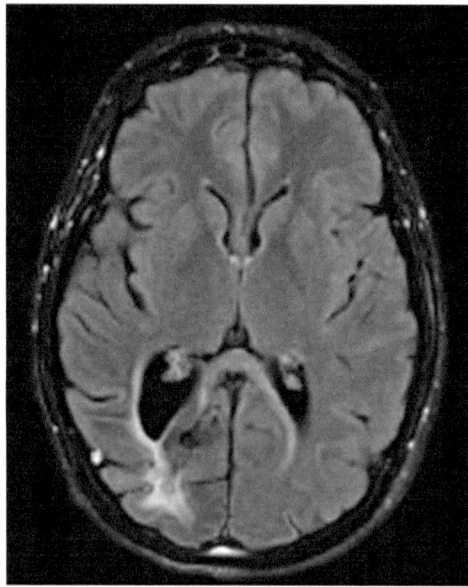

FIGURE 8.4   MRI image. 5/12/2014. After 20 months' complementary treatments, including six months of weekly *harmal* sessions, daily RSO, Pomegranate Emulsion, 18 other medicinal herbs and mushrooms, marrow washing *qigong*, intercessory prayer, ketogenic organic diet with medicinal foods, classical homeopathy, and acupuncture. The white area was scar tissue remaining from previous surgeries and radiation. The tumor appeared to have disappeared.

CASE 2. OVARIAN CARCINOMA

The second case, a married youthful grandmother of 53 years, presented herself to the physician about 14 months ago (November 2015), also to receive herbal therapy. Like the previous case, she held a license from the government to use medical *Cannabis*, and she had heard about and was interested in trying RSO. Her troubles had begun to present in 2005 when she suffered from persistent lower abdominal pains. Her family doctor found nothing amiss, and she was sent to specialists who also found nothing organically wrong. She received prescriptions for and consumed pain medications, mood elevators, and tranquilizers, but none of this seemed to be helping much or getting to the root of the problem, which for her seemed only too physical and real. Finally in 2009, some astute physician in her health maintenance organization saw that they were dealing not with functional neurasthenia, as had been until then thought, but with something deeper and more insidious. By the time the diagnosis of ovarian cancer had been made, the cancer was already advanced, having broken free of its original source and extensively invaded the abdominal wall. Surgery was scheduled promptly to remove the womb and ovaries, and as well to debulk the extensive tumor that had spread throughout the belly and affixed to one renal artery. She received a course of chemotherapy with taxol and a platinum drug and became very ill, though she had a remission. She had also taken Tamoxifen for a while, but discontinued that because of poor tolerance. The condition had been stable for some years, and then there was a recurrence. Chemotherapy was offered again but she refused. She decided she wanted to try an alternative route, had heard of RSO and natural therapeutics, and turned to the physician for herbal treatment.

On examination, the patient appeared quick, spare, bright, energetic, frightened, and desperate. She was desperate to go on living, to survive, to continue to take an active role in providing material and emotional support to her family. She was working at the time of meeting as a kindergarten teacher, which necessitated early rising and travel each day by bus to her work place. She was very emotional talking of her medical condition and her willingness to submit to any sort of regimen if it could help her. She was also adamant about not having chemotherapy again in order to avoid all the side effects and discomfort associated with it. She was very interested and eager to begin therapy with RSO, and expressed hope and expectation of good results.

After a few initial meetings, it was decided per the patient's request to begin supplementation with oral RSO, i.e., an ethanolic extract of Cannabis Flos. This was mixed 1:1 with oil of Punica Semen, and administered in drops, first one drop of the mixture per day, and then gradually ascending, per the Rick Simpson protocol, one drop every two days. The variety she was taking was *Cannabis sativa* (Midnight of Tikun Olam, Ltd.) with medium-high and equal amounts of CBD and THC.

To accelerate treatment, it was decided to employ a loading dose of the Midnight mixture. If the dose she had been taking was 50–100 mg per day, then the jump to the loading dose was to ten times that, i.e., 0.5–1 g of the mixture. Loading dose is sometimes employed in pharmacotherapeutics when the drug in question is excreted very slowly from the body. The major advantage is that a therapeutic level can be achieved *in situ* overnight, and then, because the substance is excreted so slowly, a much smaller daily amount is sufficient to maintain the therapeutic level of the slow-to-be-excreted substance.

One common example of employment of loading dose in daily medical practice was in the Emergency Department of a major metropolitan American hospital. A patient was found to have a purulent penile discharge and penicillin was employed, first as an intramuscular loading dose of 2 million units—two large shots in the buttocks—then two daily pills for 10 days as a maintenance dose. It worked for the lipophilic penicillin, which was very slowly excreted. It thus should also work for *Cannabis*, because it too is very slowly excreted. It does not apply to *Peganum harmala*, which is excreted more rapidly—the long acting effect of *P. harmala* and other indolic entheogens is due to a different mechanism to be discussed in the next chapter.

The RSO loading dose had been administered and the patient medically supervised continuously for 28 h, until she was capable to leave and was discharged. During most of that time she used eyeshades and stereophonic headphones to listen to a program of classical music aimed to facilitate positive experience in treatments with entheogenic agents (Bonny and Pahnke 1972). Aside from the length of time, and in spite of being given frequent free water to drink, she also held her bladder for the entire time. Her experience as well as the staff's during the treatment period was unremarkable. She continued to take the RSO on low-medium dose maintenance schedule without incident.

A month later, a similar session to the RSO loading dose was employed with *P. harmala*. The setting was the same as with the RSO loading dose, including supervision, eyeshades, headphones with stereo music, and the same treatment bed. She was given about 500 mg of an ethanolic *harmal* extract dissolved in hot water, which she drank without incident. The ensuing 6 h or so were uneventful, following which the patient was discharged. There were no notable sequelae in the week following that session.

Ongoing, she also had adhered to a ketogenic diet, and taken orally a daily Pomegranate Emulsion containing any number of ethanolic herbal (including mycological) extracts from this group, generally with noted antiovarian cancer activity, including *Annona squamosa, Ganoderma lucidum, Glycyrrhiza glabra, Grifola frondosa, Humulus lupulus, Ligusticum wallichii, Panax ginseng, Scutellaria baicalensis, S. barbata,* and *Tripterygium wilfordii.*

In the months following, a low dose of *P. harmala* seed ethanolic extract was often added to the mixtures of some or all of these herbs within the weekly Pomegranate Emulsion. In this way she received small daily doses of the *harmal* extract of approximately 50 mg per day over about a 3-month period.

She received regular weekly sessions of medical hypnosis and self-hypnosis training, with the objective to influence anticancer immunological reactivity through psychoneuroimmunological cascades (Benor 1996; Eccles 2007; Erwin 2011; Halley 1991; Kiecolt-Glaser and Glaser 1992; Kiecolt-Glaser et al. 2001; Kovács et al. 2008; Lansky 1982; Modlin 2004; Petry 2000; Reed 2007; Weisberg 2008; Wood and Zadeh 1999). She received regular monitoring of her abdominal status with monthly or bimonthly ultrasound examinations. A marker, CA 125, was followed a few times a year by her family doctor in the Health Maintenance Organization.

The patient underwent a lot of psychological changes and a kind of spiritual turbulence and renewal. She had to cut down on her hours teaching kindergarten and eventually quit working entirely as she became increasingly incapacitated from the illness. One of her daughters filed a report that the mother was becoming overly religious with excessive praying, and even had had a delusion that she was the Messiah.

She had been up to taking a full gram of pure RSO per day, i.e., 2 g of the RSO/pomegranate seed oil mixture. She was getting daily doses of *harmal* of approximately 50 mg per day in her Pomegranate Emulsion. Generally, if *harmal* were included in the formula, other highly cytotoxic herbs such as *Tripterygium wilfordii* were not. It was decided nevertheless, because of the messianic reports, to decrease the psychoactive elements in her regimen. The RSO was switched to another Tikun Olam Ltd. strain, i.e., Avidekel, high CBD, virtually no THC, and the dose of that lowered to 25% of what she had been taking, i.e., with a new dose of 500 mg per day RSO/pomegranate seed oil mixture. The *harmal* was completely eliminated from the formula. With this scaled-down formula, her mental state rapidly stabilized.

Meanwhile, the regular ultrasound assessments were not looking so good. In general, there had been little change for most of the year. At one point the two major tumors had consolidated into one, and there seemed to be modest reductions in tumor volume in some of the examinations. A couple times there was some ascitic fluid, but it was minor, and resolved on subsequent examinations.

At one point, there had been confusion by the patient in the reading of the report from the ultrasound examination, and the patient and also the physician erroneously believed there had been a dramatic improvement. The patient became ecstatic and went into a complete turnaround in her lifestyle and glowed as if her life had been spared and she was the recipient of a miracle. The miracle, however, was short-lived once an adequate explanation of the findings had been obtained dispelling the fleeting illusion of a cure.

The ultrasound report had shown that overall the tumor volume had been gradually but continually increasing. The CA-125 marker had increased threefold. The patient had again reaffirmed her desire to continue with complementary treatment. At that time she had been taking no new conventional medication of any kind, nor *harmal* owing to the recent cognitive lability. So, a different approach was conceptualized and begun.

As has been mentioned previously, the harmala alkaloids such as harmine are fluorophores or fluorochromes, meaning they are fluorescent and emit light. They can be photoactivated to become powerful phototoxic phytotoxins.

The penetration of *harmal* into the body through transdermal application could be theoretically facilitated with the use of pharmacological penetration enhancers. Infrared light was known to be able to penetrate tissue. Could *harmal* be combined with appropriate penetration enhancers, applied transdermally to the abdominal wall or lumbar posterior, and allowed to penetrate to the tumors? Could infrared light focused on the surface of the body correlating to the tumor mass also enter the tumor and photoactivate the *harmal*?

In order to move in the direction of assessing feasibility of this hypothesis, an ethanolic *P. harmala* seed extract was prepared and combined with an equivalent amount of DMSO (dimethylsulfoxide). These were gently heated and mixed, and then combined in a blender with about four times their total volume of glycerol. This mixture was then painted on to the surface of the skin over the tumor mass both anteriorly and posteriorly. The patient had also been complaining for 2 weeks of pain over one of her hips, which seemed like bone pain. The mixture was also painted over that area. She was given again eyeshades and headphones, allowed to recline comfortably on the treatment bed, and offered a classical program consisting of Yo Yo Ma playing the six Bach cello suites in Royal Albert Hall. After about 90 min, the infrared source, a 250-W heat lamp, was focused first on her abdomen for about 20 min, then an additional 10 min exposure given to her back. After that she was discharged, traveled home with supervision, slept well, and had no untoward effects whatsoever (Figures 8.5 and 8.6).

The following week she had been seen again. The pain in her hip had completely disappeared. She was feeling fine and eager to have another treatment. About twice the dose of the previous administration was employed, and extra DMSO added to the mixture. The mixture was applied as before, and this time the infrared light was started from the outset and allowed to shine on as the mixture was allowed to penetrate. After 2 h, the light was closed and the patient discharged with supervision to return home.

The next morning the physician received a call from the patient who was in acute distress. She had returned safely the night before and went right to bed. Unlike the previous time when she had slept over in Haifa the night before the treatment, she did not even shower before going to bed to wash off what was left of the applique. She slept soundly. In the morning she awoke not being able to rise from the bed. She vomited several times profusely, mainly the copious water she had drunk the day before. She complained about her eyes feeling out of focus. The physician spoke with her daughter and set out on the 2-h journey to her home.

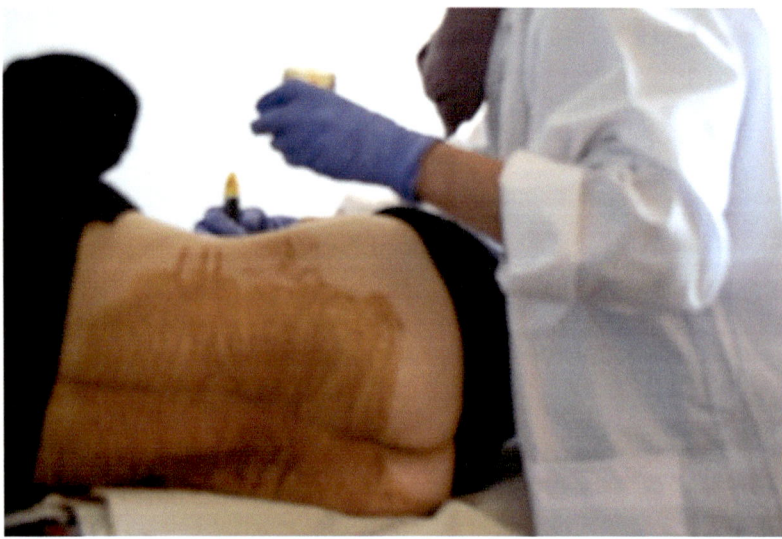

FIGURE 8.5 Applying harmal solution with penetration enhancers to the abdominal surface. Posterior surface over renal area is already covered with harmal solution.

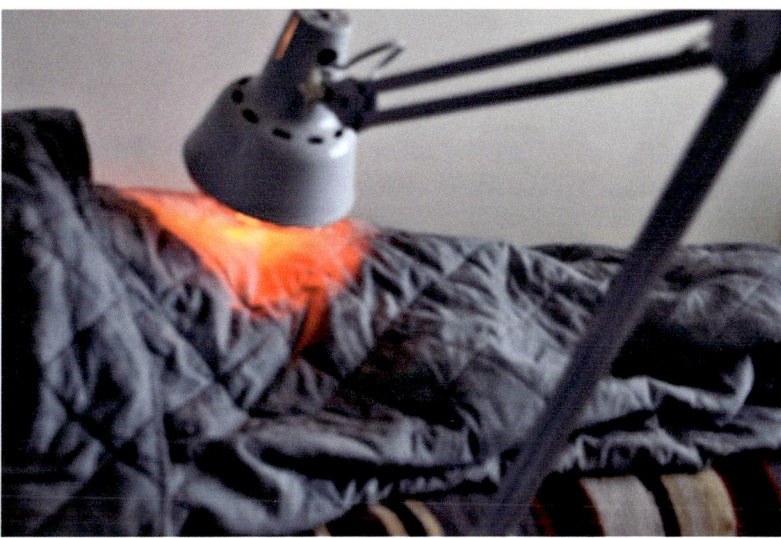

FIGURE 8.6   Activating the harmal penetrating the body with equally penetrating infrared radiation. The red color is from the near infrared red light. Infrared radiation itself is invisible to the human eye. Note that the skin is covered with the blanket. The infrared radiation, unlike visible light, was considered able to penetrate the thin opaque cloth.

She had appeared weak but had been fully oriented as she explained what was related above. After being a little baffled for a few minutes, the physician finally realized what he had been looking at, which was acute *harmal* poisoning. The daughter was instructed to promptly bathe her mother, which had been easier to accomplish than had been expected, and had probably been made a lot easier following the reassurance of a correct diagnosis of a transient and reversible phenomenon. As expected, she had felt better after the bath, and was instructed to desist from all supplements, Pomegranate Emulsion, and RSO for at least 24 h and to rest. The physician had returned to check her two hours later, and she seemed stable and on the road to recovery.

In fact, only about 48 h later, the patient had greatly surprised the physician by showing up at his clinic unannounced, ready for another treatment. She had apparently made the long trip by herself without incident.

She had appeared to be in very good shape, rested, clearer. The transdermal delivery system had proven its efficacy into at least causing vomiting and ataxia! The traditional healers using *harmal* insisted that if the patient has vomited, she had not received an adequate (therapeutic) dose. Apparently that time therapeutic dose had been obtained, and via transdermal absorption.

She was disappointed that the physician was not prepared with material for another *harmal* treatment right there on the spot. Nevertheless, she had been given the eyeshades, the infrared light focused on her belly, and a hypnotherapy session to help prepare for the next transdermal *harmal* administration with Bach and Yo Yo Ma that had been scheduled for the following week.

REFERENCES

Benor, R. 1996. Autogenic training. *Complement Ther Nurs Midwifery* 2(5): 134–8.

Bonny, H.L., and W.N. Pahnke. 1972. The use of music in psychedelic (LSD) psychotherapy. *J Music Ther* 9(2): 64–87.

Bonshtein, U. 2012. Relational hypnosis. *Intl J Clin Exper Hypn* 60: 397–415.

Eccles, R. 2007. The power of the placebo. *Curr Allergy Asthma Rep* 7(2): 100–4.

Erwin, D.M. 2011. Treatment of HPV with hypnosis—Psychodynamic considerations of psychoneuroimmunology: A brief communication. *Int J Clin Exp Hypn* 59(4): 392–8.

Halley, F.M. 1991. Self-regulation of the immune system through biobehavioral strategies. *Biofeedback Self Regul* 16(1): 55–74.

Kapoor, L.D. 2000. *Handbook of Ayurvedic Medicinal Plants*. Herbal Reference Library. Boca Raton, FL: CRC Press.

Kiecolt-Glaser, J.K., and R. Glaser. 1992. Psychoneuroimmunology: Can psychological interventions modulate immunity? *J Consult Clin Psychol* 60(4): 569–75.

Kiecolt-Glaser, J.K., P.T. Marucha, C. Atkinson, and R. Glaser. 2001. Hypnosis as a modulator of cellular immune dysregulation during acute stress. *J Consult Clin Psychol* 69(4): 674–82.

Kovács, Z.A., L.G. Puskás, A. Juhász, A. Rimanóczy, L. Hackler Jr, L. Kátay, Z. Gali, A. Vetró, Z. Janka, and J. Kálmán. 2008. Hypnosis upregulates the expression of immune-related genes in lymphocytes. *Psychother Psychosom* 77(4): 257–9.

Lansky, P. 1982. Possibility of hypnosis as an aid in cancer therapy. *Perspect Biol Med* 25(3): 496–503.

Modlin, T. 2004. Psychoneuroimmunology: Mind-brain-immune interactions. *S Afr Med J* 94(10): 797–8.

Petry, J.J. 2000. The role of the mind and emotions of patient and surgeon in the outcome of surgery. *Plast Reconstr Surg* 105(7): 2636–7.

Reed, T. 2007. Imagery in the clinical setting: A tool for healing. *Nurs Clin North Am* 42(2): 261–77.

Singh, Y., and C. Bali. 2013. Cannabis extract treatment for terminal acute lymphoblastic leukemia with a Philadelphia chromosome mutation. *Case Rep Oncol* 6(3): 585–92.

Weisberg, M.B. 2008. 50 years of hypnosis in medicine and clinical health psychology: A synthesis of cultural crosscurrents. *Am J Clin Hypn* 51(1): 13–27.

Wood, G.J., and H.H. Zadeh. 1999. Potential adjunctive applications of hypnosis in the management of periodontal diseases. *Am J Clin Hypn* 41(3): 212–25.

9 Lympho-Neuric Syncytium and the Somatodelic Hypothesis

Serotonin, or 5-hydroxytryptamine (5-HT), as has been previously noted, is a well-known neurotransmitter in the brain, especially in the Raphe nucleus, and modulation of its production, reuptake, enzymolysis, and production and activity of at least seven different proteins that serve as serotonin receptors (5-HT 1–7) and their subtypes (e.g., 5-HT2 a, 5-HT2 b, 5-HT2 c) comprise a large part of the psychopharmacologic strategies of psychiatry. Specifically, for a start, serotonin modulation underlies most antidepressants and antipsychotic drugs, the so-called "major tranquilizers." It is further well-known that receptors for serotonin occur in selected neurons of the central nervous system (CNS).

Serotonin, however, occurs not only in the CNS, but also is heavily stored in blood platelets throughout the body, and serotonin receptors bedeck not only the right neurons, but also lymphocytes, comprising the diffuse cellular matrix of immunity.

It has been observed, perhaps tongue in cheek, by zoologists that Life has a life of its own (Brooks 2000), and this could be paraphrased as the immune cells, i.e., the lymphocytes, also have a kind of "mind of their own" (Satprem 1982). Indeed, it was hypothesized by the neurobiologist-cyberneticists Vaz and Varela (1978), on mathematical grounds, that the immune system, like the central nervous system, is a *cognitive* system, specifically in the manner that it is able to differentiate the personal, immunological self from the immunological not-self. The differentiation of self from not-self is a fundament of immunologic theory, but the details are generally understood in terms of the classical clonal-selection ideas of Burnet (1959) rather than in the view of modern systems theory (Vaz 2011). Vaz and Varela, however, insisted that the immune system does have a cognitive conception of itself, and can differentiate this from not-self which it perceives as "nonsense." Such differentiation, they postulated, is the basis for executing Burnet's (1957) immunosurveillance.

Indeed, people seem to have had no trouble accepting that neural cognition, self-image, mood, and well-being could be mediated by serotonin. Why not? Because neurons, nerve cells, are involved with all that stuff, and they can be regulated by serotonin-modulating drugs. That's why.

But lymphocytes, like neurons, also have serotonin receptors on their surfaces. They do?! Yes, and those receptors can also receive serotonin or its congeners as ligands, and so serotonin can directly modulate immune function, through the immune system's innate sensitivity to this monoamine (i.e., serotonin, 5-hydroxytryptamine, 5-HT), as well as via putative psychoneuroimmunological cascades "originating" in brain.

That there are likely serotonin receptors on lymphocyte surfaces was the conclusion of Vasil'eva et al. (2011) who noted the effect of the antipsychotic drug, haloperidol, on natural killer lymphocyte activity (NKCA) in 59 first episode endogenous psychosis male patients between ages 18 and 30 years, diagnosed with acute schizophrenia or schizoaffective disorder. They found that levels of natural killer lymphocytes were depressed in half the patients, relative to controls, even before commencement of psychopharmacologic intervention. After beginning treatment with haloperidol and clozapine, levels were checked again 4 and 8 weeks into treatment. If the pretreatment levels had been low, the treatment resulted in an increase, though still not as high as controls. If the pretreatment NKCA had been high, i.e., normal, then the NKCA was reduced at the two treatment checkpoints. These effects had been reciprocally related to the maximal rate of serotonin reuptake by lymphocytes, and serotonin added to the cell culture *in vitro* had normalized the NKCA level in cultures with and without monocytes. These findings had led the authors to conclude that there had been active serotonin receptors on the surfaces of the natural killer lymphocytes of these patients.

Modulation of immune functions via platelet serotonin has been reviewed, including its role in regulating proliferation and activation of lymphocytes. Serotonin levels are increased in asthma and rheumatoid arthritis, and targeting serotonin receptors showed potential to improve survival after myocardial infarction or sepsis and to attenuate autoimmune asthmatic attacks in animal models—reinforcing the idea that lymphocytes possess surface receptors for serotonin (Lychkova and Puzikov 2014; Mauler et al. 2016).

Further suggestive evidence had been developed by Rivera-Baltanas et al. (2014) who had used immunocytochemistry to assess 5-HT2A receptor clusters in lymphocytes of patients with depression. Genetic variation in the serotonin receptor gene affected immune responses in rheumatoid arthritis (Snir et al. 2013). The serotonin-receptor 5-HT(2C) was specifically expressed in natural killer lymphocytes of patients with Alzheimer's disease (Martins et al. 2012). In mice, pharmacological stimulation of the 5-HT2A receptor led to a blocking of the immune response whereas its suppression led to immunological enhancement as measured by CD8(+)T cell counts (Davydova et al. 2010). Serotonin 3A receptors were expressed on B lymphocytes even if they had become neoplastic (Rinaldi et al. 2010). Yin et al. (2006) elucidated the role that serotonin 5-HT1B receptors on T lymphocytes had played in their proliferation.

Elena Magrini and her coworkers at the Humanitas Clinical Institute in Milan demonstrated that chemotaxis of lymphocytes to the powerful endogenous chemotactic CXCL12 protein could be "tuned" by serotonin. Using real time PCR, western blot analysis, and electrophysiological patch clamp experiments, they had been able to show that serotonin receptor 5-HT3 is functionally expressed in human primary T lymphocytes. Further, 5-HT3 receptor agonists selectively had decreased migration of lymphocytes toward the chemotactic CXCL12—indicating that the chemotaxis of the lymphocytes had been modulated, as if turning a dial, by serotonin (Magrini et al. 2011). If, in fact, lymphocytes do have serotonin receptors on their surfaces, and can be tuned by regulating their exposure to serotonin, then perhaps they as a network could behave in ways analogous to the way neurons with serotonin receptors on their surfaces behave when they are tuned by serotonin.

One interesting characteristic of the response of patients and healthy volunteers to the indolic entheogen psilocybin is the great length of time the pharmacological effects can persist, including relief of chronic pain, and even of chronic cancer pain (Schindler et al. 2015), from a single dose. Young (2013) eloquently discussed this matter from a pharmacologist's point of view and had left the mystery unclaimed, while emphasizing the importance of long-term psychological factors. Nevertheless, multiple studies have emphasized this longevity of effect from a single dose of psilocybin (Griffiths et al. 2006; Grob et al. 2011), LSD (Krebs and Johansen 2012), or Ayahuasca (Osório Fde et al. 2015; Sanches et al. 2016), and the question is why this is the case. Aside from the psychological and placebo effect that is present in every pharmacological encounter, what is the physiological reason for the long-term glow from an experience that may persist for months following a single dose?

If *harmal*, Ayahuasca, LSD, or psilocybin really do reset serotonin receptors and brain in some way to effect long-term beneficial changes in mood, it is big news. If this phenomenon associated with these compounds may also have anticancer effects, it is of very great interest indeed. Certainly anticancer effects against xenografts had been demonstrated from very high doses of LSD in rat (Scott and Stone 1959), and that effect had been attributed to the modulating effect of LSD on serotonin output.

What is the relationship between entheogenic herbs and their anticancer effects?

Cannabis sp., which to some authorities is not a hallucinogen, is nevertheless implicated in sacred rites, now in Jamaican Rastafarianism (Hickling and Griffith 1994; Legrand et al. 2015), as also in ancient times (Jiang et al. 2006, 2007; Nahas 1982; Touw 1981) (Figure 9.1).

Further, at high oral doses and also otherwise *Cannabis* sp. preparations may possess potential for extremely potent psychoactive effects. While cannabinoid pathways are not directly targeted by monoamines, and while monoaminergic pathways are not directly targeted by cannabinoids, i.e., phenyl terpenes that occur in *Cannabis* sp., indirectly the pathways *are* targeted in each case. This is due to biochemical intercommunication or "cross-talk" between cannabinoid and monoaminergic (adrenergic, dopaminergic, serotonergic) pathways, especially in that endogenous cannabinoids modulate presynaptic transmitter release of possibly all monoaminergic neurons (Chiang et al. 2013; Kreitzer 2005; Wolf et al. 2008). Thus, indirectly, sometimes psychedelic, "minor hallucinogen," and often entheogenic cannabinoids (e.g., tetrahydrocannibinol, THC) also possess potent anticancer effects making them extremely attractive leads for anticancer herbal treatment and compound-focused drug development (Caffarel et al. 2012; Gasperi et al. 2015; Guzmán 2003; Nasser et al. 2011; Pars and Howes 1977; Pisanti et al. 2013).

Yet, chemically cannabinoids are quite far removed from harmala alkaloids. If the anticancer effects of *harmal* are related to its serotonin modulation not only in the nervous system, but also on lymphocytes and even at receptors on the tumor (or neoplastic clone) itself (Fröberg et al. 2009; Henriksen et al. 2012; Oufkir and Vaillancourt 2011; Pai et al. 2009; Rinaldi et al. 2010; Soll et al. 2012), how do the cannabinoids exert their anticancer effects at serotonin receptors? First, these compounds have multiple effects which go beyond the serotonin receptor, however, just as cannabinoids can modulate serotonergic transmission through presynaptic

(a)

(b)

FIGURE 9.1 *Cannabis sativa* aerial part. (a) Full-grown plant. (b) Fresh *Cannabis* leaves have multiple potential uses. (8/10/2013, by Dāvis Mosāns, http://www.flickr.com.)

inhibition, so can the serotonergic system modulate cannabinoid signaling in the brain (Nasehi et al. 2016a,b).

Entheogenic compounds seem to provide a kind of template for a type of anticancer effect, though congeners of these entheogens that do not possess psychoactive activity may still be potent, sometimes even more potent than the parent entheogen, as anticancer agents. This was the case for example with the non-entheogenic LSD analog, d-1-methyl-lysergic acid diethylamide (MLD-41) that was an even more potent anticancer drug than LSD itself *in vivo* (Scott and Stone 1959). Further, the same may also be the case for the so-called nonpsychoactive cannabinoid cannabidiol (CBD). Not truly nonpsychoactive, but more like an order of magnitude less psychoactive than THC, CBD is increasingly recognized for its anticancer properties which may rival or be equivalent to the much more psychoactive THC. This is often the case in medicinal chemistry. The very weak estrogen 17-α estradiol is nonetheless as strong an antioxidant as the much more potently estrogenic 17-β estradiol (Lansky et al. 2005). Moreover, the stepped-up research with CBD has highlighted its possible direct interaction—and by association possibly also that of THC—with serotonin receptors (Linge et al. 2016). Cannabinoid signaling in fear conditioning is modulated by a serotonergic circuit (Nasehi et al. 2016a,b), and 5-HT3 and cannabinoid receptor type 1 (CB-1) synergistically interact in terms of memory consolidation and its deficit (Ahmadi-Mahmoodabadi et al. 2016). CBD also provides a protective effect on maintaining the blood–brain barrier as a hedge against glucose and oxygen deprivation that is mediated via serotonergic receptors (Hind et al. 2016). Thus, in short, both the "major hallucinogen-entheogens," tryptamine derivatives such as the harmala alkaloids and LSD, and the "minor hallucinogen-entheogens," non-nitrogen containing compounds such as THC, as well as the relatively non-psychoactive analogs of these compounds, MLD-41 and CBD respectively, are all potent anticancer compounds whose actions are commonly mediated, at least in part, through serotonergic tuning. The ongoing salubrious effect of their entheogenic influence on cancer and other physical diseases through psychoneuroimmunological cascades has been addressed. What is of equal interest is that the anticancer effects of these compounds may also be at least partially mediated through tuning of the serotonergic receptors on the immune system, i.e., notably the lymphocytes, as well as tuning of the serotonin receptors on the tumor cells themselves.

Drugs that suppress thought, touted as antidepressants or antipsychotic agents, e.g., selective serotonin reuptake inhibitors (SSRI's), may also interfere with immunity (Branco-de-Almeida et al. 2011; Gobin et al. 2015) and anticancer therapy (Breitbart 2011; Lash et al. 2010; Pritchard 2010). Conversely, drugs which are entheogenic may facilitate cancer resolution and enhance immunological competence (Dawood and Qubih 2012; Roberts 1999; Toghyani et al. 2015).

What can be concluded now?

First, lymphocytes collectively have an internal sense of self, and their cognitive ability to coordinate their immunological movements coincides with having serotonin receptors on their cell surfaces. Neurons, which collectively maintain through thought processes a sense of self, also may possess serotonin receptors on their cell surfaces. Like email addresses on a list, these serotonin receptors represent a pharmacological means whereby the immunological and psychological selves may be simultaneously mobilized for recognition and, if necessary, immunological or behavioral response. Because of this ability to modulate both immunological and gross behavioral reactivity with the same serotoninergic tuning agents, one can imagine that the lymphocyte-immunological network is functionally continuous with the neuron-neurological network. In fact, one could make the jump to speak of the immunological lymphocyte-populated network and the neuron-populated central nervous system as polar aspects of one network, or more graphically, one syncytium.

Unlike the fusion of thousands of skeletal muscle fibers, which constitutes a true physical syncytium, the imaginal syncytium of neurons and lymphocytes is more, like the heart, a functional syncytium. In the case of the heart tissues, they all beat as one, and can be similarly affected pharmacologically as one. The lymphic–neuric network is also a functional syncytium, because putatively it can be modulated as one by serotonin, serotonin agonists, and serotonin antagonists. And if it is truly a functional syncytium as suggested, mental impulses should be able to travel freely through the lymphocytes, i.e., a kind of "mind of the cells" (Satprem 1982), or "mind of the lymphocytes."

Syncytia (plural of syncytium) are all about their ability, whether topologically continuous or not, to contain and even facilitate chain reactions. An epileptic seizure is a synchronized discharge of the neurons comprising the syncytium, which is the brain, but imagining a wave or relaxation washing the sea of bone marrow during hypnosis or *qigong* also may transmit muscular relaxation throughout the body as a chain reaction.

Maria Lobikin and her colleagues at Tufts University in Medford, Massachusetts, studied the syncytium phenomenon as it relates to carcinogenesis and cancer progression, with the identification of "instructor cells," which are able to convey the cellular programs for de-differentiation into cancer cells and for the subsequent invasion of these cells and the corresponding metastasis of the neoplasm in terms of a bioelectrically continuous syncytium (Lobikin et al. 2012). Transmission of cancer-supportive proliferation and growth is accomplished by "instructor cells" effecting stem cell transformation through a serotonergic pathway, bioelectrical cues, and action potentials (Blackiston et al. 2011)!

The second conclusion to be drawn from the material presented above in this chapter is more of a hypothesis or even only a question. Could serotonin modulators such as the harmala alkaloids, LSD, psilocybin, dimethyltryptamines via sigma-1 receptors (Fontanilla et al. 2009; Frecska et al. 2013; Su et al. 2009; Szabo 2015; Szabo et al. 2014; van Waarde et al. 2015), and possibly also cannabinoids through presynaptic crosstalk with monoaminergic signaling, act on the lympho-neuric syncytium and, through reliance on instantaneous transmission and "instructor cells," effect a kind of "somatodelic" wave with healing action potential to retrace and reverse the whoopee made by the instructor cells to make the cancer in the first place, and could that, like the psychedelic effect of such agents on the brain, persist from a single dose for months into the future? Is that possible? Is this putative phenomenon worthy of investigation?

REFERENCES

Ahmadi-Mahmoodabadi, N., M. Nasehi, M. Emam Ghoreishi, and M.R. Zarrindast. 2016. Synergistic effect between prelimbic 5-HT3 and CB1 receptors on memory consolidation deficit in adult male Sprague-Dawley rats: An isobologram analysis. *Neuroscience* 317: 173–83.

Blackiston, D., D.S. Adams, J.M. Lemire, M. Lobikin, and M. Levin. 2011. Transmembrane potential of GlyCl-expressing instructor cells induces a neoplastic-like conversion of melanocytes via a serotonergic pathway. *Dis Model Mech* 4(1): 67–85.

Branco-de-Almeida, L.S., M. Kajiya, C.R. Cardoso, M.J. Silva, K. Ohta, P.L. Rosalen, G.C. Franco, X. Han, M.A. Taubman, and T. Kawai. 2011. Selective serotonin reuptake inhibitors attenuate the antigen presentation from dendritic cells to effector T lymphocytes. *FEMS Immunol Med Microbiol* 62(3): 283–94.

Breitbart, W. 2011. Do antidepressants reduce the effectiveness of tamoxifen? *Psychooncology* 20(1): 1–4.

Brooks, D.R. 2000. The nature of the organism: Life has a life of its own. *Ann N Y Acad Sci* 901: 257–65.

Burnet, F.M. 1959. *The Clonal Selection Theory of Acquired Immunity.* Cambridge, UK: Cambridge University Press (cited by Vaz 2011).

Burnet, M. 1957. Cancer: A biological approach. I. The processes of control. *BMJ* 1(5022): 779–86.

Caffarel, M.M., C. Andradas, E. Pérez-Gómez, M. Guzmán, and C. Sánchez. 2012. Cannabinoids: A new hope for breast cancer therapy? *Cancer Treat Rev* 38(7): 911–8.

Chiang, Y.C., Y.N. Lo, and J.C. Chen. 2013. Crosstalk between dopamine D_2 receptors and cannabinoid CB_1 receptors regulates CNR1 promoter activity via ERK1/2 signaling. *J Neurochem* 127(2): 163–76.

Davydova, S.M., M.A. Cheido, M.M. Gevorgyan, and G.V. Idova. 2010. Effects of 5-HT2A receptor stimulation and blocking on immune response. *Bull Exp Biol Med* 150(2): 219–21.

Dawood, Z.A., and T.S. Qubih. 2012. Effect of *Peganum harmala* on histological reactions after post Marek's disease vaccination in layer hens. *Iraqi J Vet Sci* 26(4): 339–46.

Fontanilla, D., M. Johannessen, A.R. Hajipour, N.V. Cozzi, M.B. Jackson, and A.E. Ruoho. 2009. The hallucinogen N,N-dimethyltryptamine (DMT) is an endogenous sigma-1 receptor regulator. *Science* 323(5916): 934–7.

Frecska, E., A. Szabo, M.J. Winkelman, L.E. Luna, and D.J. McKenna. 2013. A possibly sigma-1 receptor mediated role of dimethyltryptamine in tissue protection, regeneration, and immunity. *J Neural Transm (Vienna)* 120(9): 1295–303.

Fröberg, G.K., R. Lindberg, M. Ritter, and K. Nordlind. 2009. Expression of serotonin and its 5-HT1A receptor in canine cutaneous mast cell tumours. *J Comp Pathol* 141(2–3): 89–97.

Gasperi, V., D. Evangelista, S. Oddi, F. Florenzano, V. Chiurchiù, L. Avigliano, M.V. Catani, and M. Maccarrone. 2015. Regulation of inflammation and proliferation of human bladder carcinoma cells by type-1 and type-2 cannabinoid receptors. *Life Sci* 138: 41–51.

Gobin, V., M. De Bock, B.J. Broeckx, M. Kiselinova, W. De Spiegelaere, L. Vandekerckhove, K. Van Steendam, L. Leybaert, and D. Deforce. 2015. Fluoxetine suppresses calcium signaling in human T lymphocytes through depletion of intracellular calcium stores. *Cell Calcium* 58(3): 254–63.

Griffiths, R.R., W.A. Richards, U. McCann, and R. Jesse. 2006. Psilocybin can occasion mystical-type experiences having substantial and sustained personal meaning and spiritual significance. *Psychopharmacology (Berl)* 187(3): 268–92.

Grob, C.S., A.L. Danforth, G.S. Chopra, M. Hagerty, C.R. McKay, A.L. Halberstadt, and G.R. Greer. 2011. Pilot study of psilocybin treatment for anxiety in patients with advanced-stage cancer. *Arch Gen Psychiatry* 68(1): 71–8.

Guzmán, M. 2003. Cannabinoids: Potential anticancer agents. *Nat Rev Cancer* 3(10): 745–55.

Henriksen, R., N. Dizeyi, and P.A. Abrahamsson. 2012. Expression of serotonin receptors 5-HT1A, 5-HT1B, 5-HT2B and 5-HT4 in ovary and in ovarian tumours. *Anticancer Res* 32(4): 1361–6.

Hickling, F.W., and E.E. Griffith. 1994. Clinical perspectives on the Rastafari movement. *Hosp Community Psychiatry* 45(1): 49–53.

Hind, W.H., T.J. England, and S.E. O'Sullivan. 2016. Cannabidiol protects an *in vitro* model of the blood-brain barrier from oxygen-glucose deprivation via PPARγ and 5-HT1A receptors. *Br J Pharmacol* 173(5): 815–25.

Jiang, H.E., X. Li, Y.X. Zhao, D.K. Ferguson, F. Hueber, S. Bera, Y.F. Wang, L.C. Zhao, C.J. Liu, and C.S. Li. 2006. A new insight into *Cannabis sativa* (*Cannabaceae*) utilization from 2500-year-old Yanghai Tombs, Xinjiang, China. *J Ethnopharmacol* 108(3): 414–22.

Jiang, H.E., X. Li, D.K. Ferguson, Y.F. Wang, C.J. Liu, and C.S. Li. 2007. The discovery of *Capparis spinosa* L. (*Capparidaceae*) in the Yanghai Tombs (2800 years b.p.), NW China, and its medicinal implications. *J Ethnopharmacol* 113(3): 409–20.

Krebs, T.S., and P.Ø. Johansen. 2012. Lysergic acid diethylamide (LSD) for alcoholism: Meta-analysis of randomized controlled trials. *J Psychopharmacol* 26(7): 994–1002.

Kreitzer, A.C. 2005. Neurotransmission: Emerging roles of endocannabinoids. *Curr Biol* 15(14): R549–51.

Lansky, E.P., W. Jiang, H. Mo, L. Bravo, P. Froom, W. Yu, N.M. Harris, I. Neeman, and M.J. Campbell. 2005. Possible synergistic prostate cancer suppression by anatomically discrete pomegranate fractions. *Invest New Drugs* 23(1): 11–20.

Lash, T.L., D. Cronin-Fenton, T.P. Ahern, C.L. Rosenberg, K.L. Lunetta, R.A. Silliman, S. Hamilton-Dutoit, J.P. Garne, M. Ewertz, H.T. Sørensen, and L. Pedersen. 2010. Breast cancer recurrence risk related to concurrent use of SSRI antidepressants and tamoxifen. *Acta Oncol* 49(3): 305–12.

Legrand, M., A.B. Guttormsen, and M.M. Berger. 2015. Ten tips for managing critically ill burn patients: Follow the RASTAFARI! *Intensive Care Med* 41(6): 1107–9.

Linge, R., L. Jiménez-Sánchez, L. Campa, F. Pilar-Cuéllar, R. Vidal, A. Pazos, A. Adell, and Á. Díaz. 2016. Cannabidiol induces rapid-acting antidepressant-like effects and enhances cortical 5-HT/glutamate neurotransmission: Role of 5-HT1A receptors. *Neuropharmacology* 103: 16–26.

Lobikin, M., B. Chernet, D. Lobo, and M. Levin. 2012. Resting potential, oncogene-induced tumorigenesis, and metastasis: The bioelectric basis of cancer in vivo. *Phys Biol* 9(6): 065002.

Lychkova, A.É., and A.M. Puzikov. 2014. Serotonergic regulation of the immune system. *Usp Fiziol Nauk* 45(4): 69–88.

Magrini, E., I. Szabò, A. Doni, J. Cibella, and A. Viola. 2011. Serotonin-mediated tuning of human helper T cell responsiveness to the chemokine CXCL12. *PLoS One* 6(8): e22482.

Martins, L.C., N.P. Rocha, K.C. Torres, R.R. Dos Santos, G.S. França, E.N. de Moraes, M.A. Mukhamedyarov, A.L. Zefirov, A.A. Rizvanov, A.P. Kiyasov, L.B. Vieira, M.M. Guimarães, M.E. Yalvaç, A.L. Teixeira, M.A. Bicalho, Z. Janka, M.A. Romano-Silva, A. Palotás, and H.J. Reis. 2012. Disease-specific expression of the serotonin-receptor 5-HT(2C) in natural killer cells in Alzheimer's dementia. *J Neuroimmunol* 251(1–2): 73–9.

Mauler, M., C. Bode, and D. Duerschmied. 2016. Platelet serotonin modulates immune functions. *Hamostaseologie* 36(1): 11–6.

Nahas, G.G. 1982. Hashish in Islam 9th to 18th century. *Bull N Y Acad Med* 58(9): 814–31.

Nasehi, M., K. Davoudi, M. Ebrahimi-Ghiri, and M.R. Zarrindast. 2016a. Interplay between serotonin and cannabinoid function in the amygdala in fear conditioning. *Brain Res* 1636: 142–51.

Nasehi, M., M. Farrahizadeh, M. Ebrahimi-Ghiri, and M.R. Zarrindast. 2016b. Modulation of cannabinoid signaling by hippocampal 5-HT4 serotonergic system in fear conditioning. *J Psychopharmacol* 30(9): 936–44.

Nasser, M.W., Z. Qamri, Y.S. Deol, D. Smith, K. Shilo, X. Zou, and R.K. Ganju. 2011. Crosstalk between chemokine receptor CXCR4 and cannabinoid receptor CB2 in modulating breast cancer growth and invasion. *PLoS One* 6(9): e23901.

Osório Fde, L., R.F. Sanches, L.R. Macedo, R.G. Santos, J.P. Maia-de-Oliveira, L. Wichert-Ana, D.B. Araujo, J. Riba, J.A. Crippa, and J.E. Hallak. 2015. Antidepressant effects of a single dose of ayahuasca in patients with recurrent depression: A preliminary report. *Rev Bras Psiquiatr* 37(1): 13–20.

Oufkir, T., and C. Vaillancourt. 2011. Phosphorylation of JAK2 by serotonin 5-HT (2A) receptor activates both STAT3 and ERK1/2 pathways and increases growth of JEG-3 human placental choriocarcinoma cell. *Placenta* 32(12): 1033–40.

Pai, V.P., A.M. Marshall, L.L. Hernandez, A.R. Buckley, and N.D. Horseman. 2009. Altered serotonin physiology in human breast cancers favors paradoxical growth and cell survival. *Breast Cancer Res* 11(6): R81.

Pars, H.G., and J.F. Howes. 1977. Potential therapeutic agents derived from the cannabinoid nucleus. *Adv Drug Res* 11: 97–189.

Pisanti, S., P. Picardi, A. D'Alessandro, C. Laezza, and M. Bifulco. 2013. The endocannabinoid signaling system in cancer. *Trends Pharmacol Sci* 34(5): 273–82.

Pritchard, K.I. 2010. Do selective serotonin receptor inhibitor antidepressants reduce tamoxifen's effectiveness and increase the risk of death from breast cancer? *Breast Cancer Res* 12(Suppl 4): S18.

Rinaldi, A., A.M. Chiaravalli, M. Mian, E. Zucca, M.G. Tibiletti, C. Capella, and F. Bertoni. 2010. Serotonin receptor 3A expression in normal and neoplastic B cells. *Pathobiology* 77(3): 129–35.

Rivera-Baltanas, T., J.M. Olivares, J.R. Martinez-Villamarin, E.Y. Fenton, L.E. Kalynchuk, and H.J. Caruncho. 2014. Serotonin 2A receptor clustering in peripheral lymphocytes is altered in major depression and may be a biomarker of therapeutic efficacy. *J Affect Disord* 163: 47–55.

Roberts, T.B. 1999. Do entheogen-induced mystical experiences boost the immune system? Psychedelics, peak experience, and wellness. *Adv Mind Body Med* 15(2): 139–47.

Sanches, R.F., F. de Lima Osório, R.G. Dos Santos, L.R. Macedo, J.P. Maia-de-Oliveira, L. Wichert-Ana, D.B. de Araujo, J. Riba, J.A. Crippa, and J.E. Hallak. 2016. Antidepressant effects of a single dose of ayahuasca in patients with recurrent depression: A SPECT study. *J Clin Psychopharmacol* 36(1): 77–81.

Satprem. 1982. *The Mind of the Cells*. Trans. Francine Mahak. Pondicherry, India: Sri Aurobindo Press.

Schindler, E.A., C.H. Gottschalk, M.J. Weil, R.E. Shapiro, D.A. Wright, and R.A. Sewell. 2015. Indoleamine hallucinogens in cluster headache: Results of the clusterbusters medication use survey. *J Psychoactive Drugs* 47(5): 372–81.

Scott, K.G., and R.S. Stone. 1959. Antitumor action of lysergic acid derivatives and their serotonin blocking effect as reflected by iodine-131 distribution in rats. *Cancer Res* 19: 783–87.

Snir, O., E. Hesselberg, P. Amoudruz, L. Klareskog, I. Zarea-Ganji, A.I. Catrina, L. Padyukov, V. Malmström, and M. Seddighzadeh. 2013. Genetic variation in the serotonin receptor gene affects immune responses in rheumatoid arthritis. *Genes Immun* 14(2): 83–9.

Soll, C., M.O. Riener, C.E. Oberkofler, C. Hellerbrand, P.J. Wild, M.L. DeOliveira, and P.A. Clavien. 2012. Expression of serotonin receptors in human hepatocellular cancer. *Clin Cancer Res* 18(21): 5902–10.

Su, T.P., T. Hayashi, and D.B. Vaupel. 2009. When the endogenous hallucinogenic trace amine N,N-dimethyltryptamine meets the sigma-1 receptor. *Sci Signal* 2(61): pe12.

Szabo, A. 2015. Psychedelics and Immunomodulation: Novel approaches and therapeutic opportunities. *Front Immunol* 6: 358.

Szabo, A., A. Kovacs, E. Frecska, and E. Rajnavolgyi. 2014. Psychedelic N,N-dimethyltryptamine and 5-methoxy-N,N-dimethyltryptamine modulate innate and adaptive inflammatory responses through the sigma-1 receptor of human monocyte-derived dendritic cells. *PLoS One* 9(8): e106533.

Toghyani, M., A. Ghasemi, and S.A. Tabeidian. 2015. The effect of different levels of seed and extract of harmal (*Peganum harmala* L.) on immune responses of broiler chicks. *Int J Biol Biomolec Agric Food Biotechnol Eng* 9(1): 31–4.

Touw, M. 1981. The religious and medicinal uses of Cannabis in China, India and Tibet. *J Psychoactive Drugs* 13(1): 23–34.

van Waarde, A., A.A. Rybczynska, N.K. Ramakrishnan, K. Ishiwata, P.H. Elsinga, and R.A. Dierckx. 2015. Potential applications for sigma receptor ligands in cancer diagnosis and therapy. *Biochim Biophys Acta* 1848(10 Pt B): 2703–14.

Vasil'eva, E.F., G.I. Koliaskina, O.S. Brusov, M.I. Faktor, V.G. Kaleda, N.A. Barkhatova, and E.D. Bogdanova. 2011. The effect of serotonin on the cytotoxic activity of natural killer lymphocytes in patients with the first episode of endogenous psychosis. *Zh Nevrol Psikhiatr Im S S Korsakova* 111(4): 61–6.

Vaz, N.M. 2011. Francisco Varela and the immunological self. *Syst Res Behav Sci* 28(6): 696–703.

Vaz, N.M., and F.G. Varela. 1978. Self and nonsense: An organism-centered approach to immunology. *Med Hypotheses* 4(3): 231–67.

Wolf, S.A., S. Tauber, and O. Ullrich. 2008. CNS immune surveillance and neuroinflammation: Endocannabinoids keep control. *Curr Pharm Des* 14(23): 2266–78.

Yin, J., R.H. Albert, A.P. Tretiakova, and B.A. Jameson. 2006. 5-HT(1B) receptors play a prominent role in the proliferation of T-lymphocytes. *Neuroimmunol* 181(1–2): 68–81.

Young, S.N. 2013. Single treatments that have lasting effects: Some thoughts on the antidepressant effects of ketamine and botulinum toxin and the anxiolytic effect of psilocybin. *J Psychiatry Neurosci* 38(2): 78–83.

10 Conclusion

Much has been sung of *Peganum harmala*, and with only the greatest humility as fellow creatures under the heavens do we dare to attempt to conclude anything (Figure 10.1). How can it be concluded when it is still in progress? How can it be the end if we come back again?

It is the way of life that we return again and again to the same themes, the same needs, the same joys, and the same pains. *Harmal* comes back again and again too. Year after year, season after season, seeds fall from the capsules of the dying plant to the parched ground below, and soon they will capture moisture through the winter, and in the earliest spring send out shoots. Beautiful flowers follow, yielding to fruits, to seed, give the mother leave to pass back to the soil, as the process repeats itself for another year. Again and again. Return and return. Over and over.

Harmal has been with us a very long time, is most decidedly with us now, and with very little doubt will be with us and our seed long, long, long into the future. But, so what? How can that benefit us? And why is this important to us, and our individual and collective interests?

Harmal is, as a famous ethnobotanist once explained, to be treated and used like any other herb. That is a first principle, and from there can begin the elucidating of what about *pegano* is special, or unique.

It was noted earlier in these pages that the pomegranate, *rimon*, *Punica granatum*, according to pre-Zoroastrian and Zoroastrian sources (Chapter 3), was the "mother" of *pegano*, the "mother" of *harmal*. What is meant by this mysterious comment, and why is it relevant to the present discussion?

As revealed previously, part of the concordance between *rimon* and *harmal* is the homology of the physical forms of the two fruits. Each contains, inside a fixed space it defines, a multitude of seeds densely packed with an infinitude of complexity. The degree of complexity of the packing of a single seed is governed by its infinitely variable physical position in relation to its peers. Thus, the physical packing within the respective fruits is constrained from the point of view of the sharply faceted surfaces of the seeds of each species, but also, infinitely variable. This is very important biologically from the point of view of modulation.

Modulation is a popular term in biology, but at first glance it may seem simply synonymous with inhibition. This is only partially true, however, as modulation goes beyond mere inhibition to suggest infinite, analogic adaptation and adjustment, in a word, *tuning*. Compounds, such as those that occur in plants, enable adjustments within the mammalian body with the ability to *modulate* biological responses, and thus are known as *biological response modifiers.*

The bright red juice-encapsulated seeds (i.e., arils) of the pomegranate yield a strong red colored juice which is nearly impossible to differentiate from the liquor emerging from the tough little seeds of *Peganum*, which, after a few days, turn alcohol they are soaked in to a brilliant crimson. Both of these bright red solutions that

FIGURE 10.1 *Peganum* is not important only for humans! The green leaves of *Peganum harmala* offer insects a habitat in the middle of the barren Judaean desert. (5/12/2015, near the Inn of the Good Samaritan, Judaean Desert, Israel: by Helena Paavilainen.)

are identical to the naked eye are exceptionally rich in biological response modifiers (BRMs). Both solutions are replete with compounds that can aid the body, possibly with further aid from still other plants that contain still other BRMs, to adapt and adjust through a complex *tuning* of the neuro-lymphatic "immune system." How this is achieved, i.e., with which chemical assistants, with which BRMs, is an infinitely variable and complex matter, with however some common themes and common biological or evolutionary strategies.

The sections of a pomegranate also resemble in their packing, as along with the seeds of *harmal* inside their little capsules, brains confined within their skulls in an analogous and topologically reminiscent manner. Extracts from the fruits of the pomegranate (Ahmed et al. 2014; Morzelle et al. 2016; Pathakoti et al. 2017; Yaidikar et al. 2014; Yaidikar and Thakur 2015) and the fruit *cum* seeds of *harmal* (Biradar et al. 2013) have recently been the subjects of intensive research owing to their seemingly unlikely (to the unobservant) potency and potential as neuroprotective agents. Different biological strategies emerge in an imaginal state, homologous morphologies, different BRMs, but with uncanny potential to address the same biochemical needs of senescent, aging, brains.

Back again to the subjects of cancer, telepathy, and synchronicity, another "in real time" occurrence bears noting. As the physician was waking a little after midnight to write this Conclusion chapter, the subtle ringing of the cell phone on silent was heard, and checking the recognition screen, he noted it was the daughter of the second patient whose ongoing treatment was described in Chapter 8. The daughter was in a state of panic, because her mother had become suddenly very weak and pale looking, with irregular pulse, increased (for her) blood pressure, and very shallow and infrequent respirations. She looked, according to the daugher, very yellow, if not green. Her husband had also awoken in a panic when he saw her, and had begun to shout out in fright. She was shivering, cold, and could not stop shaking. At one point, according to her report, all were shouting out in fright. The ambulance had been called, and the paramedics were on the scene. A decision had to be made immediately whether to rush the patient to any hospital in the country—physician's choice—or to the physician himself, 2 h away—anywhere, but movement had to commence at once, or the paramedics would have to leave the scene. There were many other calls. The physician pleaded to allow the team to wait 15 min and then to reassess, but this was considered absolutely unacceptable, as was the option for the patient to be kept at home and wait the two hours needed for the physician to arrive. The patient indicated she was already feeling better and preferred to rest at home, but she was soon persuaded otherwise. She was taken down the five flights of stairs by the crew and away by ambulance to the large multi-service cancer hospital an hour's drive from her home.

With an agitated heart that gradually again became placid, the physician once more sat down to write. A couple of hours later, the phone again rung in the very wee hours for the daughter to inform the physician that the hospital staff had examined her mother, performed an electrocardiogram and found nothing wrong, everything fine, they were discharging her mother and she would soon be on her way home. No need to travel to the hospital tonight. She would have already left!

The experience of the acute anxiety which extended to all concerned made the physician again aware of the need for facilities where patients being treated with herbs such as *harmal* could come for short stays. Places where respect and a let-the-body-heal attitude could prevail. In the meanwhile, though, the system did just fine by appropriately and noninvasively checking the patient, reassuring her, and sending her home. She would live longer and have the chance for more tuning.

Harmal has many roles to play in the medicine of the modern world, of course, as a powerful drug plant, in the hands of those whom the society has assigned responsibility for such matters, i.e., physicians or their qualified substitutes. The importance of the discovery of diabetologist Andrew Stewart that harmine, alone of a screen of over 3,000 compounds—some have it as 1 of 100,000 compounds—was found capable of stimulating the regeneration of human pancreatic β cells, the insulin-producing components of the Islets of Langerhans, classically considered lacking in ability to regenerate (which is the underlying reason for the so-called incurability of Type 1 juvenile onset diabetes mellitus), is nothing short of staggering. These structures, β cells and neurons, are not supposed to be able to regenerate. Now, thanks to ancient harmala alkaloids, compounds originating in *harmal*, the impossible suddenly seems possible.

Harmal, as an entheogen, may also help to make the impossible true by enhancing the power of the deep imagination to make alterations, to *modulate* the physical structure of the body. This may sound far-fetched, but given the cascades of psychoneuroimmunology, plausible. The possible ability of *harmal* to *tune* the putative lympho-neuric syncytium through a carefully orchestrated modulation of scores or hundreds of different serotonin receptors, with even a single dose or just a few doses, is clinically appealing. Such occasional or "intercurrent" treatments with *harmal* as a tuning agent may target both the lymphatic and the neurological aspects of the syncytium in a single tuning or reset to facilitate global modulation in order to lower oxidative tension and inflammation throughout the network, to relax and help allow to heal both the psyche and soma.

At low doses, *harmal* also has potential as a daily supplement, as a nutrient, antibiotic, antiparasitic, and antiviral, or as a preservative for the foods themselves. Low doses would be one way to avoid systemic toxicity, while still having some possible sufficient antidepressant, antidiabetic, or antiparkinsonian effect. Low doses of other drugs, for example ketamine (Bauchat et al. 2011), have revealed the potential for very low doses, lower than the pharmacologically expected therapeutic ones, to stimulate clinically meaningful responses. Low doses, for example of LSD (White and Appel 1982), may provide pharmacological cues to organisms to help "remember" or ease future therapeutic interventions. Such cues may act by hormesis-based mechanisms to aid in the process of activating and tuning lympho-neuric circuits to carry and sustain healing responses in brain and soma in support of disease resistance (Calabrese et al. 2016; Raefsky and Mattson 2017). Technically, hormesis refers to a biphasic dose response of a drug with stimulation at low doses and inhibition at high doses (Calabrese et al. 2016), and it is critical to keep in mind when thinking about the phyto-metabolome (the totality of small compounds, i.e., the totality of metabolites,) of a plant. The proper study of metabolomes is metabolomics, which employs chemical analysis by nuclear magnetic resonance, chromatography, mass spectrophotometry, and related methods to help define the totality in terms of both principal components and general features, i.e., gestalts.

Drug developers can exult in *P. harmala* and the entire *Peganum* genus for the many interesting leads they provide. Certainly the race is on to synthesize the next super-harmine or super-vasicine—one that is bigger and better, more specific, potent, and well-tolerated.

Harmine and the harmala alkaloids are important not only for Parkinson's disease, diabetes, obesity, and cancer, but also in the broader context of psychiatry. The serotonin 2A receptor (5-HT2a), to which these compounds' strong binding is common with psilocybin, and which is also implicated in the action of antidepressant and antipsychotic drugs, may serve as convergence point (Howland 2016) of different streams and philosophies of psychopharmacology, and with help from within, to psychopharmacognosy as well.

Interactions of serotonin with opioid receptors, and in cross-talk with opioid pathways, is an emerging area of research as 2017 gets underway and will demand a translational approach to help enable clinical benefit from these latest findings (Graeff 2017). Indeed, antipain synergy between harmal and opium (*Papaver somniferum*) may point the way to advances in clinical dolorology and end-of-life issues (Shoaib et al. 2016).

Letting go is an important precept of most healing. Frequently we "hold on" to patterns of thought and behavior that run counter with the needs of our bodies, mind, spirit, soul, and overall health. If we want symptoms and diseases to "go away," we, the patients, must first let them go! They will never go unless we let them go.

In the end, of course, we also need to let go of life itself before we embark on the "ultimate journey" (Grof 2006) due us all. When the time comes, *harmal* also has the potential for yeoman's service in facilitating the passage.

Finally, as a practicality, *harmal* need not be used alone. By combining it with other botanicals, mushrooms, the right foods, breathing, meditation, music, exercise, surroundings, and company, the best, deepest, strongest, and smoothest results will likely be obtained (Liu et al. 2016).

REFERENCES

Ahmed, A.H., G.M. Subaiea, A. Eid, L. Li, N.P. Seeram, and N.H. Zawia. 2014. Pomegranate extract modulates processing of amyloid-β precursor protein in an aged Alzheimer's disease animal model. *Curr Alzheimer Res* 11(9): 834–43.

Bauchat, J.R., N. Higgins, K.G. Wojciechowski, R.J. McCarthy, P. Toledo, and C.A. Wong. 2011. Low-dose ketamine with multimodal postcesarean delivery analgesia: A randomized controlled trial. *Int J Obstet Anesth* 20(1): 3–9.

Biradar, S.M., H. Joshi, and K.C. Tarak. 2013. Cerebroprotective effect of isolated harmine alkaloids extracts of seeds of *Peganum harmala* L. on sodium nitrite-induced hypoxia and ethanol-induced neurodegeneration in young mice. *Pak J Biol Sci* 16(23): 1687–97.

Calabrese, E.J., G. Dhawan, R. Kapoor, I. Iavicoli, and V. Calabrese. 2016. HORMESIS: A fundamental concept with widespread biological and biomedical applications. *Gerontology* 62(5): 530–5.

Graeff, F.G. 2017. Translational approach to the pathophysiology of panic disorder: Focus on serotonin and endogenous opioids. *Neurosci Biobehav Rev* Jan 7 [Epub ahead of print].

Grof, S. 2006. *The Ultimate Journey: Consciousness and the Mystery of Death*. Santa Cruz, CA: Multidisciplinary Association for Psychedelic Studies (MAPS).

Howland, R.H. 2016. Antidepressant, antipsychotic, and hallucinogen drugs for the treatment of psychiatric disorders: A convergence at the serotonin-2A receptor. *J Psychosoc Nurs Ment Health Serv* 54(7): 21–4.

Liu, W., H. Ma, N.A. DaSilva, K.N. Rose, S.L. Johnson, L. Zhang, C. Wan, J.A. Dain, and N.P. Seeram. 2016. Development of a neuroprotective potential algorithm for medicinal plants. *Neurochem Int* 100: 164–77.

Morzelle, M.C., J.M. Salgado, M. Telles, D. Mourelle, P. Bachiega, H.S. Buck, and T.A. Viel. 2016. Neuroprotective effects of pomegranate peel extract after chronic infusion with amyloid-β peptide in mice. *PLoS One* 11(11): e0166123.

Pathakoti, K., L. Goodla, M. Manubolu, and T. Tencomnao. 2017. Metabolic alterations and the protective effect of punicalagin against glutamate-induced oxidative toxicity in HT22 cells. *Neurotox Res* Jan 9 [Epub ahead of print].

Raefsky, S.M., and M.P. Mattson. 2017. Adaptive responses of neuronal mitochondria to bioenergetic challenges: Roles in neuroplasticity and disease resistance. *Free Radic Biol Med* 102: 203–16.

Shoaib, M., S.W. Shah, N. Ali, I. Shah, S. Ullah, M. Ghias, M.N. Tahir, F. Gul, S. Akhtar, A. Ullah, W. Akbar, and A. Ullah. 2016. Scientific investigation of crude alkaloids from medicinal plants for the management of pain. *BMC Complement Altern Med* 16: 178.

White, F.J., and J.B. Appel. 1982. Training dose as a factor in LSD-saline discrimination. *Psychopharmacology (Berl)* 76(1): 20–5.

Yaidikar, L., B. Byna, and S.R. Thakur. 2014. Neuroprotective effect of punicalagin against cerebral ischemia reperfusion-induced oxidative brain injury in rats. *J Stroke Cerebrovasc Dis* 23(10): 2869–78.

Yaidikar, L., and S. Thakur. 2015. Punicalagin attenuated cerebral ischemia-reperfusion insult via inhibition of proinflammatory cytokines, up-regulation of Bcl-2, down-regulation of Bax, and caspase-3. *Mol Cell Biochem* 402(1–2): 141–8.

Epilogue

Peganum, in spite of its reputation for toxicity in the Western medical literature, may be safely and comfortably employed in medical practice so long as dosage is appropriate, galenic preparation is utilized, and proper medical correctives are included. It has wide potential for use in psychiatry. Infrequent dosing with supportive music therapy or in combination with medical hypnosis may yield beneficial effects psychically, and possibly also somatically. Potential applications include treatment of chronic pain, Parkinson's or Alzheimer's disease, diabetes, obesity, and cancer.

In the treatment of cancer, transcutaneous administration may represent a novel delivery strategy that can localize treatment to solid malignancies. Chemical penetration enhancers such as glycerol or dimethylsulfoxide may facilitate arrival to desired internal locations. The use of infrared light may both aid penetration and enable photoactivation of the fluorescent harmala alkaloids as phototoxic agents. Penetration through tissue planes including the blood–brain barrier as well as anticancer effects may be further synergized with the use of low-dose therapeutic ultrasound (Sadanala et al. 2014; Shaw et al. 2016; Su et al. 2012; Tserkovsky et al. 2012; Wang et al. 2013; Wang et al. 2015).

Very low doses (hormesis) and ultra low doses (homeopathy) of *harmal* represent additional possibilities for health maintenance programs including antibacterial, antifungal, antiprotozoal, antiviral therapies, and preventive medicine. Novel combinations of *P. harmala* with other herbs that have been shown synergistically efficacious in some settings, such as diabetes (Abedi Gaballu et al. 2015), may also generalize to efficacy in other settings, such as cancer, and are worthy of further exploration (Figure E1.1).

A new arena of medicine that might be tentatively called *harmal*-ogy, encompassing preclinical and clinical research and practice with *P. harmala* preparations, as well as other herbs that include harmala alkaloids in significant quantities—such as *Banisteriopsis caapi* in Ayahuasca, may also be in the reportage and be subjects for professional meetings and seminars. Scientists and clinicians interested in starting, or contributing, to a professional journal in this area, or in organizing or attending academic meetings addressed to this theme, are invited to contact the authors at punisyn@gmail.com or at POB 9945, Haifa, Israel.

FIGURE E1.1 Harsh desert conditions: each and every stage of the development of *Peganum harmala* is slow—but the results are worth waiting for as its seeds are holding great promise for modern medicine. (5/12/2015, Judaean Desert, by side of the Inn of the Good Samaritan, Israel: by Helena Paavilainen.)

REFERENCES

Abedi Gaballu, F., Y. Abedi Gaballu, O. Moazenzade Khyavy, A. Mardomi, K. Ghahremanzadeh, B. Shokouhi, and H. Mamandy. 2015. Effects of a triplex mixture of *Peganum harmala*, *Rhus coriaria*, and *Urtica dioica* aqueous extracts on metabolic and histological parameters in diabetic rats. *Pharm Biol* 53(8): 1104–9.

Sadanala, K.C., P.K. Chaturvedi, Y.M. Seo, J.M. Kim, Y.S. Jo, Y.K. Lee, and W.S. Ahn. 2014. Sono-photodynamic combination therapy: A review on sensitizers. *Anticancer Res* 34(9): 4657–64.

Shaw, A., E. Martin, J. Haller, and G. ter Haar. 2016. Equipment, measurement and dose: A survey for therapeutic ultrasound. *J Ther Ultrasound* 4: 7–16.

Su, X., L. Li, and P. Wang. 2012. Research progress of the anti-tumor effect of sonodynamic and photodynamic therapy. *Sheng Wu Yi Xue Gong Cheng Xue Za Zhi* 29(3): 583–7.

Tserkovsky, D.A., E.N. Alexandrova, V.N. Chalau, and Y.P. Istomin. 2012. Effects of combined sonodynamic and photodynamic therapies with photolon on a glioma C6 tumor model. *Exp Oncol* 34(4): 332–5.

Wang, H., X. Wang, P. Wang, K. Zhang, S. Yang, and Q. Liu. 2013. Ultrasound enhances the efficacy of chlorin E6-mediated photodynamic therapy in MDA-MB-231 cells. *Ultrasound Med Biol* 39(9): 1713–24.

Wang, P., C. Li, X. Wang, W. Xiong, X. Feng, Q. Liu, A.W. Leung et al. 2015. Anti-metastatic and pro-apoptotic effects elicited by combination photodynamic therapy with sonodynamic therapy on breast cancer both *in vitro* and *in vivo*. *Ultrason Sonochem* 23: 116–27.

Afterword

After finishing the preceding chapters, the authors decided to query potential contributors for a foreword or preface to this volume. The first of those potential contributors was Dr. Andrew Weil. Andrew Weil, MD, is the Lovell-Jones professor of integrative rheumatology, a professor of public health, a clinical professor of medicine, and the director of the University of Arizona Center for Integrative Medicine, University of Arizona. His influence extends worldwide as a visioner, initiator, and ongoing standard for the field of integrative medicine, which he cognized and contextuated, and of which he continues to influence, support, and direct. Programs in integrative medicine under his oversight now flourish in 44 academic medical centers including Harvard, Stanford, Yale, Johns Hopkins, Mayo Clinic, Duke University Medical Center, Children's Memorial Hospital, and the University of California-San Francisco Osher Center for Integrative Medicine. His influence also extends far back in time in the journey by which this book came to be.

As mentioned in the Foreword, this book was inspired by a Golden Guide, *Hallucinogenic Plants*, by the late Richard Schultes, PhD, who had much to do with fathering the field of ethnopharmacology, and according to Dr. Weil's Preface to this volume, it was from a lecture by Professor Schultes as his former Harvard undergraduate professor that he had first learned of *Peganum harmala* (Schultes 1976). Throughout the decades in which this book was incubating, Andrew Weil was always in the background. His book, *The Natural Mind*, contextuated the difference between straight thinking and stoned thinking with a clear and novel explication that was easy to understand and whose practicality was apparent (Weil 1972). His strong background in botany as an undergraduate was singular for a physician in 1972 as much as for today.

What *about* today? After finishing the preceding chapters, Dr. Weil was queried regarding his possible willingness to receive electronically the substantial manuscript and to possibly write a foreword or preface. He responded that he would be unable to write a foreword or preface due to time constraints, but that he agreed to being sent the document. If he liked it, he might be able to write a couple of lines of support. So, some few hours after it was roughly completed, it was sent to Dr. Weil.

The next morning, one of the authors decided to pay a visit to the downtown shipping port area of Haifa. It was an area in whose environs the company that he had cofounded and managed for 18 years had once had a "beta site" for producing moderate amounts of pomegranate medicinal extracts. Walking down the familiar main street that abutted the area of the beta, he noticed a new secondhand bookstore with a very attractive front and appealing green graphics in the large front window. Through said window was a familiar, smiling face, that of Dr. Andrew Weil from the paperback cover of his book of 22 years ago, *Spontaneous Healing* (Weil 1995). This occurrence, one might conclude, was an example of a Jungian-like synchronicity, which one might argue may have been more likely to occur, or at least to be noticed, from a non-ordinary, altered, hypnoidal, or "stoned" state of consciousness (Jung

1952). Something in the noosphere was contacted, and a meaningful coincidence occurred.

So, the physician retreated, in short, with that volume and a few others of interest, namely Piaget's *Behavior and Evolution*, three case studies of Freud, Rand's *Fountainhead*, Kubler-Ross's *Death and Dying*, and a masterpiece collection of short stories. He began of course with Dr. Weil's book, which was the most appealing, enjoying the irony that he may be reading *Spontaneous Healing* while Dr. Weil might possibly be reading *Harmal: The Genus* Peganum.

Spontaneous Healing is a fantastic, extremely useful, and valuable book strongly recommended to everyone for its brilliant exposition of a multitiered self-healing system encompassing all levels of the human being from macromolecular to emotional to spiritual. It should and may one day be a required reading for all first year medical students. Anyway, one point in the book particularly resonated with experience with the first patient from Chapter 8 who had been successful in attaining improbable healing. He, like a patient described in *Spontaneous Healing*, viewed the calamity of seemingly incurable illness as a "gift." This may be a key point in eliciting spontaneous healing responses: presenting an internal double bind (for the disease) of gratitude, which greatly seems to empower the forces of repair, resorption of sick tissue, and healing (Lansky 1979, 1982). Harmal, a plant with physiologically paradoxical properties, enabling parasympathetic nervous tuning while being simultaneously entheogenically stimulating, may help nudge the human biosystem into such an attitude.

The authors wish to express their deepest gratitude to Professor Dr. Andrew Weil for taking the time from his extremely busy schedule to write the very kind Preface to this volume. This act is another important step forward toward the return of, or *integration* of, botanical medicine, entheogenesis, and healing into the greater context of medicine at large, i.e., integrative medicine.

REFERENCES

Jung, C.G. 1952. *Synchronicity: An Acausal Connecting Principle, Vol. 8, The Collected Works of C.G. Jung.* Princeton, NJ: Princeton University Press.

Lansky, P. 1979. The somatology of paradox. *Somatics* 2 (2).

Lansky, P. 1982. Possibility of hypnosis as an aid in cancer therapy. *Persp Biol Med* 25: 496–503.

Schultes, R.E. 1976. *Hallucinogenic Plants.* Illustrated by E.W. Smith. New York: Golden Press, Racine, WI: Western Publishing Company, Inc.

Weil, A. 1972. *The Natural Mind: A New Way of Looking at Drugs and the Higher Consciousness.* Boston: Houghton-Mifflin.

Weil, A. 1995. *Spontaneous Healing: How to Discover and Embrace Your Body's Natural Ability to Maintain and Heal Itself.* New York: Ballantine Books.

Appendix 1: Laboratory Identification of Harmala and Quinazoline Alkaloids

TABLE A1.1

Methods of Identification of *Harmala* and Quinazoline Alkaloids in Laboratory Settings

Method	Alkaloid(s)	Notes	Reference
High-performance capillary electrophoresis	Vasicine, vasicinone	*Adhatoda vasica*	Avula et al. 2008
Rapid UPLC	Harmalol, harmol, harmane, harmaline, harmine	*Passiflora* sp., chemical fingerprint, simultaneous with 4 flavonoids	Avula et al. 2012
Time-resolved fluorescence in microchip electrophoresis	Harmala alkaloids	Separated in seconds, methanol extract, two-photon excited fluorescence detection	Beyreiss et al. 2013
High-performance liquid chromatography (HPLC)	Vasicine	Study of photochemical oxidation and its analogues	Chowdhury et al. 1987
High-performance liquid chromatography (HPTLC)	Vasicine, vasicinone	In *Adhatoda vasica*, different plant parts	Das et al. 2005
HPTLC	Vasicine	In a commercial cough syrup; glycyrrhizin, eugenol, and cineole also detected	Deore et al. 2014
Dissolubility property, particle size diameter	Harmine, harmaline	In different ethanol concentrations	Ding et al. 2010
UPLC/ESI-QTOF-MS	Harmane	Ultra-performance liquid chromatography with electrospray ionization quadrupole time-of-flight tandem mass spectrometry	Li et al. 2014
Ultra-performance liquid chromatography, time of flight, mass spectrophotometry UPLC/Q-TOF-MS	Vasicine	In *Adhatoda vasica* and its *in vitro* culture	Madhukar et al. 2014

(Continued)

TABLE A1.1 (CONTINUED)
Methods of Identification of *Harmala* and Quinazoline Alkaloids in Laboratory Settings

Method	Alkaloid(s)	Notes	Reference
HPTLC, densitometric	Harmine, harmaline, vasicine, vasicinone	*Peganum harmala* seeds, ethyl acetate-methanol-ammonia (7 + 1 + 0.3, v/v/v) mobile phase	Pulpati et al. 2008
Spectrophotometric, HPTLC densitometric	Vasicine, total alkaloids	*Adhatoda vasica* leaf juice	Soni et al. 2008
Capillary electrophoresis, UV	Harmine, harmaline, harmol	In aqueous infusion of *P. harmala* seeds	Tascón et al. 2016
Thin layer chromatography (TLC), HPLC	Harmine, harmaline, vasicine	Ethyl acetate–methanol–ammonia (10: 1.5: 0.5) as developing solvent	Wen et al. 2012
HPLC, ultraviolet (UV)	Vasicine, harmaline, harmine, deacetylpeganetin, peganetin	Aerial parts of *P. harmala* in China, looked at habitat, growth season, 0.1% trifluoroacetic acid and acetonitrile as mobile phase—simultaneous detection of all metabolites	Wen et al. 2014

REFERENCES

Avula, B., S. Begum, S. Ahmed, M.I. Choudhary, and I.A. Khan. 2008. Quantitative determination of vasicine and vasicinone in *Adhatoda vasica* by high performance capillary electrophoresis. *Pharmazie* 63(1): 20–2.

Avula, B., Y.H. Wang, C.S. Rumalla, T.J. Smillie, and I.A. Khan. 2012. Simultaneous determination of alkaloids and flavonoids from aerial parts of *Passiflora* species and dietary supplements using UPLC-UV-MS and HPTLC. *Nat Prod Commun* 7(9): 1177–80.

Beyreiss, R., D. Geißler, S. Ohla, S. Nagl, T.N. Posch, and D. Belder. 2013. Label-free fluorescence detection of aromatic compounds in chip electrophoresis applying two-photon excitation and time-correlated single-photon counting. *Anal Chem* 85(17): 8150–7.

Chowdhury, B.K., S.K. Hirani, and D. Ngur. 1987. High-performance liquid chromatographic study of the photochemical oxidation of vasicine and its analogues. *J Chromatogr* 390(2): 439–43.

Das, C., R. Poi, and A. Chowdhury. 2005. HPTLC determination of vasicine and vasicinone in *Adhatoda vasica*. *Phytochem Anal* 16(2): 90–2.

Deore, S.L., P.S. Jaju, and B.A. Baviskar. 2014. Simultaneous estimation of four antitussive components from herbal cough syrup by HPTLC. *Int Sch Res Notices* 976264.

Ding, K., L. Liu, X. Cheng, C. Wang, and Z. Wang. 2010. Investigation on representation methods of dissolubility property of total alkaloid extract from *Peganum harmala*. *Zhongguo Zhong Yao Za Zhi* 35(17): 2250–3.

Li, S., W. Liu, L. Teng, X. Cheng, Z. Wang, and C. Wang. 2014. Metabolites identification of harmane *in vitro*/in vivo in rats by ultra-performance liquid chromatography combined with electrospray ionization quadrupole time-of-flight tandem mass spectrometry. *Biomed Anal* 92: 53–62.

Madhukar, G., E.T. Tamboli, P. Rabea, S.H. Ansari, M.Z. Abdin, and A. Sayeed. 2014. Rapid, sensitive, and validated UPLC/Q-TOF-MS method for quantitative determination of vasicine in *Adhatoda vasica* and its *in vitro* culture. *Pharmacogn Mag* 10(Suppl 1): S198–205.

Pulpati, H., Y.S. Biradar, and M. Rajani. 2008. High-performance thin-layer chromatography densitometric method for the quantification of harmine, harmaline, vasicine, and vasicinone in *Peganum harmala. J AOAC Int* 91(5): 1179–85.

Soni, S., S. Anandjiwala, G. Patel, and M. Rajani. 2008. Validation of different methods of preparation of *Adhatoda vasica* leaf juice by quantification of total alkaloids and vasicine. *Indian J Pharm Sci* 70(1): 36–42.

Tascón, M., F. Benavente, N.M. Vizioli, and L.G. Gagliardi. 2016. A rapid and simple method for the determination of psychoactive alkaloids by CE-UV: Application to *Peganum harmala* seed infusions. *Drug Test Anal* Jul 5 [Epub ahead of print].

Wen, F., X. Cheng, W. Liu, M. Xuan, L. Zhang, X. Zhao, M. Shan et al. 2014. Chemical fingerprint and simultaneous determination of alkaloids and flavonoids in aerial parts of genus *Peganum* indigenous to China based on HPLC-UV: Application of analysis on secondary metabolites accumulation. *Biomed Chromatogr* 28(12): 1763–73.

Wen, F.F., L.M. Zheng, X.J. Li, Y. Li, L. Zhang, X.M. Cheng, C.H. Wang et al. 2012. Research on quality standards of herbs of *Peganum harmala. Zhongguo Zhong Yao Za Zhi* 37(19): 2971–6.

Appendix 2: *Peganum harmala* Inebriations/Provings

"Proving" is a word borrowed from classical homeopathy (Hahnemann 1982), whereby a substance, sometimes highly diluted, is taken by healthy individuals in order to ascertain its effects in the absence of disease symptoms. It is a prerequisite to understanding the effects of substances for their potential use as medicines. Inebriation is an appropriate term for self-experimentation with *harmal*, and is employed here after the example of Shanon (2010) in his discussions of his own and others' experiences with Ayahuasca.

Two examples, from two different male physicians, are here provided. The first includes some recollections of earlier *harmal* inebriations.

EXAMPLE 1

At 10:07 a.m., on July 31, 2016, the physician, in the presence of another medical doctor (AMD), unlocked the large blue door to the outer office, opened it, and let it close behind them on its own. Turning to the right, they, as usual, were struck by the stark, modern, artistic elegance of the entrée, which, with only one chair, was, nonetheless, the waiting room. How this works if more than one waiter (or waitress) is that collapsible chairs may be brought out from the inner office, but in general, the philosophy is, if possible, not to wait.

Inside, in the inner room, the physician, dressed for a "day off," looks out as pensively, with respect, if not trepidation, as he stirs together in freshly boiled water, 3 ml each of two ethanolic extracts (made with 95% food-grade ethanol), of (1) Semen Peganum (i.e., freshly ground seed of *P. harmala*) and (2) Correctives, whose content and preparation will now be described. First, pomegranate (*Punica granatum*) vinegar (Lansky and Newman 2007), from a fermentation of organic pomegranate juice, peels, leaves, and flowers that had descended from alcohol to acid and used to pickle wildcrafted caper (*Capparis spinosa*) (Lansky et al. 2013) buds and fruits for 15 months, was combined with naturally dried, wildcrafted aerial parts of *Acacia raddiana* boiled together on a small fire for 4 h, then allowed to stand in ambient temperature for at least 2 weeks until the liquor had largely evaporated. At that point, the residue consisting of the dried *Punica–Capparis* acidulated aqueous extract was passively extracted in 95% food-grade ethanol for an additional 3 weeks, before being filtered and concentrated to ~3%.

The aforementioned associate, AMD, was sitting in a chair off to the side, arranging knitting materials, and taking in the view from out the window. Meanwhile, the physician continued to stir. The stirring, like the *harmal*, is multipurpose. First, it is to remove, as much as possible, all the alcohol through its enhanced rate of evaporation during the stirring and constant exposure of different topological layers of the solution to air on a rotating basis. Second, it provides a focus for the mind, to concentrate and ponder upon the set for the experience to come. The stirring was not difficult, but 2 weeks earlier, the physician had suffered an injury and had a mild tension and pulling on one side of the neck that had been persistent, even prompting at one point a visit to the chiropractor. The visit had not been repeated, but the pain in the neck had persisted.

The job of AMD was to be the "sitter," that is, to sit nearby as the physician would eventually imbibe the contents of the cup, and then be encouraged to lie down and listen to the prearranged musical program through high-fidelity, stereophonic headphones. The musical program had been developed according to the work of Helen Bonny, PhD, a music therapist who had worked with Walter Pahnke, MD, PhD, and later with Stanislav Grof, MD, PhD (after Dr. Pahnke had mysteriously disappeared never to be found while diving one day off the Maine coast), in the LSD therapy research programs at Spring Grove Hospital and the Maryland Psychiatric Research Center in the 1960s and early 1970s (Bonny and Pahnke 1972). In short, Bonny groups primarily classical music selections that include both instrumental and vocal (if used, preferably in a language that the listener does not understand) according to six stages: stage 1, Waiting (there is that pesky waiting again); stage 2, Onset; stage 3, Building to Peak; stage 4, Peak; stage 5, Re-entry; and stage 6, Return. In a typical LSD psychedelic therapy session, the program could typically take about 12 h to traverse all the stages, and sufficient recorded music was available and arranged according to these stages. The prearranged agreement between the physician and the sitter was for the physician to provide a verbal or nonverbal signal of when to go, music-wise, to the next stage. As according to the Bonny–Pahnke protocols, eye-shades and headphones were available and verbal communication was to be kept to a minimum, mainly for logistical matters such as bathroom breaks and for drinking water.

After nearly 10 min of stirring, the physician finally imbibed the solution, donned the headphones and eyeshades, and lay back on the couch with a blanket in the air-conditioned room to listen to music. The first selection in the Waiting list was Pachelbel's *Canon*.

As the physician lay waiting, his thoughts wandered. It was not to be his first experience with indolic entheogens in general, nor with *harmal* specifically. Actually he was quite familiar with the power of such drugs. Once, in the company and under the medical supervision of a well-known Mexican psychiatrist, and in a group of other intrepid adventurers, he was given a dose of Dr. Albert Hofmann's famous "Problem Child" (Hofmann 1980)—lysergic acid diethylamide tartrate, aka, LSD-25, along with eyeshades to help direct his attention inward. Loud music had been playing on speakers, and at least 25 other individuals had been given the same medicine and treatment. As the substance had begun to work its pharmacological effects, the physician felt himself involuntarily begin to rock gently back and forth in his position on a foam palette on the floor. As the music carried him further within, the rocking, almost imperceptibly at first, began to slowly increase in intensity. Eventually, the rocking became stronger and faster, impossible to control. Faster and faster, stronger and stronger. The rocking had then clearly seemed to assume a life of its own as the physician-to-be felt his breathing also becoming stronger and deeper, keeping up seemingly with the intensity of the rocking, with the pounding beating of the music. He felt as if he were hurtling down a tunnel that kept getting narrower and darker. Contractions from all directions seemed to engulf him, so he was left finally gasping for breath. And then, something extraordinary happened. He felt himself going through a narrow opening and then out, free, into luminous space, relaxation, and unbelievable comfort. There was no mistaking this experience, which he seemed to remember from a long time ago. He had experienced being born!

That session went on all night, and the light of dawn seemed also like a kind of rebirth. Anyway, the physician-to-be had asked the psychiatrist if there might be a possibility to have another such experience. The psychiatrist had lamented that that would not be possible under his medical supervision, as there had been a change in the Mexican government's allowance for the procedure. Nevertheless, he said, perhaps the American might be interested to meet the psychiatrist's teacher in the art of shamanism? If so, he would make the introduction and provide the details for contacting her. He had explained that he himself had left his busy medical practice and government position for a year to follow her around Mexico as she practiced her art, and took him on as an apprentice. Still a young medical student, the physician to physician-to-be, north of the border years later, had no intention of giving up his formal training, but still, he might be able to visit her in her small native village, and perhaps, through her largesse, learn something new. So, a few weeks later, with the next short school break, the medical student set off with a small party of other future doctors to visit the legendary curandera, Maria Sabina, in Huatla.

After further adventures, including the 5 h treacherous journey on a yellow school bus up about 8 miles of impossible winding mountain roads from Teotitlan, and an additional hour's ride by four-wheel drive, they arrived just before dark to her small hut in the woods on the side of the mountain. It had just begun to pour. The small group, including two translators, one from Mazatec to Spanish, and another from Spanish to English, were along. The only sounds were those of the pounding rain on the tin roof, and the snoring of dogs and small children sleeping on the floor. The only light came from a single candle burning in a bottle cap. She asked, "Did you bring the little ones?" She was referring to the *Psilocybe* mushrooms, which had been easy to obtain in the village (Figure A2.1). They were wrapped in some kind of broad leaves. She had inspected them carefully and had seemed satisfied, in spite of the small white worms crawling on the mushrooms. When asked about the worms, she shrugged and said simply that they were part of the mushrooms and were to be respected and not worried about.

She handed the leaves back to the small coterie and instructed to chew them slowly and thoughtfully. After what seemed to be about a quarter hour of chewing, she asked in Mazatec, "Do you feel the force yet?" When the answer was negative, she answered, "Chew more." More were chewed, more waiting occurred, and again, the question, "Do you feel the force yet?" When again the answer was negative, she again responded to eat more. More mushrooms, *con* worms, were consumed, and again she asked if the force could yet be felt. Again the answer was negative, and again still more mushrooms and worms were consumed. When she then again asked if the force could be felt, it was impossible to speak. She took the silence correctly to be a yes, and blew out the candle.

FIGURE A2.1   Cultivated *Psilocybe cubensis.* The genus *Psilocybe* is widely spread throughout the world. (4/11/2007, by Dr. Brainfish, http://www.flickr.com.)

The next 10 min or so were most extraordinary. First, complex multicolored geometric patterns danced before the eyes. It was so dark; it made no difference whether the eyes were open or closed. Then, the young medical student actually felt himself die. Or at least, that was certainly what it felt like. Then the question popped into his consciousness. If he were really dead, then who was experiencing this death? And who was asking these questions? And the answer was obvious, since the experience of death was so convincing: there is no death really, at least not experientially. Consciousness still remains after death! At that realization, the curandera lit the candle again, and the session was over.

With these two experiences under his belt, as it were, being born and dying, the physician allowed himself to sink more deeply into the bed as the next selections on the Waiting list, Bach's *Suite for Orchestra No. 1 in C Major*, Haydn's *Symphony 94 in G*, Bach's *Goldberg Variations*, and Strauss' *Blue Danube Waltz*, were played. By 10:35 a.m., the physician felt tingling through his torso and limbs, and signaled to his sitter to move on to stage 2.

As noted earlier, this experience was also not the first for the physician with *harmal*. The first time the doctor was walking from his home, with full view of the 20 miles shoreline below, on the way to a house call. He had chewed a few seeds, 50 mg, on the way. That had been 10 min earlier. He felt a wave of familiar sensations that could be best ascribed to "tripping," a somatologically (Lansky 1979a) integrated totality responding to a particular class of indole alkaloids. Within the realms of human experience, there are many with which this experience overlaps, but none in nature with which the physician could recall, other than with *phenethylamines* like mescaline. It was the inward perception that something was coming on, like the feeling before getting a flu. He had felt it coming on. He had felt the *harmal* taking effect. The inward feeling was of beginning to trip. But like a feeling of flu that a shot of brandy and a good night's sleep (or a dose of a well-placed homeopathic remedy) could nip in the bud, the feeling passed. It was a moment of familiarity and it passed. But that feeling had been unmistakable.

On a different occasion, he chewed 1.8 g of seeds along with the first patient discussed in Chapter 8. It was highly instructive for the physician to be able to follow along with the patient as the effect of the drug came on. After chewing the seeds together on a balcony overlooking a large field, patient and physician then shared a smoke: hand-rolled, organic tobacco cigarettes, which, in spite of their well-known and well-accepted multi-tiered damaging effects on health, may also exert, in certain contexts, transient neuropsychiatric, including antiseizure, benefits (Rong et al. 2014; Yamazaki and Sumikawa 2017). (NB: No endorsement or encouragement of tobacco smoking is here implied!)

After the smoke and a cup of tea, the two headed outdoors to the large plot of overgrown land across the street to take a walk. At a convenient location, a small campfire was made and then extinguished as physician and patient continued on their trek. The effect of the drug, which seemed to last about 3 h, was not dramatic, though deep. It was an extremely quiet intensity with a kind of spiritual sense of well-being, which permeated the increased feeling of relaxation. In no way were executive skills impaired in either individual nor were there hallucinations of any kind. Both were able to eat afterward (salmon sashimi) and to sleep soundly the following evening.

On a different occasion, the physician had set out in the desert alone not far from where the *harmal* plants could be found *in situ* to chew carefully 4 g of seeds, along with two teaspoons of *Acacia* tincture that he had prepared. The reaction had been violent. There may have been a vague sense of inner peace, but in the context of vomiting and explosive diarrhea, it was irrelevant. There was also ataxia as the physician attempted to walk to the lavatory. The acute syndrome continued for a couple hours and then gradually abated. Probably the small amount of alcohol still remaining in the tincture contributed to the reaction. The 4 g of seeds, though, seemed to be approaching a toxic dose.

Prior to coming to the office that July 31 for the supervised, formal session, the physician had fasted the previous 12 h on just water, blueberry juice, and a single morning coffee. Then, in the familiar setting of the clinic, the feeling of familiarity began to intensify. It went beyond just coming on, to going in. There was an increasing feeling of *interiority*. The physician signaled to go on to stage 3, Building to Peak. The first selection on that list was vocal: *Tamquam Sponsus* Gregorian chants. This moved on to Mendelsohn's *Hebrides Overture*, Faure's *Pie Jesu*, and Tchaikovsky's *Swan Ballet*. Not long after that, the feeling of interiority deepened, and the acuity of eidetic imagery was enhanced. The physician signaled to move on to stage 4, Peak (Figure A2.2).

Here the feeling had mainly been extremely internal. In the sharp glow of enhanced eidetic perception, there was a luminous grouping of freshly dried dates. There was also an internal image of a small dog notable for its glow and actuality. The peak in essence was connection with a deep infinitude of quiet, soothing darkness, with the exception of a few images such as the dog and the dates.

Descent did not take long. There was a quick movement back through stage 5, Re-entry, with Tchaikovsky's *Violin Concerto in D major*, and down through stage 6, Return, with Holst's *Venus, Bringer of Peace*. And that was pretty much that. At 11:10 a.m. there was note of a drink of water, at 2:50 p.m. a feeling of being "happy," content from the excellent tolerance of the new extract and the possibilities this seemed to open for treatment integrating the somatic and supratentorial realms affected by the *harmal*. By 4 p.m. the acute experience was over, and the physician and the sitter were eating dinner in a nearby restaurant. A grilled salmon and salad were well-tolerated and sleep that evening uneventful.

FIGURE A2.2   In the physician's office.

One truly amazing aftereffect that commenced immediately upon getting up from the table was the complete disappearance of the pain in the neck that had persisted for 2 weeks up to the morning of the session. This observation is a footnote, but its importance should not be underestimated. The disappearance was total and permanent.

EXAMPLE 2

The second inebriation was in another physician, a young psychiatrist, approximately a quarter of a century younger than the first physician and known to the latter for two decades as a patient, student, and finally colleague. On August 17, 2016, the psychiatrist arrived to the clinic, and at 9:30 a.m. imbibed 3 ml of the *harmal* extract and 2.5 ml of the aforedescribed acidulated *Capparis–Punica–Acacia* "corrective." By 9:50 a.m. he felt it beginning to come on and lay down on the table with a serape, makeshift eye shades, and high-fidelity stereophonic headphones. By 10:00 a.m. he felt he was already at stage 3, Moving to Peak, and at 10:20 a.m. he felt he was at stage 4, Peak, and engulfed with feelings of love for his wife, then several months pregnant with his third child. At 10:45 a.m., in the middle of Brahms's *German Requiem*, he asked if it were possible that he was already coming down. Music was changed at that moment to Holst's *Neptune the Mystic*. At 11:05 a.m. he was resting quietly with headphones and eyeshades… peaceful, listening to Mozart's *Concertante for Violin and Viola* with Isaac Stern, and Pinchas Zukerman playing Zubin Mehta conducting. By 11:10 a.m. he made low grunts and began to move a little. At 11:17 a.m. he got up to lavatory and returned at 11:22 a.m. and 2 min later he grunted and stretched, at 11:42 a.m. coughed and made a small stretch, spoke aloud at 11:45 a.m., and at 11:47 a.m. said he was "on the ground." Skipped to stage 6, Return. He listened to *Venus* (Holst) and Gluck's *Dance of the Blessed*, at 12:05 p.m. Pachelbel's *Canon*, placed his hands behind his head, still with headphones and eyeshades. Switched to *Goldberg Variations, Nutcracker, Sugar Plums*. At 12:10 p.m. he had a big stretch followed by Bach's *Third Brandenburg Concerto* with Jean-Pierre Rampal on flute. At 12:25 p.m. he sat up and declared himself feeling "like drunk but clear." He described one experience where he was getting out of his body, but was "conscious all the time" and looking down upon himself from above. He noted at that point, however, that it had been difficult for him to focus his vision, but that he have had insights about himself, such as how he made plans in his life, and his envy toward his brother, and greediness. He related another experience as if "floating on a boat," and a transient period of macropsia where one or possibly both sitters appeared physically larger than life during his one short bathroom break. He had had visions of "ascending through a rock," a coyote, and a monster he had to fight with a piece of wood, but denied any feelings of dysphoria whatsoever throughout the entire session. When asked by the physician's assistant, after getting dressed and at table in the restaurant for the post-session salmon dinner, for which kinds of patients he thought the procedure with the drugs at that dose would be contraindicated, he replied "none."

In every way, the psychiatrist's session was tolerated excellently. He related a couple weeks post-session that although mild and certainly a lot milder than his one previous psychedelic experience with psilocybin (which had apparently been harrowing), the *harmal* session had nevertheless affected him in an extremely positive way. He felt it had not only altered his perceptions and experience of ongoing events in his life, but most poignantly, that it actually in a strange, mysterious, and occult (i.e., in the sense of occult blood, hidden, unknown) manner greatly improved the quality of those events themselves.

REFERENCES

Bonny, H.L., and W.N. Pahnke. 1972. The use of music in psychedelic (LSD) psychotherapy. *J Music Ther* 9(2): 64–87.

Hahnemann, S. 1982. *Organon of Medicine.* 6th ed., trans. J. Kunzli, A. Naude, and P. Pendleton. Los Angeles: J.P. Tarcher.

Hofmann, A. 1980. *LSD: My Problem Child.* New York: McGraw-Hill.

Lansky, E.P., and R.A. Newman. 2007. *Punica granatum* (pomegranate) and its potential for prevention and treatment of inflammation and cancer. *J Ethnopharmacol* 109(2): 177–206.

Lansky, E.P., H.M. Paavilainen, and S. Lansky. 2013. *Caper: The Genus* Capparis. Boca Raton, FL: CRC Press.

Lansky, P. 1979. The somatology of paradox. *Somatics Mag* 2(2): 31–9.

Rong, L., A.T. Frontera Jr, and S.R. Benbadis. 2014. Tobacco smoking, epilepsy, and seizures. *Epilepsy Behav* 31: 210–8.

Shanon, B. 2010. *The Antipodes of the Mind: The Phenomenology of the Ayahuasca Experience.* Oxford, UK: Oxford University Press.

Yamazaki, Y., and K. Sumikawa. 2017. Nicotine-induced neuroplasticity counteracts the effect of schizophrenia-linked neuregulin 1 signaling on NMDAR function in the rat hippocampus. *Neuropharmacology* 113(Pt A): 386–95.

Appendix 3: Hebraic Mentions

The Hebrew name for harmal is *shever lavan*, which means "white fragment." The meaning of the Hebrew name and its relation to the plant are obscure and mysterious. What are at hand are more *shvarim*, i.e., fragments, of a yet to be clearer picture.

In the Talmud (*Shabbat* 20), its seed is considered very warm, good for drinking if someone has a chill. Its unique and intense scent is believed to keep away ghosts, spirits, and demons. Bedouin are said to use it as an abortifacient, and as a treatment for joint pain, general weakness, toothache, wounds and abscesses (furuncles), skin diseases, hemorrhoids, and eye infections, and for strengthening the roots of the hair.

Folk recipes especially among Yemenite Jews made use of not only the seeds, but also the stems and leaves. In one version, leaves and stems were stuffed into a jar with olive oil and allowed to sit in the sun for a week. The medicated oil was used for anointing and for massage for arthritis and back pain, and for strengthening the scalp and hair follicles. Crushed seeds were placed on an aching tooth to relieve pain. Frying leaves or seeds in olive oil helped produce a medicinal poultice with a clean linen cloth for open wounds or abscesses. Mashing leaves and mixing in a little olive oil yielded an external treatment for hemorrhoids. A hundred grams or so of leaves and stems was decocted for half an hour in water, and the filtered liquor added to a hot bath. The patient with skin lesions was asked to soak in the mixture for half an hour. Aqueous decoctions of the plant were drunk repeatedly for hepatomegaly. Knee pain was treated by crushing plant parts, heating with cow butter, and placed on a bandage to the affected area.

Bedouin observed that goats eating the leaves or seeds would rut more readily, leading to the plant's reputation of an aphrodisiac. Similarly, its diuretic effect in camels led to its use for that purpose in human folk medicine (Krispil, 2000).

Small bags of the dried material were sewn into clothes of children and bride and groom to keep away the evil eye and to bring fertility. Throwing dried capsules onto red hot rocks would release a fragrant vapor to be used with shamanic exhortations and prayers for healing (Krispil, 2000). Maimonides described a formula from Ibn Sina for treating bites of any kind, combining 3 drams each of seeds of harmal, *Nigella sativa*, and cumin, one and a half drams each of gentian and round birthwort (*Aristolochia rotunda*), and three quarters of a gram each of white pepper and myrrh, and kneading with honey to yield half dram doses (Maimonides, 2009).

REFERENCES

Krispil, N. 2000. *Ha-Madrich ha-shalem le-tzimchei marpe be-Eretz Israel* [The Complete Guide to the Medicinal Plants of the Land of Israel]. Or Yehuda.

Maimonides, M. 2009. *On Poisons and the Protection against Lethal Drugs: A Parallel Arabic-English Edition* (Medical Works of Moses Maimonides), ed., trans. and annot. G. Bos, and M.R. McVaugh. Chicago: University of Chicago Press.

Index